C0-AKF-425

A. I. Isayev (ed.)

Modeling of Polymer Processing

PROGRESS IN POLYMER PROCESSING

Series Editor: L. A. Utracki

A. I. Isayev
Modeling of Polymer Processing

L. A. Utracki
Two-Phase Polymer Systems

A. Singh/J. Silverman
Radiation Processing of Polymers

A. I. Isayev (ed.)

Modeling of Polymer Processing

Recent Developments

With contributions from

A. Cohen · H. G. deLorenzi · N. Famili · D. Huilier · A. I. Isayev · Y. Jaluria
M. V. Karwe · G. D. Kiss · T. H. Kwon · H. F. Nied · W. I. Patterson · R. C. Progelhof
T. Sastrohartono · M. Sobhanie · Y. Wang · J. L. White

Hanser Publishers, Munich Vienna New York Barcelona

Distributed in the United States of America and in Canada by
Oxford University Press, New York

Editor:
Prof. Dr. A. I. Isayev, Institute of Polymer Engineering, The University of Akron, Akron, Ohio, USA

Distributed in USA and in Canada by
Oxford University Press
200 Madison Avenue, New York, N. Y. 10016

Distributed in all other countries by
Carl Hanser Verlag
Kolbergerstr. 22
D-8000 München 80

CIP-Titelaufnahme der Deutschen Bibliothek

Modeling of polymer processing : recent developments / A. I. Isayev (ed.). With contributions from A. Cohen ... – Munich ; Vienna ; New York ; Barcelona : Hanser, 1991
(Progress in polymer processing)
ISBN 3-446-16031-0
NE: Isayev, Avraam I. [Hrsg.]; Cohen, A.

ISBN 3-446-16031-0 Carl Hanser Verlag, Munich, Vienna, New York, Barcelona
ISBN 0-19-520865-X Oxford University Press, New York

Printed in the Federal Republic of Germany by Grafische Kunstanstalt Josef C. Huber KG, Dießen/Ammersee

PROGRESS IN POLYMER PROCESSING SERIES

Leszek A. Utracki, *Series Editor*

FOREWORD

The term "polymer processing" encompasses the whole spectrum of polymer science and engineering domains, from polymerization, compounding, forming, decorating, and assembling to properties. The polymers in question can be either natural or man-made, elastomeric, thermoset or thermoplastic. Formation method may be by extrusion (including profiles, films or fibers), molding in solid or molten state, casting, etc.

In polymer processing there is an uneasy balance between the technology and science. Most frequently the technology has led the way in opening new areas, such as in development of rubbers or thermosets, in extrusion or molding. In this case the science followed, providing understanding and tools for optimization. However, in the area of thermoplastics frequently the reverse was true; here the scientific curiosity and directed research led to synthesis of these market dominating polymers. Nowadays development of new materials or products first and foremost requires solid understanding of basic principles, followed by good engineering design of the forming process and sound knowledge of the market. The successful polymer process researcher or engineer must have a broad knowledge of fundamental principles and engineering solutions.

There are several polymer engineering societies based on the national membership promoting interests of the industry and disseminating pertinent technological information. There are also chemical and physical societies with polymer divisions taking a more basic approach to polymer science. However, searching for information to solve a specific polymer processing problem still leads to difficulties. There are hundreds of thousands of articles published annually. They provide a fragmented body of information requiring, to start with, either familiarity within the narrow domain or an encyclopedic knowledge of the field. On the other hand, with the exponential expansion of information the preparation of monographic single-authored books is a tedious and frustrating task, particularly when trying to catch up with recent and most pertinent developments.

There is also another element - the globalization of science and technology. As the development of polymer blend technology demonstrates, the center of activity within a domain may rapidly shift from one country to another, from one continent to the next. The recognition of these globalization tendencies led to the formation in 1985 of the Polymer Processing Society, PPS, the first fully international professional organization dedicated to the promotion, growth and development of scientific understanding, and innovation in polymer processing. It provides a forum for the world-wide community of engineers and scientists and publishes the Society journal, "International Polymer Processing," as well as the book series, "Progress in Polymer Processing."

Progress in Polymer Processing, PPP, was initiated by PPS in 1986 and formally established in 1988. Its aim is to provide complete, in-depth, up to date information on various aspects of polymer science and engineering. The Series Editor, with the internationally based Advisory Board, is responsible for selecting the volume topics and Volume Editor(s), as well as for general supervision of content, quality, and form of the publication.

The PPP series aims to provide multi authored monographic books on topics of current interest to the international polymer processing community. Depending on the breadth of the selected topic, the volume may have the character of either an exhaustive monograph, providing an actual, complete picture of the selected field, or of an in-depth progress report of its dominant aspects. As time progresses we hope to shift more and more toward the former character. Using the multi authored format we expect on the one hand to provide a more complete picture of the selected topic viewed from different perspectives, and on the other to shorten the production time, keeping the published information up to date. We expect to produce two to three volumes a year. Current list of titles has a dozen positions.

To accomplish these goals, we shall need the help and cooperation of the international community, as well as serious effort by everyone involved in the process, the authors, editors, members of the Advisory Board, and the publisher. It is hoped that in time we will become more accustomed to our tasks and hence more efficient, always guided by the needs of our colleagues within the polymer processing community.

L.A. Utracki
Series Editor

CONTENTS

Contributors xiv

Preface xv

1. **A Brief Overview of Modeling in Polymer Processing** 1
A.I. Isayev

ABSTRACT 2
1.1 INTRODUCTION 2
1.2 LITERATURE REVIEW 2
1.2.1 Extrusion 2
1.2.2 Mixing 4
1.2.3 Melt Spinning and Film Blowing 5
1.2.4 Coating 7
1.2.5 Calendering 8
1.2.6 Compression Molding 9
1.2.7 Injection Molding 10
1.2.8 Blow Molding and Thermoforming 10
1.3 CONCLUDING REMARKS 11
REFERENCES 11

2. **Inverse Formulation and its Application to Die Flow Simulation** 19
A. Cohen

ABSTRACT 20
2.1 THE INVERSE FORMULATION AS AN ALTERNATIVE APPROACH 20
2.2 REVIEW OF APPLICATIONS OF THE INVERSE FORMULATION 20
2.2.1 Design and Performance of Diffusers 20
2.2.2 Nonlinear Elasticity 21
2.2.3 Second Order Fluid 21
2.2.4 Ground Water Modeling 21
2.3 THE CONCEPT OF ILL-POSED AND CORRECT PROBLEMS 21
2.4 RESULTS IN THE AREA OF HEAT CONDUCTION 22
2.4.1 Identification of Temperature Dependent Thermal Properties 23
2.4.2 Identification of Boundary Conditions 24
2.4.3 Initial or Boundary Conditions 25
2.5 THE LUBRICATION APPROXIMATION OF FLOW IN AN EXTRUSION DIE 26
2.6 THE INVERSE FORMULATION OF THE DIE FLOW MODEL 30
2.7 PRESSURE DROP OPTIMIZATION AS AN ADDITIONAL CONDITION OF THE INVERSE FORMULATION 31
2.8 CONCLUSIONS 34
ACKNOWLEDGEMENTS 34
REFERENCES 34

3. **Intermeshing Co-Rotating Twin-Screw Extruders: Technology, Mechanisms and Simulation of Flow Characteristics** 37
J.L. White and Y. Wang

ABSTRACT 38
3.1 INTRODUCTION 38
3.2 TECHNOLOGY 38
3.2.1 Early Developments 38
3.2.2 Colombo and LMP Development (1937-1959) 41
3.2.3 Erdmenger, Meskat, I.G. Farbenindustrie-Bayer Early Twin-Screw Development 42
3.2.4 Ellermann's Continuous Mixing Machine 48
3.2.5 Erdmenger-Bayer Four-Screw Machine 50
3.2.6 Readco-Baker Perkins Development 50
3.2.7 Throttling Devices 52
3.2.8 Recent Bayer Developments 52
3.2.9 Hermann Berstorff Developments 53
3.2.10 Recent Werner and Pfleiderer Developments 53
3.2.11 Newer Machine Manufacturers 53
3.3 APPLICATIONS 53
3.4 FLOW SIMULATION 54
3.4.1 General 54
3.4.2 Right-Handed Screw Elements 54
3.4.3 Left-Handed Screw Elements 63
3.4.4 Kneading Disc Element 64
3.4.5 Modular Machine 70
ACKNOWLEDGEMENT 72
NOMENCLATURE 72
REFERENCES 72

4. **Numerical Simulation of the Transport Processes in a Twin-Screw Polymer Extruder** 77
T.H. Kwon, Y. Jaluria, M.V. Karve and T. Sastrohartono

ABSTRACT 78
4.1 INTRODUCTION 78
4.2 LITERATURE REVIEW 78
4.3 THE PRESENT PROBLEM 80
4.4 ANALYSIS 82
4.4.1 Mathematical Model for the Flow and Heat Transfer in a Single-Screw Extruder 82
4.4.2 Mathematical Model for the Flow in the Intermeshing Zone of a Twin-Screw Extruder 87
4.5 NUMERICAL SOLUTION 88
4.6 NUMERICAL RESULTS AND DISCUSSION 91
4.6.1 Results on the Translation Zone 91
4.6.2 Choice of Parameter Values 101

	4.6.3 Results for Screw Moving Case	102
	4.6.4 Results for the Intermeshing Zone Based on the FEM Model	103
4.7	MODELING OF THE TRANSPORT IN A TWIN-SCREW EXTRUDER	111
4.8	CONCLUSIONS	111
	ACKNOWLEDGEMENTS	112
	NOMENCLATURE	112
	REFERENCES	114

5.	**Finite Element Simulation of Thermoforming and Blow Molding** ***H.G. deLorenzi and H.F. Nied***	117
	ABSTRACT	118
5.1	INTRODUCTION	118
5.2	FINITE ELEMENT FORMULATION	123
	5.2.1 Theoretical Background	123
	5.2.2 The Finite Element Concept	126
	5.2.3 The Equilibrium Equations	127
	5.2.4 The Solution Procedure	133
5.3	MATERIAL BEHAVIOR	133
	5.3.1 Background and Experimental Observations	133
	5.3.2 Nonlinear Elastic Constitutive Models	134
	5.3.3 Experimental Techniques for Obtaining Material Data	138
5.4	THERMOFORMING EXAMPLES	145
	5.4.1 Free Inflation of a Flat Membrane	146
	5.4.2 Comparison with Experimental Results for a Deep Vacuum Formed Cylinder	150
	5.4.3 Forming a Deep Cylinder With a Male Mold	154
	5.4.4 Thermoforming a 3-D Box	156
5.5	BLOW MOLDING EXAMPLES	157
	5.5.1 Blow Molded Jar	157
	5.5.2 Blow Molding of a Rectangular Box	158
5.6	DESIGN METHODOLOGY	159
	5.6.1 Design Philosophy	159
	5.6.2 Example of Iterative Design Procedure	160
5.7	SOME UNRESOLVED ISSUES	164
5.8	CONCLUSIONS	168
	ACKNOWLEDGEMENT	169
	REFERENCES	169

6.	**Simulation of Packing and Cooling Phases of Thermoplastics Injection Molding** ***D. Huilier and I. Patterson***	173
	ABSTRACT	174
6.1	INTRODUCTION	174
6.2	MODELING THE INJECTION MOLDING PROCESS	174
	6.2.1 Filling Modules in CAD/CAM/CAE Systems	175
	6.2.2 Packing-Cooling in CAD/CAM/CAE Systems	175

	6.2.3	Literature Review of Packing-Cooling Modeling	176
	6.2.4	Problem Formulation	177
	6.2.5	Definition of the Packing and Cooling Phases	177
6.3	A DIFFERENT MODELING APPROACH: PACK1 AND PACK2		178
	6.3.1	Pack1	180
	6.3.2	Computational Results: Pack1	183
	6.3.3	Pack2: A Modification of Pack1	186
	6.3.4	Computational Results: Pack2	190
	6.3.5	Summary	196
6.4	EXTENSIONS AND DEVELOPMENTS		197
	6.4.1	Physics and Thermodynamics	197
	6.4.2	Modeling and Industrial Practice	198
6.5	CONCLUSION		198
	ACKNOWLEDGEMENTS		199
	NOMENCLATURE		199
	REFERENCES		200
	APPENDIX		202

7.	**Simulation of Injection Molding of Rubber Compounds** *M. Sobhanie and A.I. Isayev*		205
	ABSTRACT		206
7.1	INTRODUCTION		206
7.2	THEORITICAL		209
	7.2.1	Governing Equations	209
	7.2.2	Viscous Heating	211
	7.2.3	Curing Kinetic Model	212
7.3	NUMERICAL SOLUTION METHOD		214
	7.3.1	Pressure Profile	214
	7.3.2	Temperature Profile	224
	7.3.3	Avdancement of the Melt Front	225
	7.3.4	Treatment of the Melt Front	226
	7.3.5	Solution Algorithm	227
	7.3.6	Comparison of the Techniques	227
7.4	NUMERICAL EXAMPLE		232
	7.4.1	Processing Conditions and Cavity Dimensions	232
	7.4.2	Physical Properties of Rubber	233
	7.4.3	Rheological Constants	233
	7.4.4	Curing Kinetic Constants	233
	7.4.5	Predictions for Filling Stage	235
	7.4.6	Predictions for Post-Filling Stage	239
7.5	CONCLUDING REMARKS		241
	ACKNOWLEDGEMENT		242
	NOMENCLATURE		242
	REFERENCES		243
	APPENDIX		246

8. **Viscoelastic Modeling of Injection Molding of a Strip and a Center Gated Disk Cavity** 247
N. Famili and A.I. Isayev

ABSTRACT 248
8.1 INTRODUCTION 248
8.2 THEORETICAL ANALYSIS 249
8.2.1 Governing Equations 249
8.2.2 Problem Setup 250
8.3 NUMERICAL SOLUTION 253
8.4 NUMERICAL RESULTS 256
8.5 CONCLUSIONS 274
NOMENCLATURE 274
REFERENCES 275

9. **An Interactive PC-Based Data Acquisition System for Injection Molding** 277
G.D. Kiss and R.C. Progelhof

ABSTRACT 278
9.1 INTRODUCTION 278
9.1.1 History 278
9.1.2 Field Signals 279
9.1.3 Computer Interface: A/D Converter 280
9.1.4 Computer Characteristics 281
9.2 DATA ACQUISITION SYSTEM 281
9.2.1 System Preface 281
9.2.2 System Goals 281
9.2.3 Hardware 282
9.2.4 Software 283
9.2.5 Examples of Application 291
9.3 SUMMARY AND CONCLUSIONS 299
REFERENCES 300

10. **Subject Index** 303

CONTRIBUTORS

A. Cohen, The Dow Chemical Company, Central Research, Materials Science & Development Laboratory, 1702 Building, Midland, Michigan 48674, U.S.A.

H.G. deLorenzi, General Electric Corporate Research and Development, Building K-1, Room 3 B19, P. O. Box 8, Schenectady, New York 12301, U.S.A.

N. Famili, Institute of Polymer Engineering, The University of Akron, Akron, Ohio 44325-0301, U.S.A.

D. Huilier, Ecole d'Application des Hauts Polymeres, Institut Charles Sadron Universite Louis Pasteur, 4 rue Boussingault, 6700 Strasbourg, FRANCE.

A.I. Isayev, Institute of Polymer Engineering, The University of Akron, Akron, Ohio 44325-0301, U.S.A.

Y. Jaluria, Department of Mechanical and Aerospace Engineering, Rutgers University, New Brunswick, New Jersey 08903, U.S.A.

M.V. Karwe, Department of Mechanical and Aerospace Engineering, Rutgers University, New Brunswick, New Jersey 08903, U.S.A.

G.D. Kiss, Bell Communications Research, 445 South St., Morristown, New Jersey 07962-1910, U.S.A.

T.H. Kwon, Department of Mechanical Engineering, Pohang Institute of Science and Technology, P.O. Box 125, Pohang, KOREA

H.F. Nied, General Electric Corporate Research and Development, Building K-1, Room 3 B19, P.O. Box 8, Schenectady, New York 12301, U.S.A.

W.I. Patterson, Department of Chemical Engineering, McGill University, Montreal, Quebec H3A 2A7, CANADA.

R.C. Progelhof, University of South Carolina, Mechanical Engineering, Columbia, South Carolina 29208, U.S.A.

T. Sastrohartono, Department of Mechanical and Aerospace Engineering, Rutgers University, New Brunswick, New Jersey 08903, U.S.A.

M. Sobhanie, Institute of Polymer Engineering, The University of Akron, Akron, Ohio 44325-0301, U.S.A.

Y. Wang, Institute of Polymer Engineering, The University of Akron, Akron, Ohio 44325-0301, U.S.A.

J.L. White, Institute of Polymer Engineering, The University of Akron, Akron, Ohio 44325-0301, U.S.A.

PREFACE

Computers are widely used in modern technology to analyze transport phenomena and to control various processes. Highly sophisticated computer-aided design (CAD) and computer-aided manufacturing (CAM) systems are commercially available. A large number of conferences covering the development of CAD and CAM systems are held every year in various parts of the world and the term "computer-aided engineering" (CAE) has become a buzzword. The number of publications dealing with research and development in this area continues to increase. The use of computers in polymer processing provides many benefits, including an increase in productivity which otherwise could not be achieved. The main aim of this volume is to cover some recent advances in various areas of polymer processing, with particular attention to modeling. Major emphasis is given to simulation of kinematics and dynamics of flow processes in various polymer processing operations. This volume will be helpful to those involved in research and development of polymer processing.

In general, the volume can be divided into five sections. The first section (Chapter 1) summarizes recent advances in modeling of various polymer processing operations. It mainly serves as a source of references and gives a brief overview of the subject. It also discusses the complex problems involved in modeling and outlines the major future thrust.

The second section, which includes Chapters 2 to 4, describes some recent efforts in modeling of extrusion processes. In particular, Chapter 2 reviews the application of the inverse formulation for solution of a broad range of engineering problems. Extension of the inverse formulation to simulation of flow through a extrusion die is the main subject of this chapter. The advantages and difficulties of its application to the extrusion die design are considered. Chapters 3 and 4 are related to modeling of flow in twin screw extruders. Chapter 3 gives a critical review of the development of technology and modeling of flow in intermeshing co-rotating twin screw extruders. The development of twin-screw technology starting from the beginning of the century is described with extensive references given to patent literature. Simulation of isothermal flow of Newtonian and non-Newtonian fluids through screw and kneading disc elements is analyzed by both the finite difference method and the flow analysis network technique. Chapter 4 includes the results of the numerical simulation of the non-isothermal non-Newtonian flow in co-rotating and counter-rotating twin screw extruders. The flow regions are divided into the translation region where flow is similar to that in a single-screw channel, and the intermeshing region where mixing takes place. The finite difference method and the finite element method are employed for modeling of flow and temperature fields in the translation and mixing region, respectively.

The third section, which comprises Chapter 5, is concerned with the finite element simulation of thermoforming and blow molding processes. Use of finite element analysis in these processes is relatively new and beneficial for their optimal design. The chapter combines a comprehensive review of the state-of-the-art together with theoretical formulation of the problem along with a description of material behavior relevant to thermoforming and blow molding. A number of highly illustrative examples of application of the developed approaches to the process simulation is given. Design methodology and procedures are outlined. Limitations of the modeling efforts and some suggestions for future work are presented. Hopefully, this will lead to significant further developments in both the thermoforming and the blow molding industries.

The fourth section, which consists of Chapters 6 to 8, deals with simulation of the injection molding of thermoplastics and rubber compounds. Chapter 6 describes approaches available for the treatment of packing and cooling stages of injection molding of thermoplastics. In particular, modeling of the non-isothermal compressible flow in the packing stage is based upon the generalized Newtonian fluid model and Tait's equation of state. The effect of varying the physical properties and processing conditions on kinematics and dynamics of flow during the packing stage and on shrinkage in the moldings is elucidated for the cases of simple cavities. Chapter 7 deals with injection molding of rubber compounds. A general computer code for the non-isothermal vulcanization and filling of a thin cavity having an arbitrary three-dimensional geometry is described. In particular, this program employs a control-volume finite-element method for solution of the continuity and momentum equations and a finite-difference method for the solution of the energy equation. The simulation can accommodate any type of generalized Newtonian fluid model and vulcanization kinetics. A detailed example of the application of this program is presented for molding a one-quarter disk mold with a runner system and two gates. Pressure, velocity, temperature, induction time, melt front, weld-line formation and the evolution of state of cure in the mold are evaluated. Some comparisons of the finite element techniques based on the Galerkin and control-volume methods are also made. Chapter 8 deals with non-isothermal viscoelastic simulation of injection molding of both a rectangular strip and a center-gated disk. Both filling and cooling stages are considered. Time and spatial development of velocity, shear rate, shear and normal stresses, birefringence and thickness of the thermal boundary layer are determined. Results of simulation are compared with some data available for frozen-in birefringence in molded products.

The fifth section, which consists of Chapter 9, describes a personal computer-based data acquisition system developed for the injection molding process. This chapter extensively discusses types of field signals encountered in applications and describes computer interfaces used to convert the signals into a form suitable for input to the personal computer. A detailed description of a data acquisition system developed to monitor the critical processing parameters of a closed loop controlled injection molding machine for high precision parts is given. Several examples of the use of this data acquisition system are presented.

Finally, it is noted that the mathematical models with corresponding simulation codes, discussed in the present volume, for various polymer processing operations are not unique due to the variety of the processes under consideration and due to the underlying assumptions made in each particular case. The latter brings some natural restrictions for their engineering applications. Users should be aware of that fact and carefully analyze their specific problems. In view of the absence of a general model for any polymer processing operation, the modeling efforts presented in this volume are a further small step toward the solution of the global problem, namely, the development of science-based polymer technology.

A.I. Isayev
Akron, 1990

CHAPTER 1

A BRIEF OVERVIEW OF MODELING IN POLYMER PROCESSING

by A.I. Isayev

Institute of Polymer Engineering
The University of Akron
Akron, Ohio 44325
U.S.A.

ABSTRACT
1.1 INTRODUCTION
1.2 LITERATURE REVIEW
1.2.1 Extrusion
1.2.2 Mixing
1.2.3 Melt Spinning and Film Blowing
1.2.4 Coating
1.2.5 Calendering
1.2.6 Compression Molding
1.2.7 Injection Molding
1.2.8 Blow Molding and Thermoforming
1.3 CONCLUDING REMARKS
REFERENCES

ABSTRACT

The chapter provides an overview of modeling of such polymer processing operations as: mixing, extrusion, melt spinning, film blowing, coating, calendering, compression molding, injection molding, blow molding, and thermoforming. The main reason for writing this chapter was to provide the reader with a complete source of references on the subject. Furthermore, the chapter explains the principal physical phenomena occuring in the process, provides the basic principles for modeling, including formulation of the equations of continuity, motion and energy, as well as the rheological constitutive equations and the equations of state. The numerical methods to solve these equations are also briefly discussed.

1.1 INTRODUCTION

In recent years great efforts have been made in modeling of various polymer processing operations. There are several driving forces to model the processes. First, modeling helps in better understanding the effects of processing conditions, raw material properties and equipment size on process behavior. Secondly, the developed model assists in determining an optimal processing operation design without the necessity of going through many trial and error attempts. Furthermore, the model may eventually lead to the development of quantitative relationships between processing conditions, structure and properties of the final product. Evidently, polymer processing operations are quite complicated, since many physical phenomena occur during processing. However, even simple modeling efforts with some seemingly crude underlying assumptions are quite useful in developing a hierarchy of models to gradually build up a scientific basis for modern polymer technology. Schenkel [1] provided a general overview of trends and highlights in polymer processing over the last 50 years.

1.2 LITERATURE REVIEW

1.2.1 Extrusion

Extrusion is the most common method of processing of polymers, which has been extensively studied. Many comprehensive books [2-25] have been written on extrusion. A recent book by Rauwendaal [25] gives a detailed description of extrusion machinery design, including single and twin screw extruders and some consideration of die design. The subject of modeling extrusion processes is considered by Fenner [26] and by Pearson [27]. A comprehensive review of the extrusion processes including some aspects of modeling was recently published by Stevenson [28].

Simulation of a single screw extruder usually includes a solid bed conveying region, a melting region, and a metering region. Available simulation attempts of single screw extruders are based upon flow of an inelastic fluid disregarding the time-dependent behavior of polymeriĉ materials. The majority of earlier models for the metering zone were based on isothermal fully developed flow and included pressure and drag flow [see, for example, 25,28-30]. Non-isothermal simulations gradually evolved by inclusion of heat conduction in the direction of screw channel thickness with viscous heat generation [31], then heat conduction in the channel width direction [32] and heat convection in the down channel direction [33]. Numerical techniques utilized were the finite difference method and the finite element method. The former method has been applied since the initial stage of development of the non-isothermal models [12,26,34,35], while the latter method is being used in the more recent simulations [26,36-40]. Even three-dimensional finite element simulations of screw extrusion are in an advanced stage of development [41]. Recently, modeling of two-dimensional isothermal extrusion of a Newtonian fluid with slip boundary conditions was also performed [42]. The flow analysis network method has been used for modeling of flow in pin-barrel screw extruders [43].

Melt pumping analysis of extrusion is suitable for both rubber compounds and thermoplastic melts. However, in thermoplastic melts two other stages before pumping are important. They are the solids conveying region and the plasticating or melting region. Analysis of the solids conveying region was originally proposed by Darnell and Mol [44], based on an elastic plug frictional motion under isotropic pressure along screw channel and barrel. Broyer and Tadmor have included nonisotropic stress distribution [45], and non-isothermal effects in solids conveying [46]. Effects of gravity, centrifugal forces and two-dimensional character of plug flow have been considered by Lovegrove and Williams [47,48]. They have also made a preliminary study of the stresses and deformations in elastic plugs [49]. Fenner [50], Tadmor and Gogos [29], Rauwendaal [25] and Stevenson [28] presented extensive discussions on this subject.

The melting zone in an extruder is located between the solids conveying and metering zones. Analysis of polymer behavior in the melting zone was based on screw push-out experiments as described by Maddock [51] and by Street [52] about three decades ago. Tadmor and his coworkers [6,53-55] continued these experiments and proposed a model now known as the "Tadmor melting model." The analysis included momentum, energy and mass balances on the melt film at the barrel surface, energy and mass balance on the solid bed and energy balance at the melt-solid bed interface. Various modifications have been added to the Tadmor model. The subject was recently reviewed by Fenner [26], Rauwendaal [25] and Stevenson [28]. Two modifications are worth mentioning namely, the five-zone model [27] by Shapiro, Halmos, Pearson and Trottnow [56-58] and the three-layer model by Lindt and Elbirli [59-62] and by Dekker [63]. The five-zone model (in addition to the solid bed and the barrel film) includes the melt pool, screw shank film and screw flight film. The three-layer model includes the barrel film, screw film, and the solid bed between them. A new approach to model melting in single screw extruder was proposed by Viriyayuthakorn and Kassahun [41]. Their approach, based on a finite element technique, eliminates the need to assume a melting model. However, the calculations require substantial computational efforts using a Cray computer.

A die system is an important part of any extrusion operation. It generally consists of the die preform, distributing and streamlining polymer melt, and the die lip producing a product of the required geometry. Several books [3,25,50,64,65] have been published related to all aspects of extrusion die design, including interaction between material properties and die design, modeling of flow, thermal control of extrusion dies and manufacturing and computer-aided design. Our main goal here is to give an overview of the literature on modeling of flow in extrusion dies. In particular, one method of die design using the inverse formulation is considered in Chapter 2 of this book.

Die flow analysis is usually carried out with the objective either to calculate the pressure, stress, velocity and temperature field as well as dimensions of the extruded product based on die geometry or to establish a die geometry that will ascertain the required extrudate shape. In most modeling efforts it was customary to employ the lubrication approximation. Numerical methods employed in simulation of flow in extrusion dies are finite differences [66] or finite elements [67]. Due to more than sixty years of refinement the finite difference method has now reached a high degree of sophistication [66]. However, even with this refinement, the method suffers from inherent difficulties when used in modeling of flow in dies of complex geometry. These arise while specifying the finite difference operators for Neumann boundary conditions on complex boundaries and in employing nonuniform and nonrectangular meshes in the computational domain. Thus, in recent years there has been an increasing trend to employ finite element methods in flow modeling in extrusion dies, based on the primitive variables formulation [68,69] and penalty function formulation [70,71].

Stevenson [28] extensively reviewed the literature related to extrusion die geometries and their flow analysis. However, several papers which appeared just recently are worth mentioning. In particular, the problem of scale up of extrusion dies has been considered [72,73]. Models for process control of profile extrusion based on the heat penetration through the die wall [74-76] and single-roller die extrusion [77] have been proposed. Further complication of flow analysis of extrusion dies due to the presence of yield stress in filled thermoplastics and rubbers was also analyzed [78,79]. Significant efforts have been made to incorporate the effect of viscoelasticity into isothermal and non-isothermal modeling of flow in slit dies [80,81] and spiral mandrel dies [82]. Recent achievements in modeling of planar and axisymmetric contraction and expansion flow using viscoelastic constitutive equations have also been reviewed [83-86].

Extrudate swell is an important characteristic of polymers, since it determines the geometrical dimensions of the final extruded product; the most recent finite element analysis of the extrudate swell has been reported [84,87,88]. One of these papers [87] also included results of calculation of the slip effect. Furthermore, some modeling efforts have been directed toward inclusion of reactions on flow of thermosetting resins through a coat hanger die [89] and flow induced fiber orientation in an expanding channel tubing die [90].

Analysis of coextrusion flow is an important part of the extrusion operation. Most recent efforts in modeling of coextrusion flow have been reported [91-96]. In particular, they consider flow of two polymers between parallel plates [91-93], in a tube [93] and a die of arbitrary cross section [94]. Further, a multilayer extrusion has also been considered in slit and annular dies [95] and multimanifold vane flat die and feed block geometries [96].

1.2.2 Mixing

Mixing is an important step in polymer processing technology since most polymers are used in the form of compounds. The compounds may include blends of two or more homopolymers, and polymers filled with fibrous or particulate solids and various other ingredients. Performance characteristics of the products made from the various compounds are highly dependent upon the quality of mixing. Thus, there have been significant efforts noted in the literature to model mixing processes. In particular, Chapters 3 and 4 describe some specifics of recent developments in the area of flow modeling in twin screw extruders. A general review on the subject of mixing was recently published [97]. A fundamental discussion of compounding machinery and the mechanism of mixing are also described there. Basic steps involved in mixing are convective redistribution of materials, shear-induced extension of material surfaces, stress-induced dispersion and diffusion of mass and heat [27]. In real mixing processes all these steps may occur simultaneously. Accordingly, modeling of mixing requires a formidable level of sophistication. However, modeling can be simplified by considering one or two steps at a time which are considered dominant in a particular mixing process. Visualization of a mixing process can be done by incorporating colored ingredients into materials to be mixed and simulating the process, and by using numerical techniques to solve transport equations governing the mixing. Due to the complexity of the problems considered in mixing, both routes are important.

A dispersive mixing model was recently described for the case where separation of the fragments following rupture of agglomerates determines the dynamics of the process [98]. The relative motion of fragments including the effective van der Waals and hydrodynamic forces were included. The effect of simple shear flow, pure elongational flow, uniaxial extensional flow and biaxial extensional flow on the dynamics of agglomerate size distribution was elucidated. It was found that pure elongational flow was the most efficient in particle separation. Process analysis of a laboratory internal mixer has been performed and a model for the variation of the internal wall temperature has been proposed [99]. Flow dynamics in internal mixer has been studied in papers [100,101] using lubrication approximation in conjunction with an isothermal power-law model. The finite element method was employed to determine the flow field in the intermeshing region of a counterrotating twin-screw extruder [102]. A vortex formation near the converging boundary was detected which reduced in size due to shear thinning. It was noted, that the predicted pressure drop based on finite element calculations and that base on lubrication approximation were comparable. Modeling of the intermeshing corotating twin-screw extruder has also been performed based on the non-isothermal non-Newtonian fluid model [103]. The developed model allows calculation of the energy, specific energy and melt temperature rise during processing as a function of the melt viscosity as well as the screw speed and screw geometry (including location and number of transport elements, kneading disks, pitch, screw clearance and flight width). The model has been verified in several experiments.

1.2.3 Melt Spinning and Film Blowing

Melt spinning and film blowing are the most widely used processing operations for manufacturing fibers and films, respectively. Both processes are continuous and non-isothermal. There have been numerous studies on modeling of these processes. In particular, books by Ziabicki [104], Petrie [105], Ziabicki and Kawai [106] and Pearson [27] consider important aspects of modeling of fiber spinning with pertinent previous literature. In addition, reviews by Denn [107,108] and White [109] describe various approaches to modeling.

In a simple modeling of melt spinning, the process can be treated as one-dimensional extensional flow with all variables considered to be constant through cross section of the jet and dependent only along the jet. The process is usually assumed to be at steady state. Thus, most of the works on modeling seek a steady-state solution of the melt spinning process based on elongational flow of a Newtonian or power-law fluid. However, it is noted that the transient processes are also important due to the fact that there is a distribution of properties through the cross section of fibers especially near the surface. This was recognized long ago in view of observed distribution of the frozen-in birefringence in fibers [110,111]. The effect can only be understood in terms of non-isothermal fiber spinning of viscoelastic fluid. Other transient characteristics of the fiber spinning process are also important. These include breakage of filaments and draw resonance.

In recent years, more attention has been paid to numerical simulation of the melt spinning of viscoelastic fluids. Most of the literature describes modeling efforts based on differential viscoelastic constitutive equations. Denn and co-workers [112], Fisher and Denn [113], and Sridhar and Gupta [114] studied dynamics of isothermal steady state spinning with a Maxwell model. In particular, Sridhar and Gupta [114] have observed that at low strain rates velocity profiles along the spinline can be predicted by the Newtonian model, and at higher rates by the Jeffrey and Maxwell models. However, at strain rates higher than 100 s^{-1} none of the simple models have been found to be adequate. The fractional increase in tensile stress along the fiber has been shown to be a unique function of the total strain. Next, Gupta and co-workers [115,116] considered steady state spinning of the Oldroyd fluid B. The predictions of the Oldroyd fluid have been found to be quite different from those of the upper convected Maxwell model. In particular, the fluid behaved in a Newtonian manner for both very low and very high Deborah numbers. The results have been compared with experiments for the Boger fluid and excellent agreement has been obtained for velocity profiles and spinline stresses at various draw ratios and spinline lengths. Zhiganov et al. [117-120] developed a mathematical model for unsteady non-isothermal spinning of Maxwell fluid. They were able to describe the draw resonance of jet drawing by taking into account the increase in the shear modulus of the jet with increasing strain rate and decreasing temperature. Upadhyay and Isayev [121] modeled non-isothermal flow using the Leonov differential constitutive equation. They were able to predict the experimental results of Bogue and co-workers for the stress growth at constant elongational strain rate, constant pulling speed and stress relaxation after elongation under various thermal histories including constant cooling rate and a sudden jump or drop in temperature.

Several simulation attempts have been made to model fiber spinning using integral constitutive equations. In particular, Malkus [122] has done the finite element simulation of isothermal fiber spinning using a Maxwell integral model. The author was able to carry out computations up to a Deborah number of about one. At higher Deborah numbers the numerical scheme failed to converge. Larson [123] studied the elongational flow using the Doi-Edwards model. He obtained solutions at low draw ratio and for short spinlines. Papanastasiou et al. [124,125] presented the finite element simulation of the fiber spinning using a BKZ-type integral constitutive equation and the Curtiss-Bird model. Solutions were obtained at high values of elasticity, draw ratio, and drawing force. The predictions have been found to agree with experimental data from the literature on melt spinning of polystyrene and polypropylene.

Recently, Zieminski and Spruiell [126] developed a mathematical model to describe the high-speed melt-spinning behavior of crystallizable polymers. This model includes the effects of acceleration, gravity, and air friction on the kinematics of the process. In addition, the effects of temperature and molecular orientation on the crystallization kinetics, and those of molecular weight and crystallinity on the elongational viscosity have been incorporated into the model. The model compared well with the experimental data on online diameter, birefringence and temperature profiles at take up speeds ranging from 2.8 to 6.6 km/min.

Usually, the process of fiber spinning is carried out on bundles of jets coming from a spinneret. In this case, the cooling conditions for each filament can be different, due to the assymmetric location of the quenching air source. A variation of quenching conditions would lead to changes in molecular orientation. Hence, the properties of spun fibers may vary from row to row. This effect has been simulated by Dutta [127] for steady state non-isothermal multifilament spinning of a Newtonian fluid with temperature dependent viscosity.

Tubular film blowing process involves extrusion of melt through a tubular die with subsequent simultaneous stretching and inflation of the tube until solidification. The process is highly non-isothermal and unsteady. In a majority of earlier modeling attempts, the process was considered to be at steady state and isothermal. The kinematics of tubular film blowing was first analyzed by Alfrey [128], Pearson [129] and Pearson and Petrie [130-132]. Books by Pearşon [27], Han [133], Middleman [134] and Tadmor and Gogos [29] describe details of the process modeling. Recent review papers by Petrie [135] and by White and Cakmak [136] give the latest account of available literature on kinematics and dynamics of the process as well as on heat transfer and structure development. Modeling efforts by Kanai and White [137-139] and by Yamane and White [140] give local kinematics and heat transfer rates as well as bubble stability and crystallization effects. In particular, the approach of Kanai and White [137-139] was based upon a non-isothermal Newtonian model, while the formulation of Yamane and White [140] employs a power-law model under non-isothermal conditions. They looked at the sensitivity of bubble shape to the power-law index and activation energy of viscous flow. Their modeling efforts are similar to that of Han and Park [141,142] but include experimental heat transfer coefficient and crystallization. Based on their modeling results, Yamane and White [140] have concluded that for materials having high activation energy, the temperature dependence of rheological properties dominates in formation of bubble shapes at fixed drawdown ratios, blowup ratios, and frostline heights.

There have also been some efforts, notably by Wagner [143] and by Luo and Tanner [144], in modeling of the effect of viscoelasticity on bubble shape under non-isothermal conditions. Furthermore, a model to describe elastic strain in film blowing process has been formulated by Lohse and Marinow [145,146]. The model considered melt flow in the film blow head, and the deformation in the tube formation zone including elastic strain. Using this model it was possible to determine the latent elastic strain in the longitudinal and cross directions of high density polyethylene films. More recently, Campbell and Cao [147-148] extended simulation of blown film extrusion beyond the freeze line to include the actual frost line. Their two-phase model incorporates viscoplasticity and crystallinity of the polymer melt. The prediction of bubble shape, temperature and velocity from the die to the frost line agreed well with experimental data. Thus, a possibility now exists that further improvement of the constitutive equation used in the simulation would lead to predictions of properties of polymeric films from molecular and processing parameters.

1.2.4 Coating

Coating is a process of application of a polymeric fluid to the surface of sheets or wires. Coating is usually applied for mechanical strength and environmental protection. The quality of coating is determined by flow and heat transfer phenomena. Analysis of the process has been presented in books by Tadmor and Gogos [29], Pearson [27], Fenner [12] and Middleman [134]. Recently, Mitsoulis [149] reviewed fluid flow and heat transfer in wire coating. The earliest modeling efforts were based on the lubrication approximation of a Newtonian and power-law fluid under isothermal conditions [150-152]. The analysis of flow in an annulus with the inner cylinder moving in the axial direction has been considered. Later, Fenner and Williams [153] carried out an analysis for a tapering die. Kasajima and Ito [154] have analyzed the drag flow of the polymer coating and heat transfer during cooling of the coating. Winter [155] has treated heat transfer problems inside and outside of a coating die. Analytical treatment of the process has been given by Han and Rao [156] including experimental and theoretical studies of the coextrusion wire-coating process. Analysis of the isothermal and non-isothermal wire-coating coextrusion has been considered by LeNir [157] and Basu [158], respectively. In addition, Han and Rao [159] conducted a theoretical and experimental study of isothermal wire-coating coextrusion based on a power-law fluid.

Modeling of isothermal flow in the wire coating dies without use of the lubrication approximation has been done by Caswell and Tanner [160] by means of the finite element method. Non-isothermal analysis has been carried out by Carley et al. [161] indicating the effect of viscous dissipation on temperature rise inside wire-coating dies. Recently, Chung [162] reported results of modeling including the effect of melt compressibility. Based on the Criminale-Ericksen-Filbey constitutive equation, Tadmor and Bird [163] have considered the effect of viscoelasticity on the eccentricity of the wire. Furthermore, Mitsoulis et al. [164-166]

analyzed the effect of viscoelasticity, slip and thermal effects in high-speed wire coating using the finite element method. They also used the finite element method to simulate wire-coating coextrusion [167,168] especially to locate the interface between two layers.

The spin coating deposition process has found wide use in polymer processing of thin films for microelectronic applications. A comprehensive theoretical study of the fundamental physical mechanisms of polymer film formation onto substrates has been described by Zenekhe [169-171] and by Lawrence [172]. These mechanisms include centrifugal and viscous forces, solute diffusion and solvent evaporation. A model was proposed that incorporates some of these mechanisms to predict the film thickness [169-173].

1.2.5 Calendering

Calendering is a continuous process for manufacturing films and sheets by pressing a thermoplastic melt or elastomer between a pair of counter-rotating heated rolls. The earlier attempts of modeling of the process initiated by Gaskell several decades ago [174] were based on the lubrication approximation and are described in the textbooks of McKelvey [151], Middleman [134], Tadmor and Gogos [29], Torner [175] and Tanner [176]. These modeling attempts analyze the symmetrical calendering of Newtonian or power-law fluid between rolls of equal diameters rotating at the same speed. Kamal et al. [177] recently reviewed this development in calendering in their paper on the film embossing process. Takserman-Krozer et al. [178] and Daud [179] investigated the unsymmetric case of calendering of Newtonian and power-law fluids, respectively. In their case, rolls of unequal diameter were rotating at different speeds.

Calendering analysis without the lubrication approximation has been reported by Agassant and Espy [180], Seeger et al. [181], Dimitrijew and Sporjagin [182] and Mitsoulis et al. [183]. These two-dimensional analyses predict a formation of vortex patterns in the melt bank far away from the nip region as observed in various experimental studies [184-188].

Several attempts have been made to extend theory based on the lubrication approximation to calendering of viscoelastic fluid [189-197]. However, this was done by neglecting the normal stress terms in the momentum equation. Thus, at present, the contribution of viscoelasticity in calendering is not clear. The full equations of motion have to be solved in order to clarify this point.

The finite element method is most suitable for solution of the calendering problem. Mitsoulis and coworkers [183,195] used this technique extensively in simulation of the calendering process. The results obtained included predictions of the shape and location of the free surface, vortex patterns, temperature and pressure distributions, roll separating force, torque and power consumption. Some comparisons of these predictions with experimental data have also been made. Recently, Zheng and Tanner [197] investigated calendering of a viscoelastic fluid under isothermal conditions using both analytical and numerical methods. In particular, they considered a perturbation analysis of calendering of Phan-Thien-Tanner (PTT) viscoelastic fluid using the lubrication approximation and evaluated normal stresses by assuming a known velocity field. In addition, the two-dimensional flow problem in calendering using the boundary integral method without the lubrication approximation for the Newtonian, power-law and PTT models was simulated. This investigation for the case of calendering of the viscoelastic fluid was limited to a single relaxation process. Nevertheless, even with one relaxation mode formulation the differences between the inelastic and viscoelastic simulations of calendering were detected.

1.2.6 Compression Molding

Compression molding is one of the oldest and most widely used processes for manufacturing rubber, thermoset and plastic products. Recently, interest in compression molding has increased due to use of the process for making continuous composites. The subject of simulation of compression molding has been discussed very briefly in the textbooks [29,198]. Recent reviews on the subject were given by Tucker [199,200] and by Melby and Castro [201].

There are several recent references on modeling of molding of continuous composites worth mentioning. In particular, Gutowski et al. [202-204] have described a general model for three-dimensional flow and one-dimensional consolidation of the polymer composites. The fibers were considered to make up a deformable, nonlinear elastic network through which the resin flows according to Darci's law. The model has been applied to compression molding and bleeder ply molding. The finite element method for the simulation of isothermal mold filling in compression molding of random and unidirectional long fiber reinforced composites has been described by Hojo et al. [205]. The model has been found to accurately predict the mold-filling pattern observed in experiments. In particular, the flow front of unidirectional composites was found to expand in the direction perpendicular to the fiber axis, leading to a high anisotropy.

A two-dimensional finite element simulation of flow and heat transfer in compression mold filling of reactive polymers and fiber-reinforced composites has been developed by Tucker and coworkers [206,207]. It was detected that a thin layer of warm low-viscosity fluid formed at the mold surface acts as a lubricant. The latter leads to mostly planar extensional deformation in the rest of the charge during non-isothermal compression molding. This approach is consistent with a model for the flow of a chopped fiber reinforced polymer compound in compression molding proposed by Barone and Caulk [208]. In this paper, compression molding was modeled as a two-dimensional membrane-like sheet which extends uniformly through the cavity thickness with slip at the mold surface. In a recent paper, Barone and Osswald [209] used a boundary element method for analyzing the flow of sheet molding compounds during compression molding. In contrast to earlier modeling based on lubrication models, this approach showed that the flow front propagation depends on the charge thickness. In particular, the flow front progression for elliptical, rectangular and L-shaped charges of various thicknesses was predicted.

The importance of viscoelastic effects in compression molding was considered by Isayev and Azari [210], who experimentally and theoretically studied shear-free squeezing flow of elastomers. It was found that under constant mold closing velocity the simulation based on an inelastic fluid showed higher closing force than those based on a viscoelastic fluid, especially at the initial stage of the process. This indicates that the transient response in compression molding is important. In contrast, Shenoy and Saini [211] who studied the compression molding of ultrahigh-molecular weight polyethylene stated that the effect of elastic forces can be neglected. The latter led to the conclusion that the entire mechanics of compression molding are dependent on geometric parameters and a melt flow index.

A simplified analysis of the cure of sheet molding compounds in molds of various geometries was also proposed by Hutchinson and Nixon [212,213].

Recently, there have been substantial efforts to develop a mold heating analysis program for compression molding. The most notable contribution in this area was made by Upadhyay [214]. His theoretical analysis has been based on the boundary element method to calculate the temperature field in the mold in order to optimize the location and power of the heaters in the mold. The experimental validation of the developed analysis was carried out on a specially designed mold under various conditions. The experimentally determined heat transfer coefficients for a given mold were employed in the calculations.

1.2.7 Injection Molding

Injection molding is the most widely employed polymer processing operation. Several books give a detailed description of the process [2,27,29,133,134,175,200]. During the last decade, significant progress has been made in the development of science-based technology of the injection molding of thermoplastics, elastomers, thermosets and other reactive systems. These efforts have been extensively reviewed in various books [215-218]. Development of the science-based technology became possible due to modeling and experimental efforts of various groups, notably by Austin, Menges, Kamal, Wang and their coworkers. Chapters 5-7 describe some recent efforts in modeling of injection molding of thermoplastics and elastomers, where relevant references are given. To that end, one should also add several other investigations related to inelastic modeling of thermoplastics made by Richardson [219-225], Agassant and coworkers [226-228], and a review by Bowers [229]. Furthermore, Blanc et al. [230] also reported some results of a study of injection molding of reinforced thermosets, including the observation of the fiber orientation and simulation of the process. Some recent efforts on injection molding of elastomers have been reported by Bowers [231] and Isayev [232]. The present status of research in reaction injection molding has been reviewed by Macosko [218] and Lee [233]. Application of computer-aided engineering to injection molding was recently published by Menges and Recker [234]. Injection molding of continuous fiber-reinforced polymers using the preplaced mats in the mold is still under development. A model for mold filling simulations of such composites has been described by Chan and Hwang [235] where the non-isothermal non-Newtonian flow through porous fabric mat has been considered.

There have also been some efforts in modeling of heat transfer on mold surface between the polymer and the mold and in the mold itself [236-238] in order to detect the nonuniformity of the mold temperature. In particular, Kwon [237,238] has developed a computer code for design of a mold cooling system based on the boundary integral method.

Despite the rapid progress in understanding the injection molding process, further efforts in simulation of injection molding are needed. In particular, the treatment of the fountain flow region during cavity filling, packing stage and viscoelastic modeling of injection molding including shrinkage, orientation and stress development in the molded products are still in an embryonic stage.

1.2.8 Blow Molding and Thermoforming

Blow molding and thermoforming are processes in which a preformed tube (parison) or sheet of softened polymer is inflated or pushed against a cold mold, where it solidifies to the final shape. These processes are described in several books [27,29,239]. State-of-the-art modeling efforts have been described in Chapter 8. Thus, further discussion of this subject is not required in the present chapter. However, several recent studies (not included in Chapter 8) should be mentioned. These are modeling efforts by Menges and Weinand [240], Potente et al. [241], Allard et al. [242] and Cohen et al. [243]. These investigators have considered the stress and thickness evolution during thermoforming and presented some comparisons with experiments. In addition, an extensive review on analysis of thermoforming crystallizing polyethylene terephthalate was published by Throne [244] together with modeling of non-isothermal crystallization kinetics relevant to thermoforming. A review on blow molding was recently prepared by Belcher [245].

1.3 CONCLUDING REMARKS

The present review of the literature on modeling of polymer processing indicates that during the last decade development of science-based polymer processing technology using computers has made major advances. These include development of models of the various processes and their application to design processes in the industrial environment. The main goal of modeling of the process is to learn how, by knowing the material parameters and product requirements, one can choose optimal processing conditions and processing machinery to get the acceptable product. In many cases a solution of this problem is only partially achieved. This is due to the fact most of the simulations consider an individual stage of the process or one particular phenomenon, while in a real process there is interaction between the various stages which influence the behavior of the process as a whole. The latter leads to considerable restrictions on the predictive capability of the simulation. The other major deficiency of the many present modeling efforts is that only a few of the developed models have earnestly been tested in experiments. Accordingly, their real utility is not immediately understood. Thus, it is desirable that any future modeling efforts go along with experimental verification. Otherwise, the real possibility exists for erroneous predictions of the process behavior.

Several problem areas are also evident from this review. The first problem is related to a lack of a material characterization data bank. In particular, any successful modeling requires various data on material characterization including rheological, thermal, kinetic and physical properties in a wide range of strain rates, temperatures, pressures and cooling conditions. The second problem is the absence of a reliable unified constitutive rheological model which can be used in all physical states of polymers including the solid, rubbery and liquid state, through which the polymer passes during processing. The third problem is the fact that the modeling efforts of various research groups lead to development of computer programs using different approaches. In particular, numerous commercial computer packages have been developed. In many cases, these programs are proprietary and their content is rarely revealed to users. Thus, comparisons are not made between various programs intended to be used for the same purpose. Usually, the computer code users are polymer processing practioners who have an access to one specific code. They check the predictive capability of the particular code used by their company, but these efforts usually go unreported. Thus, resolving these major issues is a challenging problem facing researchers involved in the modeling of various polymer processing operations.

REFERENCES

1. Schenkel, G., *Intl. Polym. Process.*, *3*, 3 (1988).
2. Bernhardt, E.C., Ed., "Processing of Thermoplastic Materials," Van Nostrand Reinhold, New York (1959).
3. Schenkel, G., "Plastics Extrusion Technology and Theory," Illiffe Books Ltd., London (1966).
4. Schenkel, G., "Kunststoff Extruder-Technik," Carl Hanser Verlag, Munich (1963).
5. Schenkel, G., "Schneckenpressen für Kunststoffe," Carl Hanser Verlag, Munich (1959).
6. Tadmor, Z., Klein, I., "Engineering Principles of Plasticating Extrusion," Van Nostrand Reinhold, New York (1970).
7. Simonds, H.R., Weith, A.J., Schack, W., "Extrusion of Rubber Plastics and Metals," Van Nostrand Reinhold, New York (1952).

8. Fisher, E.G., "Extrusion of Plastics," Illiffe Books Ltd., London (1954).
9. Jacobi, H.R., "Grundlagen der Extrudertechnik," Rudolf Zechner Verlag, Speyer am Rhein (1963).
10. Mink, W., "Grundzuege der Extrudertechnik," Rudolf Zechner Verlag, Speyer am Rhein (1963).
11. Griff, A.L., Ed., "Plastics Extrusion Technology," Van Nostrand Reinhold, New York (1968).
12. Fenner, R.T., "Extruder Screw Design," Illiffe Books Ltd., London (1970).
13. Bikales, N.M., Ed., "Extrusion and Other Plastics Operations," John Wiley & Sons, New York (1971).
14. Richardson, P.N., "Introduction to Extrusion," Society of Plastics Engineers, Inc. (1974).
15. Janssen, L.P.B.M., "Twin Screw Extrusion," Elsevier, Amsterdam (1978).
16. Brydson, J.A., Peacock, D.G., "Principles of Plastics Extrusion," Applied Science Publishers Ltd., London (1973).
17. Martelli, F.G., "Twin Screw Extrusion, A Basic Understanding," Van Nostrand Reinhold, New York, (1983).
18. "Kunststoff-Verarbeitung im Gespraech, 2 Extrusion," BASF, Ludwigshafen (1971).
19. "Der Extruder als Plastifiziereinheit," VDI-Verlag, Duesseldorf (1977).
20. Levy, S., "Plastics Extrusion Technology Handbook," Industrial Press Inc., New York (1981).
21. Potente, H., "Auslegen von Schneckenmaschinen-Baureihen, Modellgesetze und ihre Anwendung," Carl Hanser Verlag, Munich (1981).
22. Herrmann, H., "Schneckenmaschinen in der Verfahrenstechnik," Springer-Verlag, Berlin (1972).
23. Dalhoff, W., "Systematische Extruder-Konstruktion," Krausskopf-Verlag, Mainz (1974).
24. Harms, E., "Kautschuk Extruder, Aufbau und Eisatz aus Verfahrenstechnischer Sicht," Krausskopf-Verlag, Mainz, Bd.2, Buchreihe Kunststoffetechnik (1974).
25. Rauwendaal, C., "Polymer Extrusion," Hanser Publishers, Munich (1986).
26. Fenner, R.T., in "Computational Analysis of Polymer Processing," Pearson, J.R.A., Richardson, S.M., Eds., Applied Science, London (1983), p.83.
27. Pearson, J.R.A.,"Mechanics of Polymer Processing," Elsevier, London (1986).
28. Stevenson, J.F., in "Comprehensive Polymer Science," Aggarwal, S.L., Ed., Pergamon Press, Oxford (1989), p.303.
29. Tadmor, Z., Gogos, C.E., "Principles of Polymer Processing," John Wiley & Sons, New York (1979).
30. Middleman, S., "Fundamentals of Polymer Processing," McGraw-Hill, New York (1977).
31. Colwell, R.E., Nickolls, K.R., *Ind. Eng. Chem.*, *51*, 841 (1959).
32. Griffith, R.M., *Ind. Eng. Chem. Fundam.*, *1*, 180 (1962).
33. Fenner, R.T., *Polymer*, *16*, 298 (1975).
34. Kuhnle, H., *J. Polym. Eng.*, *6*, 51, (1986).
35. Kuhnle, H., *Kunststoffe*, *72*, 267 (1982).
36. Hami, M.L., Pitman, J.F.T., *Polym. Eng. Sci.*, *20*, 339 (1980).
37. Roylance, D., *Polym. Eng. Sci.*, *20*, 1029 (1980).
38. Choo, K.P., Hami, M.L., Pitman, J.F.T., *Polym. Eng. Sci.*, *21*, 100 (1981).
39. Agur, E.E., Vlachopoulos, J., *Polym. Eng. Sci.*, *22*, 1084 (1982).
40. Pittman, J.F.T., Raschid, K., *J. Polym. Eng.*, *5*, 1 (1985).

41. Viriyayuthakorn, M., Kassahun, B., *SPE Tech. Papers*, *30*, 81 (1984).
42. Meijer, H.E.H., Verbraak, C.P.J.M., *Polym. Eng. Sci.*, *28*, 758 (1988).
43. Brzoskowski, R., White, J.L., Szydlowski, W., Nakajima, N., Min, K., *Intl. Polym. Process.*, *3*, 134 (1988).
44. Darnell, W.H., Mol, E.A.J., *SPE J.*, *12*, 20, April (1956).
45. Broyer, E., Tadmor, Z., *Polym. Eng. Sci.*, *12*, 12 (1972).
46. Tadmor, Z., Broyer, E., *Polym. Eng. Sci.*, *12*, 378 (1972).
47. Lovegrove, J.G.A., Williams, J.G., *J. Mech. Eng. Sci.*, *14*, 114 (1973).
48. Lovegrove, J.G.A., Williams, J.G., *Polym. Eng. Sci.*, *14*, 589 (1974).
49. Lovegrove, J.G.A., Williams, J.G., *J. Mech. Eng. Sci.*, *16*, 271 (1974).
50. Fenner, R.T., "Principles of Polymer Processing," MacMillan, London (1979).
51. Maddock, B.H., *SPE J.*, *15*, 383 (1959).
52. Street, L.F., *Int. Plast. Eng.*, *1*, 289 (1961).
53. Tadmor, Z., Duvdevani, I.J., Klein, I., *Polym. Eng. Sci.*, *7*, 198 (1967).
54. Klein, I., Tadmor, Z., *Polym. Eng. Sci.*, *9*, 11 (1969).
55. Tadmor, Z., *Polym. Eng. Sci.*, *6*, 185 (1966).
56. Shapiro, J., Halmos, A.L., Pearson, J.R.A., *Polymer*, *17*, 905 (1976).
57. Shapiro, J., Halmos, A.L., Pearson, J.R.A., *Polymer*, *17*, 912 (1976).
58. Halmos, A.L., Pearson, J.R.A., Trottnow, *Polymer*, *19*, 1199 (1978).
59. Lindt, J.T., *Polym. Eng. Sci.*, *16*, 284 (1976).
60. Lindt, J.T., *Polym. Eng. Sci.*, *21*, 1162 (1981).
61. Elbirli, B., Lindt, J. T., Gottgetreu, S.R., Babe, S.M., *Polym. Eng. Sci.*, *24*, 988 (1984).
62. Lindt, J.T., Elbirli, B., *Polym. Eng. Sci.*, *25*, 412 (1985).
63. Dekker, J., *Kunststoffe*, *66*, 130 (1976).
64. Michaeli, W., "Extrusion Dies," Hanser Publishers, Munich, (1984).
65. Rao, N.S., "Designing Machines and Dies for Polymer Processing with Computer Programs," Hanser Publishers, Munich (1981).
66. Roache, P.J., "Computational Fluid Dynamics," Revised Edition, Hermosa, Albuquerque (1982).
67. Zienkiewicz, O.C., "The Finite Element Methods," 3rd Ed., McGraw-Hill, London (1977).
68. Crochet, J.M., Walters, K., Davies, R.J., "Numerical Simulation of Non-Newtonian Flow," Elsevier (1984).
69. Hughes, T.J.R., Liu, W.K., Brook, A., *J. Comput. Phys.*, *30*, 1 (1979).
70. Hughes, T.J.R., Brooks, A., in "Finite Elements in Fluids," Gallagher, R.M., Norrie, D.H., Oden, J.T., Zienkiewicz, O.C., Eds., John Wiley & Sons, New York (1982).
71. Zienkiewicz O.C., Godbole, P.N., in "Finite Elements in Fluids," Gallagher, R.H., Norrie, D.H., Oden, J.T., Zienkiewicz, O.C., Eds., John Wiley & Sons, New York (1982).
72. Kaiser, H., *Polym. Process. Eng.*, *5*, 1 (1987).
73. Kanai, T., Kimura, M., Asano, Y., *J. Plast. Film Sheeting*, *2*, 224 (1986).
74. Yang, B., Lee, J.L., *Polym. Eng. Sci.*, *28*, 697 (1988).
75. Yang, B., Lee, L.J., *Polym. Eng. Sci.*, *28*, 708 (1988).
76. Lee, W.K., Lee, L.J., in "Comprehensive Polymer Science," Aggarwal, S.L., Ed., Pergamon Press, Oxford (1989).
77. Tang, S.H., Griffith, R.M., *Rubber Chem. Technol.*, *59*, 826 (1986).
78. White, J.L., Wang, Y., Isayev A.I., Nakajima, N., Weissert, F.C., Min, K., *Rubber Chem. Technol.*, *60*, 337 (1987).

79. Poslinski, A.J., Ryan, M.E., Gupta, R.K., Seshadri, S.G., Frechette, F.J., *Polym. Eng. Sci.*, *28*, 453 (1988).
80. Isayev, A.I., Famili, N., *J. Plast. Film Sheeting*, *2*, 269 (1986).
81. Aldhouse, S.T.E., Mackley, M.R., Moore, I.P.T., *J. Non–Newtonian Fluid Mech.*, *21*, 359 (1986).
82. Kalyon, D.M., Yu, J.S., Du, C.C., *Polym. Process. Eng.*, *5*, 179 (1987).
83. Isayev, A.I., Upadhyay, R.K., in "Injection and Compression Molding Fundamentals," Isayev, A.I., Ed., Marcel Dekker, New York, (1987).
84. Luo, X.L., Tanner, R.I., *Rheol. Acta*, *26*, 499 (1987).
85. Mitsoulis, E., *Polym. Eng. Sci.*, *26*, 1552 (1988).
86. Keunings, R., in "Fundamentals of Computer Modelling for Polymer Processing," Tucker, C.L., Ed., Hanser Publishers, Munich (1989).
87. Phan Thein, N., *J. Non–Newtonian Fluid Mech.*, *26*, 327 (1988).
88. McClelland, M.A., Finlayson, B.A., *J. Non–Newtonian Fluid Mech.*, *27*, 363 (1988).
89. Debry, H.G., Charbonneaux, T.G., Macosko, C.W., *Polym. Process. Eng.*, *3*, 151, (1986).
90. Doshi, S.R., Dealy, J.M., Charrier, J.M., *Polym. Eng. Sci.*, *26*, 468 (1986).
91. Sornberger, G., Vergnes, B., Agassant, J.F., *Polym. Eng. Sci.*, *26*, 455 (1986).
92. Sornberger, G., Vergnes, B., Agassant, J.F., *Polym. Eng. Sci.*, *26*, 682 (1986).
93. Mitsoulis, E., Heng, F.L., *J. Appl. Polym. Sci.*, *34*, 1713 (1987).
94. Karagiannis, A., Mavridis, H., Hrymak, A.N., Vlachopoulos, J., *Polym. Eng. Sci.*, *28*, 982 (1988).
95. Nordberg, M.E. III, Winter, H.H., *Polym. Eng. Sci.*, *28*, 444 (1988).
96. Mitsoulis, E., *Adv. Polym. Technol.*, *8*, 225 (1988).
97. White, J.L., Min, K., Mixing of Polymers, in "Comprehensive Polymer Science," v.7, Aggarwal, S.L., Ed., Pergamon Press, Oxford (1989).
98. Manas-Zloczower, I., Feke, D.L., *Intl. Polym. Process.*, *2*, 185 (1988).
99. Menges, G., Grajewski, *Intl. Polym. Process.*, *3*, 74 (1988).
100. Freakley, P.K., Patel, S.R., *Polym. Eng. Sci.*, *27*, 1358 (1987).
101. Freakley, P.K., Patel, S.R., *Rubber Chem. Technol.*, *58*, 751 (1985).
102. Speur, J.A., Mavridis, H., Vlachopoulus, J., Jansen, L.P.B.M., *Adv. Polym. Technol.*, *7*, 39 (1987).
103. Meyer, H.E.H., Elemans, P.H.M., *Polym. Eng. Sci.*, *28*, 275 (1988).
104. Ziabicki, A., "Fundamentals of Fiber Formation," John Wiley & Sons, London, (1976).
105. Petrie, C.J.S., "Elongational Flows," Pitman, London (1979).
106. Ziabicki, A., Kawai, H., "High Speed Spinning," John Wiley & Sons, New York (1985).
107. Denn, M.M., *Ann. Rev. Fluid. Mech.*, *12*, 365 (1980).
108. Denn, M.M., in "Computational Analysis of Polymer Processing," Pearson, J.R.A., Richardson, S.M., Eds., Applied Science, London, p.179 (1983).
109. White, J.L., *Polym. Eng. Rev.*, *1*, 297 (1982).
110. Andrews, R.D., Rudd, J.F., *J. Appl. Phys.*, *27*, 990 (1956).
111. Kang, H.J., White, J.L., *Intl. Polym. Process.*, *1*, 12 (1986).
112. Denn, M.M., Petrie, C.J.S., Avenas, P., *AIChEJ.*, *21*, 91 (1975).
113. Fisher, J.F., Denn, M.M., *AIChEJ.*, *22*, 236 (1976).
114. Sridhar, T., Gupta, R.K., *J. Non–Newtonian Fluid Mech.*, *27*, 349 (1988).
115. Sridhar, T., Gupta, R.K., Boher, D.V., Binnington, R., *J. Non–Newtonian Fluid Mech.*, *21*, 115 (1986).
116. Gupta, R.K., Puszynski, J., Sridhar, *J. Non–Newtonian Fluid Mech.*, *21*, 99 (1986).

117. Zhiganov, N.K., Yankov, V.I., Krasnov, E.P., *Khim. Volokna*, *4*, 45 (1986).
118. Zhiganov, N.K., Nekrasov, Yu. P., Smirnov, A.V., Mezhirov, M.S., *Khim. Volokna*, *2*, 7, (1988).
119. Zhiganov, N.K., Makshakov, S.P., Pavlov, V.A., *Khim. Volokna*, *2*, 9, (1988).
120. Zhiganov, N.K., Yankov, V.I., Alekseev, E.N., Genis, A.V., *Khim. Volokna*, *4*, 18 (1988).
121. Upadhyay, R.K., Isayev, A.I., *J. Rheol.*, *28*, 281 (1984).
122. Malkus, D.S., *J. Non–Newtonian Fluid Mech.*, *8*, 223 (1981).
123. Larson, R.C., *J. Rheol.*, *27*, 475 (1983).
124. Papanastasiou, A.C., Macosko, C.W., Scriven, L.E., Chen, Z., *AIChEJ.*, *33*, 834 (1987).
125. Chen, Z., Papanastasiou, A.C., *Intl. Polym. Process.*, *2*, 33 (1987).
126. Zieminski, K.F., Spruiell, J.E., *J. Appl. Polym. Sci.*, *35*, 2223 (1988).
127. Dutta, A., *Polym. Eng. Sci.*, *27*, 1050 (1987); *Text. Res. J.*, *57*, 13 (1987).
128. Alfrey, T., *SPE Trans.*, *5*, 68 (1965).
129. Pearson, J.R.A., "Mechanical Principles of Polymer Melt Processing," Pergamon, Oxford (1966).
130. Pearson, J.R.A., Petrie, C.J.S., *J. Fluid Mech.*, *40*, 1 (1970).
131. Pearson, J.R.A., Petrie, C.J.S., *J. Fluid Mech.*, *42*, 609 (1970).
132. Pearson, J.R.A., Petrie, C.J.S., *Plast. Polym.*, *38*, 85 (1970).
133. Han, C.D., "Rheology in Polymer Processing," Academic Press, New York (1976).
134. Middleman, S., "Fundamentals of Polymer Processing," McGraw-Hill, New York (1977).
135. Petrie, C.J.S., in "Computational Analysis of Polymer Processing," Pearson, J.R.A., Richardson, S.M., Eds., Applied Science, London (1983), p.217.
136. White, J.L., Cakmak, M., *Adv. Polym. Technol.*, *8*, 27, (1988).
137. Kanai, T., White, J.L., *Polym. Eng. Sci.*, *24*, 1185 (1984).
138. Kanai, T., White, J.L., *J. Polym. Eng.*, *5*, 135 (1985).
139. Kanai, T., *Intl. Polym. Process.*, *1*, 137 (1987).
140. Yamane, H., White, J.L., *Intl. Polym. Process.*, *2*, 107 (1987).
141. Han, C.D., Park, J.Y., *J. Appl. Polym. Sci.*, *19*, 3257 (1975).
142. Han, C.D., Park, J.Y., *J. Appl. Polym. Sci.*, *19*, 3291 (1975).
143. Wagner, M.H., *Rheol. Acta*, *15*, 40 (1976).
144. Luo, X.L., Tanner, R.I., *Polym. Eng. Sci.*, *25*, 620 (1985).
145. Lohse, G., Marinow, S., *Plast. Kautsch.*, *33*, 148 (1986).
146. Lohse, G., Marinow, S., *Plast. Kautsch.*, *33*, 221 (1986).
147. Campbell, G.A., Cao, B., *J. Plast. Film Sheeting*, *3*, 158 (1987).
148. Campbell, G.A., Cao, B., *Tappi J.*, *70*, 41 (1987).
149. Mitsoulis, E., *Adv. Polym. Technol.*, *6*, 467 (1986).
150. Carley, J.F., in "Processing of Thermoplastics Materials," Bernhardt, E.C., Ed., Van Nostrand Reinhold, New York (1959).
151. McKelvey, J.M., "Polymer Processing," John Wiley & Sons, New York (1962).
152. Bagley, E.B., Storey, S.H., *Wire Prod.*, *38*, 1104 (1963).
153. Fenner, R.T. and Williams, J.G., *Trans. J. Plast. Inst.*, *35*, 701 (1967).
154. Kasajima, M., Ito, K., *Appl. Polym. Symp.*, *20*, 221 (1972).
155. Winter, H.H., *Adv. Heat Trans.*, *13*, 205 (1977).
156. Han, C.D., Rao, D., *Polym. Eng. Sci.*, *18*, 1019 (1978).
157. LeNir, V., *Wire J.*, *7*, 59, (1974).
158. Basu, S., *Polym. Eng. Sci.*, *21*, 1128 (1981).

159. Han, C.D., Rao, D.A., *Polym. Eng. Sci.*, *29*, 128 (1980).
160. Caswell, B., Tanner, R.I., *Polym. Eng. Sci.*, *18*, 416 (1978).
161. Carley, J.F., Endo, T., Krantz, W.B., *Polym. Eng. Sci.*, *19*, 1178 (1979).
162. Chung, T.S., *Polym. Eng. Sci.*, *26*, 410 (1986).
163. Tadmor, Z., Bird, R.B., *Polym. Eng. Sci.*, *14*, 124 (1974).
164. Mitsoulis, E., *Polym. Eng. Sci.*, *26*, 171 (1986).
165. Wagner, R., Mitsoulis E., *Adv. Polym. Technol.*, *5*, 305 (1985).
166. Mitsoulis, E., Wagner, R., Heng, F.L., *Polym. Eng. Sci.*, *28*, 291 (1987).
167. Mitsoulis, E., Wagner, R., in Proceedings of 3rd World Congress, Chemical Engineering, Tokyo, Japan, 1986, Vol. IV, p.534.
168. Mitsoulis, E., *J. Rheol.*, *30*, 523 (1986).
169. Zenekhe, S.A., *Polym. Mater. Sci. Eng.*, *55*, 99 (1986).
170. Zenekhe, S.A., *Ind. Eng. Chem. Fund.*, *23*, 425 (1984); *Polym. Eng. Sci.*, *23*, 830 (1983).
171. Zenekhe, S.A., ACS Symposium Ser., *346*, 261 (1987).
172. Lawrence, C.J., *Phys. Fluids*, *31*, 2786 (1988).
173. Shimoji, S., Japan *J. Appl. Phys.*, Part 2, *26*, L905 (1987).
174. Gaskell, R.E., *J. Appl. Mech*, *17*, 334 (1950).
175. Torner, R.V., "Grundprozesse der Verarbeitung Von Polymeren," VEB Deutscher Verlag fur Grund Stoffindustrie, Leipzig, G.D.R. (1974).
176. Tanner, R.I., "Engineering Rheology," Oxford University Press, New York (1985).
177. Kamal, M.R., Haber, A., Ryan, M., *Polym. Eng. Sci.*, *25*, 698 (1985).
178. Takserman-Krozer, R., Schenkel, G. and Ehrmann, G., *Rheol. Acta*, *14*, 1066 (1975).
179. Daud, W.R.W., *J. Appl. Polym. Sci.*, *31*, 2457 (1986).
180. Agassant, J.F., Espy, M., *Polym. Eng. Sci.*, *25*, 118 (1985).
181. Seeger, R., Schnabel, R., Reher, E.O., *Plast. Kautsch.*, *29*, 406 (1982).
182. Dimitrijew, J.G., Sporjagin, E.A., *Plast. Kautsch.*, *24*, 484 (1977).
183. Mitsoulis, E., Vlachopoulos, J., Mirza, F.A., *Polym. Eng. Sci.*, *25*, 6 (1985).
184. Bergen, J.T., Scott, G.W., *J. Appl. Mech.*, *18*, 191 (1951).
185. Agassant, J.F, Avenes, P., *J. Macromol. Sci., Phys.*, *14*, 345 (1977).
186. Bourgeois, J.L., Agassant, J.F., *J. Macromol. Sci., Phys.*, *14*, 367 (1977).
187. Hartzmann, G., Herner, M., Muller, G., *Angew. Macromol. Chem.*, *47*, 257 (1975).
188. Vlachopoulos, J., Hrymak, A.N., *Polym. Eng. Sci.*, *20*, 725 (1980).
189. Tanner, R.I., *ASLE Trans.*, *8*, 179 (1965).
190. Davenport, T.C., "The Rheology of Lubricants," Applied Science, London (1973).
191. Paslay, P.R., *J. Appl. Mech.*, *24*, 602 (1957).
192. Tokita, N., White, J.L., *J. Appl. Polym. Sci.*, *10*, 1011 (1966).
193. Chong, J.S., *J. Appl. Polym. Sci.*, *12*, 191 (1968).
194. Tanner, R.I., Austr. *J. Appl. Sci.*, *14*, 129 (1963).
195. Mitsoulis, E., Vlachopolos J., Mirza, E.A., *Polym. Eng. Sci.*, *24*, 707 (1984).
196. Lipp, R., *Plast. Kautsch.*, *33*, 4 (1986).
197. Zheng, R., Tanner, R.I., *J. Non-Newtonian Fluid Mech.*, *28*, 149 (1988).
198. Denn, M.M., "Process Fluid Mechanics," Prentice Hall, Englewood Cliffs (1989).
199. Tucker, C.L., in "Injection and Compression Molding Fundamentals," Isayev, A.I., Ed., Marcel Dekker, New York, 1987, Chapter 7, p.481.
200. Tucker, C.L., Ed., "Fundamentals of Computer Modelling for Polymer Processing," Hanser Publishers, Munich (1989).
201. Melby, E.G., Castro, J.M., in "Comprehensive Polymer Science," Aggarwal, S.L., Ed., Vol. 7, Pergamon Press, Oxford, 1989, p.51.

202. Gutowski, T.G., Morigaki, T., Cai, Z., *J. Compos. Mater.*, *21*, 172 (1987).
203. Gutowski, T.G., Cai, Z., Bauer, S., Boucher, D., Kingery, J., Wineman, S., *J. Compos. Mater.*, *21*, 650 (1987).
204. Rogowski, A.H., Gutowski, T.G., Upadhyay, R.K., Intl. *SAMPE Symp. Exhib.*, *33*, 1448 (1988).
205. Hojo, H., Yaguchi, H., Onodera, T., Kim, E.G., *Intl. Polym. Process.*, *3*, 54 (1988).
206. Jackson, W.C., Advani, S.G., Tucker, C.L., *J. Compos. Mater.*, *20*, 539 (1986).
207. Lee, C.C., Tucker, C.L., *J. Non–Newtonian Fluid Mech.*, *24*, 245 (1987).
208. Barone, M.R., Caulk, D.A., *J. Appl. Mech.*, *53*, 361 (1986).
209. Barone, M.R., Osswald, T.A., *Polym. Compos.*, *9*, 158 (1988).
210. Isayev, A.I., Azari, A.D., *Rubber Chem. Technol.*, *59*, 868 (1986).
211. Shenoy, A.V., Saini, D.R., *Plast. Rubber Process. Appl.*, *5*, 313 (1985).
212. Nixon, J.A., Hutchinson, J.M., *Plast. Rubber Process. Appl.*, *5*, 349 (1985).
213. Hutchinson, J.M., Nixon, J.A., *Plast. Rubber Process. Appl.*, *5*, 359 (1985).
214. Upadhyay, R.K., *Adv. Polym. Technol.*, *8*, 243 (1988).
215. Bernhardt, E.C., Ed., "Computer Aided Engineering for Injection Molding," Hanser Publishers, Munich (1983).
216. Manzione, L.T., Ed., "Application of Computer Aided Engineering in Injection Molding," Hanser Publishers, Munich (1987).
217. Isayev, A.I., Ed., "Injection and Compression Molding Fundamentals," Marcel Dekker, New York (1987).
218. Macosko, C., "Reaction Injection Molding," Hanser Publishers, Munich (1989).
219. Richardson, S.M., in "Computational Analysis of Polymer Processing," Pearson, J.R.A., Richardson, S.M., Eds., Applied Science Publishers, London, Chapter 5, p.183 (1983).
220. Richardson, S.M., *Rheol. Acta*, *24*, 497 (1985).
221. Richardson, S.M., *Rheol. Acta*, *24*, 509 (1985).
222. Richardson, S.M., *Rheol. Acta*, *25*, 180 (1986).
223. Richardson, S.M., *Rheol. Acta*, *25*, 308 (1986).
224. Richardson, S.M., *Rheol. Acta*, *25*, 372 (1986).
225. Richardson, S.M., *Rheol. Acta*, *26*, 102 (1987).
226. Alles, H., Philipon, S., Agassant, J.F., Vincent, M., Dehay, G., Lerebours, P., *Polym. Process. Eng.*, *4*, 71 (1986).
227. Agassant, J.F., Alles, H., Philipon, S., Vincent, M., *Polym. Eng. Sci.*, *28*, 460 (1988).
228. Alles, H., Agassant, J.F., Vincent, M., Dehay, G., Lerebours, P., Gingliner, B., *Intl. Polym. Process.*, *2*, 101 (1988).
229. Bowers, S., *Plast. Rubber Process. Appl.*, *7*, 191 (1987).
230. Blanc, R., Philipon, S., Vincent, M., Agassant, J.F., Alglave, H., Mueller, R., Froelich, D., *Intl. Polym. Process.*, *2*, 21 (1987).
231. Bowers, S., *Prog. Rubber Plast. Technol.*, *2*, 23 (1986).
232. Isayev, A.I., in "Comprehensive Polymer Science," Aggarwal, S.L., Eds., Pergamon Press, Oxford, (1989), Chapter 11, p.355.
233. Lee, L.J., in "Comprehensive Polymer Science," Aggarwal, S.L., Eds., Pergamon Press, Oxford, (1989), Chapter 12, p.379.
234. Menges, G., Recker, H., in "Comprehensive Polymer Science," Aggarwal, S.L., Eds., Pergamon Press, Oxford, (1989), Chapter 14, p.451.
235. Chan, A.W., Hwang, S.T., *Polym. Eng. Sci.*, *28*, 333 (1988).
236. Frutiger, R.L., *Polym. Eng. Sci.*, *26*, 243 (1986).
237. Opolski, S.W., Kwon, T.H., *SPE Tech. Papers*, *33*, 264 (1987).

238. Kwon, T.H., *J. Eng. Ind.*, *110*, 384 (1986).
239. Throne, J.L., "Thermoforming," Hanser Publishers, Munich (1987).
240. Menges, G., Weinand, D., *Kunststoffe*, *78*, 456 (1988).
241. Potente, H., Michel, P., Liu, J.P., *Kautsch. Gummi Kunstst.*, *41*, 466 (1988).
242. Allard, R., Charrier, J.M., Ghosh, A., Marangou, M., Ryan, M.E., Shrivastava, S., Wu, R., *J. Polym. Eng.*, *6*, 363 (1986).
243. Cohen, A., Arends, C.B., Dibbs, M.G., Seitz, J.T., *Polym. Mater. Sci. Eng.*, *58*, 142 (1988).
244. Throne, J.L., *Adv.Polym. Technol.*, *8*, 131 (1988).
245. Belcher, S.L., in "Comprehensive Polymer Science," Aggarwal, S.L., Ed., Pergamon Press, Oxford, (1989), Chapter 15, p.489.

CHAPTER 2

THE CONCEPT OF INVERSE FORMULATION AND ITS APPLICATION TO DIE FLOW SIMULATION

by A. Cohen

The Dow Chemical Company
Central Research
Materials Science & Development Laboratory
1702 Building
Midland, Michigan 48674
U.S.A.

ABSTRACT
2.1 THE INVERSE FORMULATION AS AN ALTERNATIVE APPROACH
2.2 REVIEW OF APPLICATIONS OF THE INVERSE FORMULATION
2.2.1 Design and Performance of Diffusers
2.2.2 Nonlinear Elasticity
2.2.3 Second Order Fluid
2.2.4 Ground Water Modeling
2.3 THE CONCEPT OF ILL-POSED AND CORRECT PROBLEMS
2.4 RESULTS IN THE AREA OF HEAT CONDUCTION
2.4.1 Identification of Temperature Dependent Thermal Properties
2.4.2 Identification of Boundary Conditions
2.4.3 Initial or Boundary Conditions
2.5 THE LUBRICATION APPROXIMATION OF FLOW IN AN EXTRUSION DIE
2.6 THE INVERSE FORMULATION OF THE DIE FLOW MODEL
2.7 PRESSURE DROP OPTIMIZATION AS AN ADDITIONAL CONDITION OF THE INVERSE FORMULATION
2.8 CONCLUSIONS
ACKNOWLEDGEMENTS
REFERENCES

ABSTRACT

This paper reviews basic ideas of the inverse formulation and its application to a broad range of engineering problems. The advantages and intrinsic difficulties of this method are elucidated. The general discussion is extended into an analysis of die flow in the lubrication approximation for generalized Newtonian fluid. The suggested mathematical treatment is based on the introduction of a new variable that combines material and geometrical parameters. Using the method of characteristics, the solution for die thickness, i.e. gap distribution, is obtained over the defined two-dimensional flow domain.

2.1 THE INVERSE FORMULATION AS AN ALTERNATIVE APPROACH

The main reason for mathematical modeling of polymer processing is to identify the conditions leading to acceptable performance of a specific piece of machinery. These conditions might include geometrical variables, e.g. thickness distribution in extrusion die as well as physical conditions, such as preheating of a preform in the injection-blow molding process.

When a conventional direct formulation is followed, once the governing equations are solved within a specified domain with known boundary conditions, the actual methodology of problem solving is no different from the Edisonian approach of trial and error: the processing conditions are guessed and coded into the model, and the results of the simulation are studied. The problem is considered solved once the conditions leading to acceptable performance are identified. The difference between the empirical method and numerical modeling is not in logic or method but rather in the use of the computer technology to reduce the cost of mistakes in terms of time and resources.

The identification of the necessary processing conditions requires extrapolation of the causes from the known effects. The inverse formulation can be utilized to avoid the often inefficient loop of intuitive guesses and adjustments. The formulation is inverse in the sense that while in the conventional direct problem the "effect" results from the "cause", in the inverse problem the "cause" formally results from the "effect".

Representing the modeling problem mathematically, [1], by an operator equation

$$Au=f, \quad u\subset U, \quad f\subset F \tag{1}$$

with a given operator $A: U \rightarrow F$ and known element f, the solution u of the inverse formulation is found by inverting the operator A.

Depending upon what causal feature is the subject of an investigation, the general theory of modeling, [2], identifies three types of problems as (i) inductive, (ii) inverse, and (iii) converse. The corresponding formulations are aimed at the identification of the mathematical model, the parameters of the process, and the boundary conditions, respectively.

2.2 REVIEW OF APPLICATIONS OF THE INVERSE FORMULATION

The inverse approach was successfully applied in many areas of mechanics and engineering:

2.2.1 Design and Performance of Diffusers

Design and performance of diffusers were analyzed by Nelson, Yang and Hudson [3]; Laidler and Myring developed a duct design algorithm by which duct wall shapes were calculated from prescribed wall friction velocity distribution [4]; Ahmed and Myring [5] presented the inverse integral boundary-layer solution for optimal design of axisymmetric ducts; Zanetti [6] used a time-dependent numerical method to solve the inverse problem for inviscid compressible subsonic or transonic flows; a classical work by Wu and Brown on the mean-stream method [7] published more than 30 years ago, led to an analytical solution of the inverse problem for the axisymmetric flow channels.

2.2.2 Nonlinear Elasticity

Applications of inverse methods in the nonlinear elasticity was summarized by Ericksen [8]. Cherepanov provided exact solutions for a number of inverse problems of the plane elasticity [9]; Franek, Kratochvil and Travnicek addressed the problem of identification of the constitutive equation in finite elasticity [10]. A linear approximation was used by Volkova [11] for analysis of an inverse problem for the theory of elasticity of anisotropic media. A variety of inverse problems of shape optimization of two- and three-dimensional elasticity were solved by Vigdergauz [12-14].

2.2.3 Second Order Fluid

Kaloni and Huschilt discussed solutions for an incompressible second order fluid [15]. They commented on a very limited application of inverse methods to viscometric flows and provide a short review of available results.

2.2.4 Ground Water Modeling

Kuiper [16] compared three different approaches developed for the solution of the inverse problem in two-dimensional steady state groundwater modeling. Velikina and Krass [17] identified seepage parameters of the geological field using the variational method. Galanin and Tikhonov [18] investigated inverse sorption problem using the asymptotic form of the solution.

2.3 THE CONCEPT OF ILL-POSED AND CORRECT PROBLEMS

The main difficulty of inverse problems is their violation of one or more of the Hadamart conditions of correctness [19]:

1. A solution u, belonging to the space U, exists for any element f of the space F, see (1).
2. The solution u is stable for small perturbation of f.
3. The solution u is unique.

The ill-posed nature of these problems originates from the fact that they present a mathematical model of physically irreversible processes during which the uniqueness of the relationship between cause and effect is obliterated.

Hadamart's example [20] of the Cauchy problem for the Laplace equation in two dimensions $\Delta u(x,y) = 0$, $-\infty < x < \infty$, where Δ is the Laplacian operator, with boundary conditions: $u(x,0) = f(x)$, $\partial u/\partial y(x,0) = \phi(x)$, demonstrates the instability of the solution to small perturbations of boundary conditions. For $f_1(x) \equiv 0$ and $\phi_1(x) = 1/a \sin(ax)$, the solution of the Cauchy problem is $u_1 = 1/a^2 \sin(ax) \sinh(ay)$, for $a > 0$. On the other hand, for $f_2(x) = \phi_2(x) \equiv 0$, the solution of the Cauchy problem becomes $u_2(x) \equiv 0$.

Using C measure to evaluate the distance, ρ_C, between two sets of the boundary conditions and the solutions, we obtain

$$\rho(f_1, f_2) = \sup_x\{\text{abs}(f_1(x) - f_2(x))\} = 0,$$

$$\rho_c(\phi_1, \phi_2) = \sup_x\{\text{abs}(\phi_1(x) - \phi_2(x))\} = 1/a,$$

for the boundary conditions, and

$$\rho_c(u_1, u_2) = \sup_x\{\text{abs}(u_1(x) - u_2(x))\} =$$

$$\sup_x\{\text{abs}(1/a^2 \sin(ax)\sinh(ay))\} = 1/a^2 \sinh(ay)$$

for the solutions, where "abs" stands for the absolute value and "sup" stands for the least upper bound. By selecting sufficiently large values for the parameter a, the distance between functions ϕ_1 and ϕ_2 can be made smaller than any predetermined number, i.e.

$$\rho_c(\phi_1, \phi_2) \to 0 \quad \text{at } a \to \infty,$$

while the distance between solutions u_1 and u_2 can be made arbitrarily large for any fixed and positive value of y, i.e.

$$\rho_c(u_1, u_2) \to \infty \quad \text{at } a \to \infty \text{ and } y = \text{const.} > 0.$$

Thus, the Cauchy problem is unstable and, hence, is ill-posed.

Tikhonov [21] revised the Hadamart criteria of correctness and suggested that the solution space U should be narrowed in order to achieve stability of the solution. The physical justification for such an approach comes from the observation that, as a rule, not all information known about a solution is incorporated into the mathematical formulation [22]. A detailed analysis of ill-posed problems has been provided by Tikhonov and Arsenin [23].

2.4 RESULTS IN THE AREA OF HEAT CONDUCTION

Inverse methods found their most extensive application in the areas of heat and mass transfer, especially in heat conduction. For the sake of convenience, the inverse problems can be divided into three groups according to specifics of their applications:

2.4.1 Identification of Temperature Dependent Thermal Properties

Identification of temperature dependent thermal properties, i.e. heat capacity, thermal conductivity and intensity of heat sources, based upon measured temperature values inside a sample [24-33]. The developed methodology of experiment design and data reduction can be readily applied to polymer processing. Aleksashenko [28] presents three criteria for data reduction procedure in order to optimize the amount of temperature measurements: 1. The temperature intervals used in a model should coincide with the experimental interval; 2. The accuracy of temperature measurements should be coordinated with the total number of data points in order to assure the prescribed accuracy of the thermal characteristics of the inverse formulation; 3. The thermal characteristics should be monotonic functions of temperature in order to assure their uniqueness.

The following equation was used in [26] in order to relate the total error β with the required number of measurements Π of the non-dimensional field variable $0 \leq \Theta \leq 1$:

$$\Theta(p) = (1-\beta)^{p-1}/(1+\beta)^{p} \quad , \text{ where } p = 1,2,\ldots,\Pi.$$

The values $\Theta(p)$ are the centers of the intervals $(\Theta(p)*(1 - \beta),\Theta(p)*(1 + \beta))$ covering the segment [0,1]. The edges of the intervals are:

$$\Theta_e(p) = (1-\beta)^{p} / (1+\beta)^{p}.$$

The last data point, i.e. $p = \Pi$, is selected in such a way that an uncovered portion of the interval [0,1] results in a relative error not larger than β. Thus for $\beta = 0.3$ (30%), $\Pi = 6$; for $\beta = 0.2$, $\Pi = 10$; and for $\beta = 0.1$, $\Pi = 30$.

Several methods have been developed to solve inverse heat conduction problems [34]. For special cases, it is possible to develop analytical solutions, including the use of self-similar solutions. Although these methods are mostly applicable only in the case of constant material characteristics, averaging the properties over separate intervals extends their range of validity [26].

The majority of currently used methods are based on computer technology due to the large volume of required calculations. As noted by Alifanov [1], there are five prime algorithms providing stable methods of solution: (i) direct approximate-analytic method, (ii) direct numerical method, (iii) method of iteration regularization, (iv) regularized algebraic method, and (v) regularized numerical method.

The method of iteration regularization, based on a gradient algorithm, is considered to be the most universal. The essence of the method is a minimization of the norm $J(u)$ of the deviation of the left-hand side of the Equation (1), $A\,u$, from the right hand-side, f, in the metric of the square integrable space $F \subset L_2$:

$$J(u) = \| Au - f \|_F ,$$

e.g. this metric can be presented as

$$J(u) = \sum_i \int_{x_i} [Au - f]^2 \, d\underline{x}.$$

Using the method of steepest descent, an iteration process was used in [29] to construct a minimization sequence $\{u^i\}$:

$$u^{i+1} = u^i + \alpha^i G(J'^{(r)}), i = 0,1,...I$$

where u^0 is initial approximation; α^i is the depth of descent, obtained from the condition

$$J^{i+1} = \min_{\alpha} J(u^i + \alpha^i G(J'^{(i)})),$$

$G(J')$ is the direction of descent; and I is the number of final iteration, defined from the accuracy condition, $J(u^I) \cong \varepsilon$, where ε is the known integral error of experimental data.

2.4.2 Identification of Boundary Conditions

Another important area of applications is the identification of boundary conditions. An analytical solution of the problem was developed by Tsoi and Yusupov [35] for pipes, flat and spherical shells. The direct problem of heat transfer was solved by expanding the Laplace image of the temperature over a certain class of base functions. The coefficients of the expansion were defined from an orthogonalization procedure applied to the residual of the heat transfer equation. The inverse Laplace transform, involving the convolution theorem, led to an approximate solution for the temperature. The inverse problem was then formulated as a solution of the Volterra integral equation with a convolute kernel:

$$\phi(t) - C(Bi) \int_0^t \exp[1 - A(Bi)(t-\tau)] \, \phi(\tau) \, d\tau = N(Bi) \, \Phi(t),$$

where $\phi(t)$ is the heat flux at time t, $\Phi(t)$ is the known temperature difference; and A, C and N are defined functions of the Biot number Bi.

Matsevity et al. [36] proposed a method of spectral functions of boundary effects W_{ij}. The temperature $T(x,y)$ was approximated as

$$T(x,y) = \sum_{i=1}^{m} \sum_{j=0}^{n_i} a_{ij} W_{ij}(x,y),$$

where W_{ij} is a solution of the Laplace equation for the disturbance function ξ^j applied over i-th portion of the boundary, with ξ being the distance measured along this portion of the boundary. The parameters a_{ij} are determined from a system of linear algebraic equations

$$T(x_1,y_1) = \sum_{i=1}^{m} \sum_{j=0}^{n_i} a_{ij} W_{ij}(x_1,y_1),$$

$$T(x_{mn}, y_{mn}) = \sum_{i=1}^{m} \sum_{j=0}^{n_i} a_{ij} W_{ij}(x_{mn}, y_{mn}),$$

where $T(x_1,y_1), \ldots, T(x_{mn},y_{mn})$ are measured values of the temperature in specified locations (x_i,y_j). If the total number of the measurements N was insufficient in order to find a_{ij} from the algebraic system, minimization of functional I by the least square method was used to find these values :

$$I = \sum_{s=1}^{N} [T(x_s, y_s) - \sum_{i=1}^{m} \sum_{j=0}^{n_i} a_{ij} W_{ij}(x_s, y_s)]^2 .$$

2.4.3 Initial or Boundary Conditions

The last group of problems is related to the recovery of initial or boundary conditions for stationary boundaries or to the finding of a position of moving boundary. The work of Alnajem and Ozisik [37] provided an analytical approach to the inverse problem involving abrupt changes of the surface conditions. Hsieh and Lin [38] developed approximate methods for inverse problems with unknown initial conditions.

Bratuta, Matsevity and Multanovsky [39] provided an algorithm resulting in the determination of the boundary conditions in surface boiling. Hsieh and Kassab introduced an auxiliary problem in the solution of the inverse heat conduction problem with geometries not fully specified [40]. Malakhov [41] investigated the identification methods of equivalent boundary conditions for a system of bodies exchanging heat.

Extensive review of publications directed to determination of external heating conditions using inverse methods can be found in [42,43].

The Stefan problem, named after J. Stefan [44], consists of finding a position of a moving boundary where some defined process takes place, e.g. the absorption or release of latent heat. For example, finding a location of the solidification front during injection molding process can be treated as Stefan problem.

Bobula and Twardowska [45] have proven the existence and uniqueness theorem for the inverse Stefan problem for the parabolic partial differential equation. Szczurek has obtained a series approximation for the transient temperature field and applied it to one-dimensional Stefan problems [46]. These series, which express the temperature field **T** in terms of a prescribed interface position **y**, are useful for treating direct and inverse problems of phase change. For example, the temperature T(x) of a plate is expressed as

$$T(x) = \sum_{k=1}^{\infty} f_k(y) (x - y)^k$$

with

$$f_k = (-1)^k / k! \sum_{j=[k/2+1]}^{j=k} (Ja)^{j-1} j! \, \Sigma^* P_j,$$

$$\text{where } P_j \equiv \prod_{i=1}^{j} (y^{(i)})^{\beta_i} / \beta_i! \, (i!)^{\beta_i},$$

Σ^* denotes the summation over all sequences of non-negative integers $\beta_1, \beta_2, ...\beta_j$, which satisfy the system of the Diofantic equations for given values of k and j :

$$\beta_1 + ... + \beta_j = 2j - k \text{ and } 1\beta_1 + ... + j\beta_j = j,$$

$y^{(i)}$ stands for the i-th derivative of y with respect to the time t, $(Ja) = c(To - T_S)/h$ denotes the Jacob number, c is the heat capacity per unit volume, h is the latent heat of fusion, To is the reference temperature and T_S is the fusion temperature.

2.5 THE LUBRICATION APPROXIMATION OF FLOW IN AN EXTRUSION DIE

The direct approach to the modeling of extrusion process underwent remarkable progress during the last twenty years. Tanner probably provided one of the first insights into the use of finite element methods for die flow simulation [47]. Tadmor et al. [48] developed the flow analysis network method to model two-dimensional steady flow in extrusion dies. Extensive work on entry and exit flows [49-51], die swell analyses [52] and die flow optimization [53] were conducted. The review by Crochet [54] of the general area of numerical simulation of flow processes provides a short introduction into the physics and history of the modeling together with an interesting collection of modeling examples.

Standard computer programs (see for example [55]) can be used for die flow simulation in the direct formulation. Also, though lacking some rheological and numerical features of specialized research programs, they provide two- and three-dimensional modeling capabilities and, in general, require substantial computer hardware capabilities.

In order to apply the concept of the inverse formulation to the modeling of extrusion die performance, we will start from the formulation of conservation laws taking into account specific geometry and boundary conditions of the flow in the die.

The geometry of extrusion dies is characterized by the smallness of one of their dimensions as compared with two others: $h << L$, $h << Y$, where h is the characteristic gap, or thickness, say in the direction z; and L and Y are characteristic dimensions in the directions x and y, respectively.

The Reynolds number, defined as a ratio between inertia and viscous forces, is expressed as $Re = \rho U h^2 / L\eta$. In the extrusion of polymeric materials, the ranges of variations for the characteristic velocity U, density ρ, and viscosity η are such that $Re << 1$.

This dominance of the viscous forces combined with the smallness of one of the characteristic dimensions allows the use of the lubrication approximation [56] for analysis of die performance:

$$(h^3/(12\eta)P,_x),_x + (h^3/(12\eta)P,_y),_y = 0, \quad (2)$$

where the subscripts x and y stand for partial derivatives over the variables.

Equation (2), known as Reynolds' equation, establishes the pressure distribution P(x,y) caused by motion of material in a flow domain with z spacing equal to h(x,y). The x-y plane of the domain is illustrated in Figure 2.1.

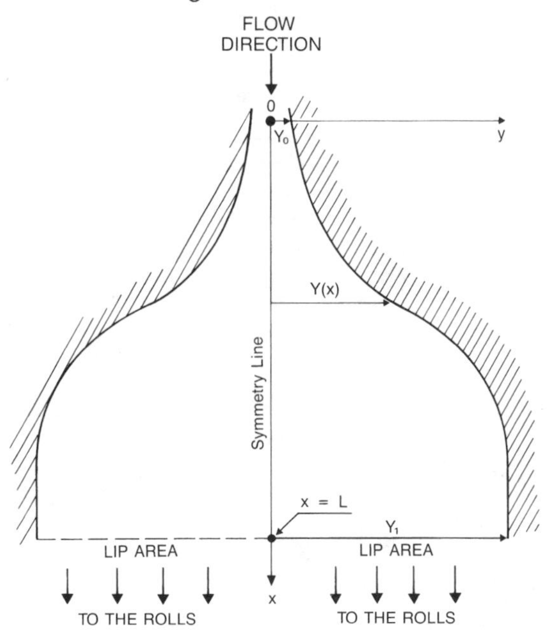

Figure 2.1 Schematic diagram of sheet die.

The shown half of the symmetric die is located between the symmetry line, $y = 0$, and the rigid border, $y = Y(x)$. The flow from the extruder enters the die at $x = 0$, and the formed sheet leaves it through the lips, i.e. the line $x = L$.

The similarity between Reynolds' equation and Laplace's Equation (3), describing a steady-state temperature distribution in a medium with the variable conductivity $k = k(T)$, is obvious:

$$(kT,_x),_x + (kT,_y),_y = 0 \tag{3}$$

Since a numerical solution of Laplace's equation can be readily obtained from any one of a large number of computer codes, the analysis of die performance can easily be conducted. The only work that has to be done to a program solving of Laplace's equation is to modify the algorithm that computes the conductivity.

To accommodate a strain-rate sensitivity of the viscosity, the effective "conductivity" of Reynolds' equation should be expressed through a gradient of the main variable - "temperature" - i.e. the pressure. Problems of stability, appearing as a result of these changes, can be resolved by a selection of the appropriate step of computation or by methods of numerical relaxation.

The dependence of the viscosity $\eta = \eta(P,_i)$ on the pressure gradient leads to an important distinction between Laplaces's and Reynolds' equations. While for the Laplace equation, with the temperature dependent conductivity $k(T)$, there exists a function F defined as

$$F = \int_{T_0}^{T} k(\Theta)/k(T_0)\, d\Theta$$

such that Equation (3) can be rewritten as $F_{,xx} + F_{,yy} = 0$, no similar function can be defined for Reynolds' equation without imposing an additional condition on the pressure field. This distinction, originating from the rheological sensitivity of polymeric materials, obviously disappears once one assumes a constant viscosity for the material.

Interpreting the results of the numerical simulation, it is possible to study flow uniformity, viscous dissipation, and the presence of stagnant areas. Figure 2.2 shows an example of the finite element mesh of the left half of the extrusion die, which was analyzed using the modifications enhanced on ABAQUS - a proprietary finite-element code from HKS, Inc.

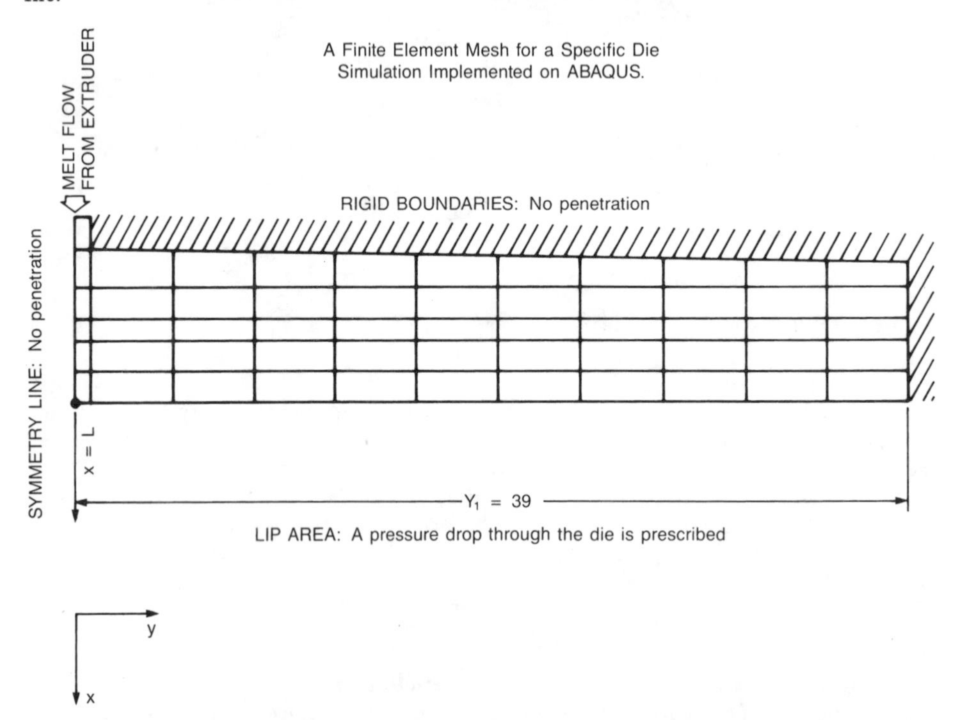

Figure 2.2 A finite element mesh for a specific die simulation implemented on ABAQUS.

Due to the symmetry, the "no penetration" boundary condition was imposed at the center line, as well as at the rigid border. The second boundary condition, "no slip," was not applied. The introduced error rapidly diminishes outside of a narrow strip along the rigid border, due to the elliptic nature of the governing Eq (2). The width of the strip is proportional to the gap of a die, i.e. it is much smaller than the characteristic dimensions of a die.

The viscosity was represented by the Spencer-Dillon model generalized as [57]:

$$\eta = \eta_0 \exp\{-\alpha \tau\} \tag{4}$$

where η_0 is the "zero stress" viscosity, τ is the shear stress, and α is the pressure sensitivity coefficient.

Distributions of flux across the lips, i.e. along the line $x = L$, are presented in Figure 2.3 for a number of pressure drops across the die. As can be seen from the results, somewhere around 1,000 psi, i.e. 7,000 kPa, the flow uniformity starts to deteriorate rapidly.

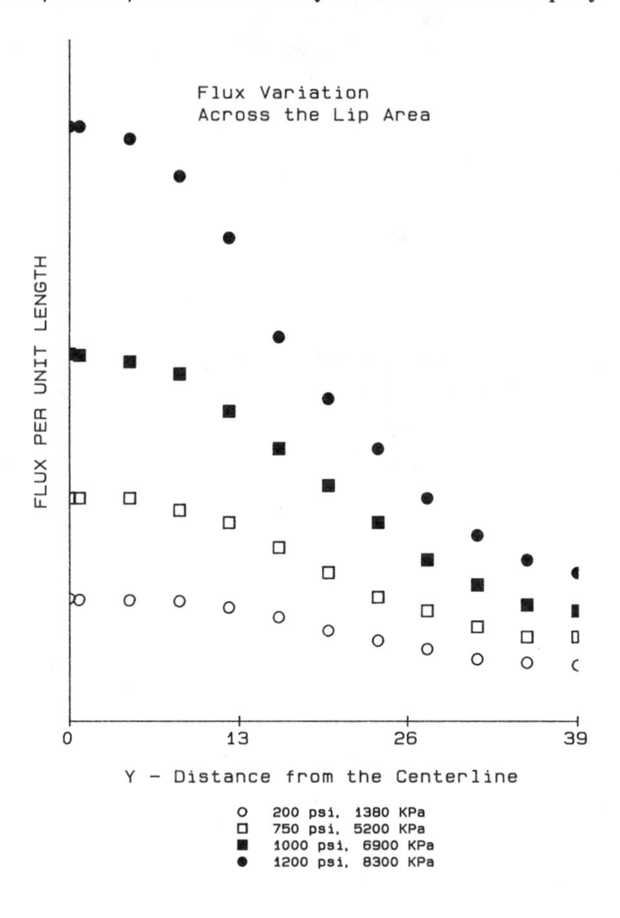

Figure 2.3 Flux variation across the lip area.

This nonlinearity of the performance, originating form the material properties of the extrudate, imposes an upper limit on the total flux, or maximum admissible pressure drops, if a defined uniformity has to be maintained.

To achieve an improved quality of die performance, the new shape of the die, h(x,y), should be found. That can be achieved using the above approach. The work consists of successive changes in the discrete arrays of h, e.g. one value per element. This sequence of adjustments continues until a desirable performance is predicted. The success and required effort depends upon the experience and intuition of the analyst.

2.6 THE INVERSE FORMULATION OF THE DIE FLOW MODEL

For the case of die flow simulation, the inverse approach enables one to start from a consequential description, i.e. the pressure distribution providing a total flux of acceptable uniformity, and proceed to the determination of the causal features, i.e. the gap distribution h(x,y), corresponding to the prescribed performance.

We shall introduce a new variable Ψ, defined as

$$\Psi = \ln (h^3 / 12\eta) . \tag{5}$$

Ψ combines the unknown geometrical parameter h and the material parameter η : These parameters are rendered non-dimensional by the appropriate scaling factors of, say, h_0 and η_0. Using the new variable, Reynolds' Equation (2) can be written as

$$\Psi,_x P,_x + \Psi,_y P,_y = - \Delta P, \tag{6}$$

where Δ stands for the Laplacian, i.e. $\Delta P = P,_{xx} + P,_{yy}$.

The left side of Eq (6) can be presented as a total derivative of Ψ along the line s:

$$d\Psi/ds = - \Delta P , \tag{6.1}$$

with the slope of s between the x and y axes coinciding with the components of the pressure gradient:

$$\underline{i} s,_x + \underline{j} s,_y = \underline{\nabla} P = \underline{i} P,_x + \underline{j} P,_y . \tag{7}$$

Thus, selecting s as a new variable, one may transform the partial differential Eq (6) into the ordinary differential Eq (6.1). Lines s passing through every point (x,y) of the flow domain are called characteristics of the equation.

It follows from Eq (7) that the slope of the characteristics is expressed as

$$(dy/dx)_s = P,_y / P,_x .$$

Along these characteristics, the full derivative of Ψ over x is expressed as

$$(d\Psi/dx)_s = \Psi,_x + \Psi,_y (dy/dx)_s = - \Delta P/P,_x . \tag{8}$$

For a known pressure distribution P(x,y), Eqs (7) and (8) enable construction of Ψ(x,y) over the total flow domain, as done for Cauchy-type boundary conditions [58]:

$$\Psi_o = \ln (Q / (2 * Y_o * (dP/dx)_o))$$

or

$$\Psi_1 = \ln (Q / (2 * Y_1 * (dP/dx)_1)),$$

where the subscripts 0 and 1 correspond to $x = 0$ and $x = L$, respectively; Q is the total flux through the die, and Y stands for the half width of the die as shown in Figure 2.1.

If the effects of the extrudate swell, caused by the elastic deformations during extrusion, are neglected, the gap h has a constant value on the line $x = L$. At the entry, $x = 0$, the value of h can be defined as a constant without loss of generality:

$$h(0,y) = h_o, \quad h(L,y) = h_1. \tag{9}$$

The values of the pressure gradient $(dP/dx)_0$ and $(dP/dx)_1$ are selected for a material with viscosity $\eta(\tau)$ to provide a required total flux Q at $x = 0$, and $x = L$, with the corresponding gap values defined in Eq (9).

The definition of the function $P(x,y)$ as $P(ax^2 + y)$, where

$$a(P)x^2 + y = C(P)$$

is the equation for the isobars, assures "no penetration" at $y = 0$ and $y = Y(x)$. The value $a(P)$ is calculated at the intersection of the isobar $C(P)$ and the outline of the rigid boundary $y = Y(x)$:

$$a(P) = Y'/2Y. \tag{10}$$

The functional dependence $P(ax^2 + y)$ can be presented, for example, as

$$P = A \exp[-(ax^2 + y)/D] \tag{11}$$

with constants A and D found from the selected values of the pressure gradient at $x = 0$ and $x = L$. The uniformity of the flow for such a pressure distribution will be assured by the selection of $Y' = 0$ in the vicinity of the lip region.

The selection of the function for $C(P)$ is not unique. The specific selection $C(P) = a(P)x^2 + y$ presents the simplest possible form of the dependence since it has a pressure dependent constant $a(P)$ and, being an even function, it assures that the "no penetration" condition on the center line is satisfied. Approximating $C(P)$ by a polynomial of higher order, one opens up the possibility to impose additional constraints on the solution, for example, providing for minimal residence time in a die or optimizing the total pressure drop.

2.7. PRESSURE DROP OPTIMIZATION AS AN ADDITIONAL CONDITION OF THE INVERSE FORMULATION

The non-uniqueness of the possible pressure distribution can be resolved by the introduction of additional conditions. As an example of such a condition, the creeping one-dimensional flow is analyzed here in conjunction with the requirement of a minimal pressure drop providing a prescribed total flux. This problem was selected because it also has an analytical solution which can be used to check the numerical result.

The relative ease of the mathematical treatment is a consequence of the simplified form of the mass conservation law for one-dimensional flow:

$$(h^3/12\eta)(dP/dx) = Q = \text{const}. \tag{12}$$

The inverse problem is defined by the selection of prescribed values for the function h(x) at the end of the interval [0,L]:

$$h(0)=h_o,\ h(L)=h_1. \tag{13}$$

The viscosity η is described as a function of the apparent shear rate $\dot{\gamma} = 6Q / h^2$ as

$$\eta = \eta_o\, \dot{\gamma}^{-k/2} \tag{14}$$

Instead of guessing P(x) and calculating h(x) from Eq (11), mass conservation is supplemented by the requirement that the resistance to the flow be found from the minimization of a functional D defined as

$$D = \int_o^L \eta h^{-2} (1+h'^2)^{1/2}\, dx. \tag{15}$$

The functional D is derived by integration of an apparent shear stress

$$\tau = \eta\dot{\gamma} = 6\,\eta\, Q/h^2$$

over the wetted area of the channel; the increment of the area is expressed as

$$ds = (1 + h'^2)^{1/2}\, dx.$$

The notations ' and '' , here and in Eq (16), denote first and second derivatives by x respectively, e.g. $h' = dh/dx$.

A necessary condition for the existence of an extremum for the functional D is provided by the Euler-Lagrange equation [58]:

$$h h'' + b(1 + h'^2) = 0, \tag{16}$$

where $b = 2 - k$.

Eq (16) can be solved using the substitution

$$h'(x) = s(h), \tag{17}$$

suggested in [59] as

$$x = x^{*} - H \int_{h(x)/H}^{1} \frac{y^{b}\,dy}{(1 - y^{2b})^{1/2}}, \quad \text{for } 0 < x < x^{*};$$

$$x = x^{*} + H \int_{h(x)/H}^{1} \frac{y^{b}\,dy}{(1 - y^{2b})^{1/2}}, \quad \text{for } x^{*} < x < L,$$

in which H and x* are constants of integration that are found from conditions Eq (13).

Since for most fluids, $d\tau/d\dot{\gamma} \geq 0$ the coefficient b in Eq (16) can have only non-negative values. In this case, $h'' \leq 0$, i.e. the optimal function h(x) has a convex shape with the maximum H reached at x*: $h(x^*) = H$, as illustrated in Figure 2.4.

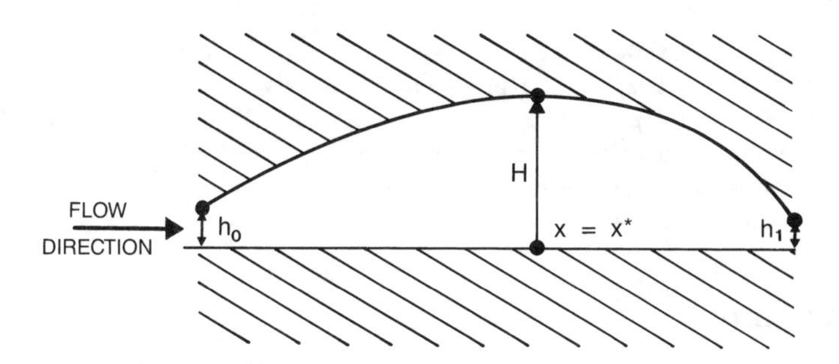

Figure 2.4 Illustration of the optimal shape for a one-dimension channel.

For $b \cong 1$, i.e. for Newtonian and generalized Newtonian fluids, H can be found from the approximation

$$L = H\left[K(1/b) - \frac{\alpha_0^{1+b} + \alpha_1^{1+b}}{1 + b}\right], \tag{18}$$

where $\alpha_0 = h_0/H$, $\alpha_1 = h_1/H$,

$$K(1/b) = 1/b\, \frac{\Gamma(1/2)\ \Gamma(1/2\,(1+1/b))}{\Gamma(1/2(2+1/b))},$$

and Γ is the Gamma function, [58]

For $b \to 0$, i.e. for an ideal plastic fluid, H is approximated as

$$H = h_0 \left(1 + b \frac{L^2}{2\,(h_1^2 + h_o^2)} \right).$$

2.8 CONCLUSIONS

Numerical solutions based on the direct approach to modeling provide a consistent interpretation of the defined model, the parameters of the process, and the boundary or initial conditions. It is necessary to proceed through a sequence of iterations in order to arrive at a viable engineering solution. The changes or adjustments made in the model, parameters, or boundary conditions during the iterations are based on a comparison of the available results with the engineering requirements. The efficiency of this iterative process depends upon the complexity of the problem and the capabilities of the analyst.

The inverse approach opens up the possibility to address the requirements explicitly. The problems of uniqueness and correctness arise and these represent a specific difficulty of this formulation.

The problem of die flow can be treated successfully in the inverse formulation. Using the method of characteristics, the die profile h(x,y) satisfying the requirements of flow uniformity and providing a prescribed total flux can be found in a single pass. The non-uniqueness of the pressure distribution can be used to introduce an additional requirement or criterion. The solution presented for one-dimensional flow coupled with a minimal pressure drop requirement illustrates this possibility.

ACKNOWLEDGEMENT

I wish to thank Prof. J.O. Wilkes (Department of Chemical Engineering, University of Michigan) and Dr. J. Bicerano (Central Research, Dow Chemical) for reading the manuscript and providing valuable discussion of its contents.

REFERENCES

1. Alifanov, O.M., *Inzh.–Fiz. Zh.*, *45*, 742 (1983).
2. Vennikov, V.A., "Similarity Theory and Modeling," Vysshaya Shkola, Moscow (1966).
3. Nelson, C.D., Yang, T., Hudson, W.G., *J. Eng. Power*, Jan, 125 (1975).
4. Laidler, P., Myring, D.F., *Aeron. J.*, *88*, 38 (1984).
5. Ahmed, N.M.A., Myring, D.F., *Intern. J. Numer. Meth. Engn.*, *22*, 377 (1986).
6. Zanetti, L., *AIAA J.*, *18*, 754 (1979).
7. Wu, Chung-hua, Brown, C.A., J.A.S., 19, 3 (1952).
8. Ericksen, J.L., in "Finite Elasticity," AMD 27, ASME, New York (1977).
9. Cherepanov, G.P., *Prikl. Mat. Mekh.*, *38*, 963 (1974).
10. Franek, A., Kratochvil J., Travnicek L., *J. Elast.*, *14*, 363 (1984).
11. Volkova, E.A., *Collect. Sci. Works*, *62*, Novosibirsk (1982).

12. Vigdergauz, S.G., *PMM USSR*, *47*, 523 (1983).
13. Vigdergauz, S.G., *PMM USSR*, *41*, 902 (1977).
14. Vigdergauz, S.G., *Mech. Solid.*, *18*, 90 (1983).
15. Kaloni, P.N., Huschilt, K., *Int. J. Non-Lin. Mech.*, *19*, 373 (1984).
16. Kuiper, L.K., *Wat. Resou. Res.*, *22*, 705 (1986).
17. Velikina, G.M., Krass, M.S., *Wat. Resou.*, *5*, 846 (1978).
18. Galanin, M.P., Tikhonov, N.A., *Vest. Mosk. Univers.*, *Ser. XV*, *3*, 70 (1985).
19. Hadamart, J., "Sur les problemes aux derivees partielles et leur signification physique," *Bull. Univ. Princeton*, *13* (1902).
20. Hadamart, J., "Le probleme de Cauchy et les equations aux derivees partielles lineaires hyperboliques," Hermann, Paris (1932).
21. Tikhonov, A.N., *Dokl. Akad. Nauk USSR*, *39*, 5 (1943).
22. Buchgeim, A.L., "Volterra Equations and Inverse Problems," Nauka, Novosibirsk (1983).
23. Tikhonov, A.N., Arsenin, V.Ya., "Solutions of Ill-Posed Problems," Halsted Press (1977).
24. Bogomolova, I.A., Glasko, V.B., Kalner, V.D., Kulik, N.I., Kondorskaya, E.E., Tikhonov, A.N., Yurasov, S.A. *Inzh.-Fiz. Zh.*, *53*, 835 (1987).
25. Alifanov, O.M., Rumyantsev, S.V., *Inzh.-Fiz. Zh.*, *53*, 843 (1987).
26. Aleksashenko, A.A., *Teor. Osn. Khim. Tekhnol.*, *13*, 188 (1979); *18*, 115 (1984).
27. Tikhonov, A.I., *Inzh.-Fiz. Zh.*, *29*, 1 (1975).
28. Aleksashenko, A.A., *Izv. Akad. Nauk USSR*, *Energ. Transp.*, *16*, 104 (1978); *20*, 122 (1982); *20*, 127 (1982); *21*, 149 (1983).
29. Artyukhin, E.A., Nenarokomov, A.V., *Inzh.-Fiz. Zh.*, *53*, 474 (1987).
30. Aleksashenko A.A., *Inzh.-Fiz. Zh.*, *33*, 1484 (1977).
31. Aleksashenko A.A., *Izv. Akad. Nauk USSR*, *Energ. Transp.*, *20*, 148 (1982); *23*, 111 (1985).
32. Frolov, V.V., *Inzh.-Fiz. Zh.*, *30*, 363 (1976).
33. Iskenderov, A.D., Dzhafarov, D.F., Gumbatov, O.A., *Inzh.-Fiz. Zh.*, *41*, 142 (1981).
34. Matsevity, Yu.M., *Inzh.-Fiz. Zh.*, *53*, 486 (1987).
35. Tsoi, P.V., Yusupov, S.Yu., *Izv. Akad. Nauk USSR*, *Energ. Transp.*, *18*, 170 (1980).
36. Matsevity, Yu.M., A.P. Slesarenko and O.S. Tsukanyan, *Inzh.-Fiz. Zh.*, *53*, 480 (1987).
37. Alnajem, N.M., Ozisik, M.N., *J. Heat Trans.*, *107*, 700 (1985).
38. Hsieh, C.K., Lin, J-f, Proc. 8-th Intern. Heat Trans. Conf. (IHTC) (1986).
39. Bratuta, E.G., Matsevity, Yu.M., Multanovsky, A.V., *Inzh.-Fiz. Zh.*, *49*, 486 (1985).
40. Hsieh, C.K., Kassab, A.J., *Int. J. Heat Mass Transf.*, *29*, 47 (1986).
41. Malakhov, N.M., *Inzh.-Fiz. Zh.*, *33*, 1062 (1977).
42. Pfahl, R.C., Mitchel, B.J., *Int. J. Heat Mass Transfer*, *13*, 275 (1970).
43. Temkin, A.G., "Inverse Methods of Heat Conduction," Energiya, Moscow (1973).
44. Stefan, J., *Ber. Akad. Wiss. Wien.*, (1890).
45. Bobula, E., Twardowska, K., *Bull. Pol. Acad. Sci*, *Tech. Ser.*, *33*, 359 (1985).
46. Szczurek, J., *Bull. Pol. Acad. Sci*, *Tech. Ser.*, *27*, 11 (1979).
47. Tanner, R.I., Proc. 7th Intern. Congr. Rheol., Gothenburg, 140 (1978).
48. Tadmor, Z., Broyer, E., Gutfinger, C., *Polym. Eng. Sci.*, *14*, 660 (1974).
49. Viriyayuthakorn, M., Caswell, B., *J. Non-Newtonian Fluid Mech.*, *8*, 95 (1979).
50. Upadhyay, R.K., Isayev, A.I., *Rheol. Acta*, *25*, 80 (1986).
51. Crochet, M.J., Proc. Xth Intern. Congr. Rheol., Sydney (1988).
52. Tanner, R.I., *J. Non-Newtonian Fluid Mech.*, *6*, 289 (1980).

53. Pearson, J.R.A., *J. Plast. Inst.*, *32*, 239 (1964).
54. Crochet, M.J., *Chem. Engn. Sci.*, *42*, 5 (1987).
55. FIDAP, Fluid Dynamics International, Inc., Evanston, Ill.
56. Reynolds, O., *Phil. Trans. Roy. Soc.*, *177*, 157 (1886).
57. Spencer, R.S., Dillon, R.E., *J. Colloidal Sci.*, *4*, 241 (1949).
58. Korn, G.A., Korn, T.M., "Mathematical Handbook," McGraw-Hill, New York (1968).
59. Gambier, B., *Acta Math.*, *33* (1910).

CHAPTER 3

INTERMESHING CO-ROTATING TWIN SCREW EXTRUDERS: TECHNOLOGY, MECHANISMS, AND SIMULATION OF FLOW CHARACTERISTICS

by J.L. White and Y. Wang

Institute of Polymer Engineering
The University of Akron
Akron, Ohio, 44325
U.S.A.

ABSTRACT
3.1 INTRODUCTION
3.2 TECHNOLOGY
3.2.1 Early Developments
3.2.2 Colombo and LMP Development (1937-1959)
3.2.3 Erdmenger, Meskat, I.G. Farbenindustrie-Bayer Early Twin-Screw Development
3.2.4 Ellermann's Continuous Mixing Machine
3.2.5 Erdmenger-Bayer Four-Screw Machine
3.2.6 Readco-Baker Perkins Development
3.2.7 Throttling Devices
3.2.8 Recent Bayer Developments
3.2.9 Hermann Berstorff Developments
3.2.10 Recent Werner and Pfleiderer Developments
3.2.11 Newer Machine Manufacturers
3.3 APPLICATIONS
3.4 FLOW SIMULATION
3.4.1 General
3.4.2 Right-Handed Screw Elements
3.4.3 Left-Handed Screw Elements
3.4.4 Kneading Disc Element
3.4.5 Modular Machine
ACKNOWLEDGEMENT
NOMENCLATURE
REFERENCES

ABSTRACT

A critical review is presented of the development of technology and modeling of flow in intermeshing co-rotating twin screw extruders. The development of technology from the beginning of the century through the modern modular intermeshing machine is described.

The simulation of flow of Newtonian and non-Newtonian fluids through screw and kneading disc elements is described. Problems of modular machines and starvation effects are considered.

3.1 INTRODUCTION

Intermeshing co-rotating twin screw extruders play an important role in polymer processing technology. They are widely used for blending, compounding of particulates, devolatilizing, and reactive extrusion. The intermeshing co-rotating design is only one of three major variations of twin screw extruders. The other two classes are intermeshing counter-rotating and non-intermeshing (often) tangential twin screw machines. These machines serve many of the same applications. Researchers and engineers dealing with twin screw machines often seem at a disadvantage because of lack of background and overview of earlier work. There are earlier reviews of note. A 1972 monograph by Herrmann [1] has given an overview of the different types of twin screw technology. A more updated review is given by White et al. [2] which covers both technology and fundamental studies of flow mechanisms. Both of these articles give extensive references. There have also been many papers through the years giving the present status of twin screw technology. These papers (most of them in German) usually have no bibliographies and most are now out of date. Two of the better ones in English are by Schutz [3] and by Prause [4].

It is our purpose in the present paper to present a critical review of the technology and efforts at modeling of intermeshing co-rotating twin screw extruders, especially modular twin screw machines. In carrying through this task, the present paper would seem to be much more extensive, detailed, and updated than our earlier review [2].

3.2 TECHNOLOGY

3.2.1 Early Developments

Intermeshing co-rotating twin screw extruders date to the 19th century. An 1869 U.S. patent of Francois Coignet [5] of Paris, France, describes a fully-intermeshing twin screw device which is shown in Figure 3.1. Coignet calls this machine a malaxator and uses it to pump artificial stone paste.

In the first decade of the 20th century, we find an intermeshing co-rotating twin screw device in a German patent by Adolf Wünsche [6] of Charlottenburg (near Berlin). This is shown in Figure 3.2.

A 1916 British patent application of Roland William Easton [7] also describes an intermeshing co-rotating twin screw machine (Fig. 3.3). It is proposed for primary use as a gravel pump, but other applications are mentioned. Easton argues that the intermeshing self-wiping character prevents backflow. Easton [8] later filed a U.S. patent which describes a more detailed view of the technology. He notes specifically that the advantage of the twin screw machine are the self-wiping screws, keeping each other clean.

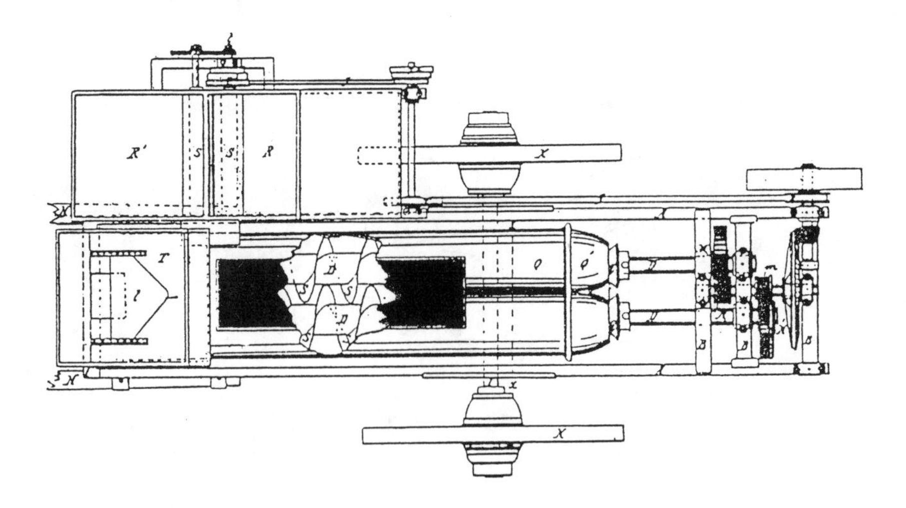

Figure 3.1 Coignet's [5] intermeshing co-rotating twin screw machine.

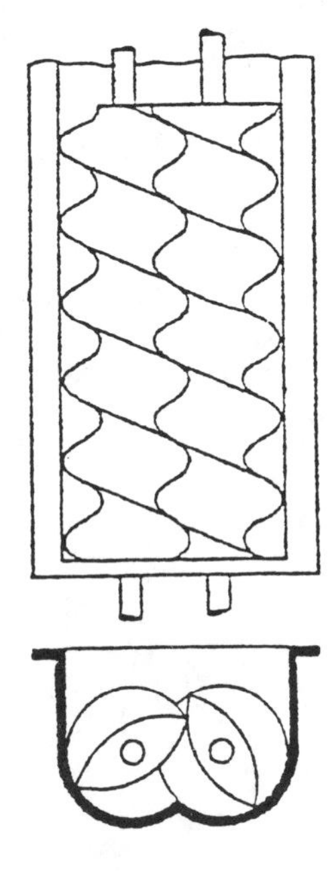

Figure 3.2 Wünsche's [6] intermeshing co-rotating twin screw machine.

A screw extrusion machine with six co-rotating screws arranged in a circle is described in a 1920 patent by H. La Casse [9]. This is represented as a mixing machine for plastic materials. The screws are not fully self-wiping.

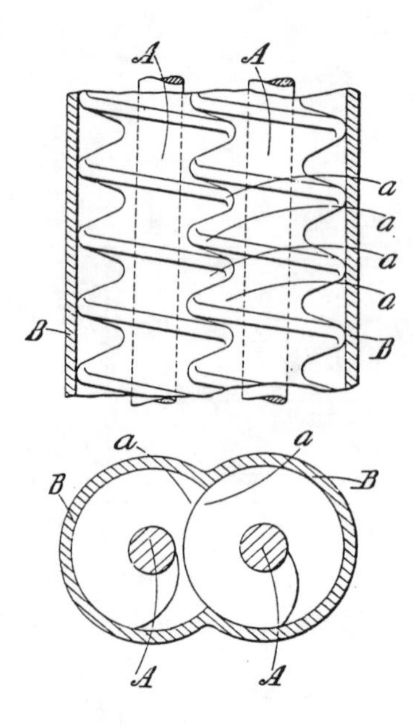

Figure 3.3 Easton's [7] intermeshing co-rotating twin screw extruder.

A 1932 American patent by William K. Nelson [10] of Chicago assigned to the Universal Gypsum and Lime Company, describes an improved method to produce gypsum wall board which uses, in part, a machine with two co-rotating double-tipped intermeshing "paddles" which wipe and scrape each other clean (Fig. 3.4). This is mentioned in the patent claims.

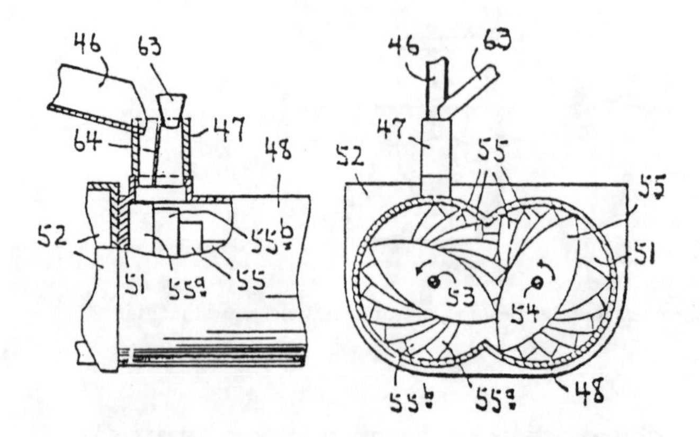

Figure 3.4 Nelson's [10] intermeshing co-rotating mixing paddle machine.

The fully-intermeshing co-rotating twin screw extruder is again described in a 1936 patent by Fred Forest Pease [11] of Lever Brothers in Cambridge, MA (Fig. 3.5). There is considerably more engineering detail in the Pease patent than in the Coignet, Wünsche, or Easton patents.

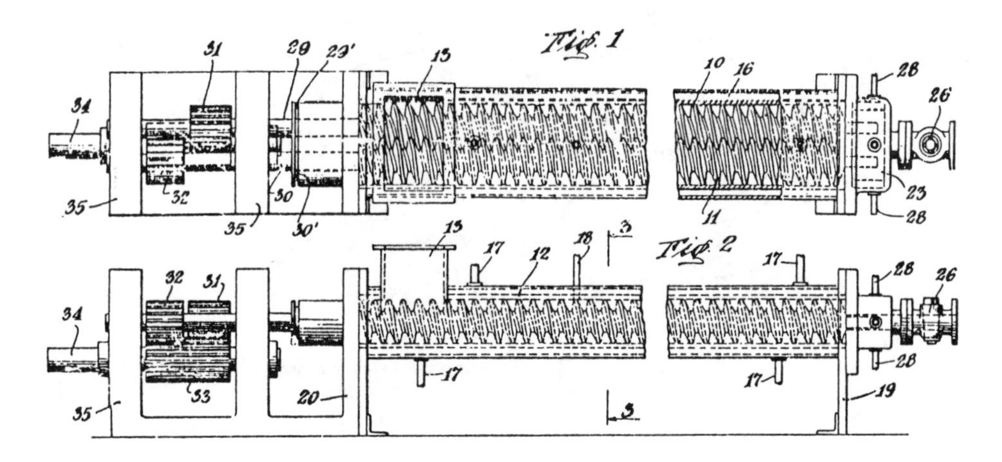

Figure 3.5 Pease's [11] intermeshing co-rotating twin screw extruders.

3.2.2 Colombo and LMP Development (1937-1959)

In 1937 Roberto Colombo had been one of the principals in the founding of Lavorazione Materie Plastische (LMP) in Turin, Italy [12]. The company was formed to produce thermosetting plastics for the Italian automotive industry. Colombo developed an intermeshing co-rotating twin screw extruder in the course of these activities and it was decided to commercialize it. The first LMP twin screw extruder was built in 1938 for the extrusion of polyvinyl chloride. In 1939, LMP sold its first lot of machines to the I.G. Farbenindustrie in Germany. Colombo [13] filed for patents in Italy in February 1939, in France in February 1940 and in Switzerland in January 1941 (Fig. 3.6). The Italian and Swiss patents describe twin screw and multiple [13] screw designs. During World War II, the twin screw extrusion activities LMP were disrupted.

LMP does not indicate what part of the I.G. Farbenindustrie (which included BASF, Bayer, Hoechst, and other firms) purchased the early twin screw machines. However, it would appear that this was the Ludwigshafen (BASF) facility. In 1943, the I.G. Farbenindustrie filed the first twin screw reactive extrusion patent [14] for addition (sodium polymerization of butadiene) and condensation (hexane diisocyanate butanediol) polymerization. This used a Colombo-type machine. The patent was not issued until 1953 when it was assigned to BASF.

In the post-war period, Colombo and LMP filed patents throughout the world. A British patent was filed in 1946 and issued to LMP in 1949 [15]. A US patent was filed by Colombo [16] in 1947 and issued in 1951. A Canadian patent was filed in 1949 and issued in 1956 [17]. These patents describe twin and multiple-screw intermeshing self-wiping machines. The screws have multiple modular but machined sections to gradually compress the material

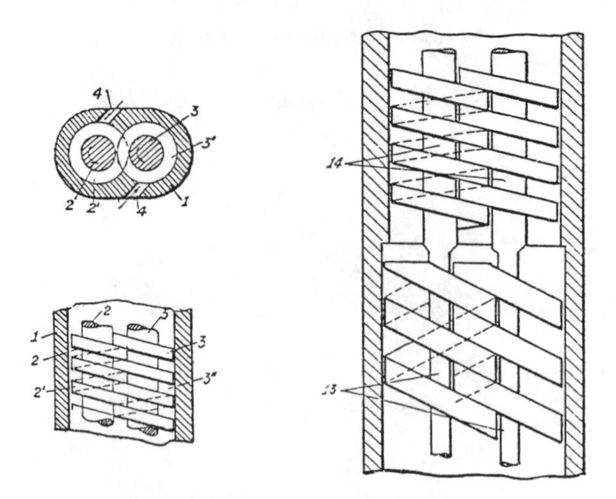

Figure 3.6a Colombo's [13] intermeshing co-rotating twin screw machine.

as it moves along the screw. The barrel diameter also decreases as one moves from section to section. The purpose was to build up pressure along the screw.

The LMP Company licensed R.H. Windsor Ltd. in Britain, C.A.F.L. in France, and Ikegai Iron Works in Japan to manufacture Colombo designed intermeshing co-rotating twin screw extruders [12]. The R.H. Windsor Company marketed their machines in the USA through the F.J. Stokes Company.

1953 papers by S.H. Greenwood [18,19] describe the "Stokes-Windsor" intermeshing co-rotating twin screw extruder and show photographs of machined modular screws. The machine is recommended by Greenwood for extrusion of rigid polyvinyl chloride. It is erroneously described as a constant volume or positive displacement pump. This is the case for fully-intermeshing counter-rotating but not co-rotating machines.

Several papers in *Kunststoffe* in the 1950s describe the Colombo-LMP-Windsor machine for German language audiences [20-22]. A 1956 paper by Baigent [23] described the machine for British audiences.

In 1957 K.H. Baigent [24] of R.H. Windsor, an LMP licensee, developed an intermeshing co-rotating twin screw injection molding machine. This involved non-modular screws in a constant cross-sectional barrel. The patent was issued in 1959.

3.2.3 Erdmenger, Meskat, I.G. Farbenindustrie-Bayer Early Twin Screw Development

A more far reaching program on intermeshing co-rotating twin screw extrusion was initiated by the I.G. Farbenindustrie Wolfen Works (near Bitterfeld). This program is associated with the career of Rudolf Erdmenger [25], who had worked with the firm of Maschinenfabrik Paul Leistritz in Nuremburg in 1937-39 before joining the I.G. Farbenindustrie. At Leistritz, engineers were working with the I.G. Farbenindustrie in Hoechst to develop an intermeshing counter-rotating twin screw kneading machine [26]. Such counter-rotating twin screw machines were widely known as positive displacement screw pumps. However, at Paul

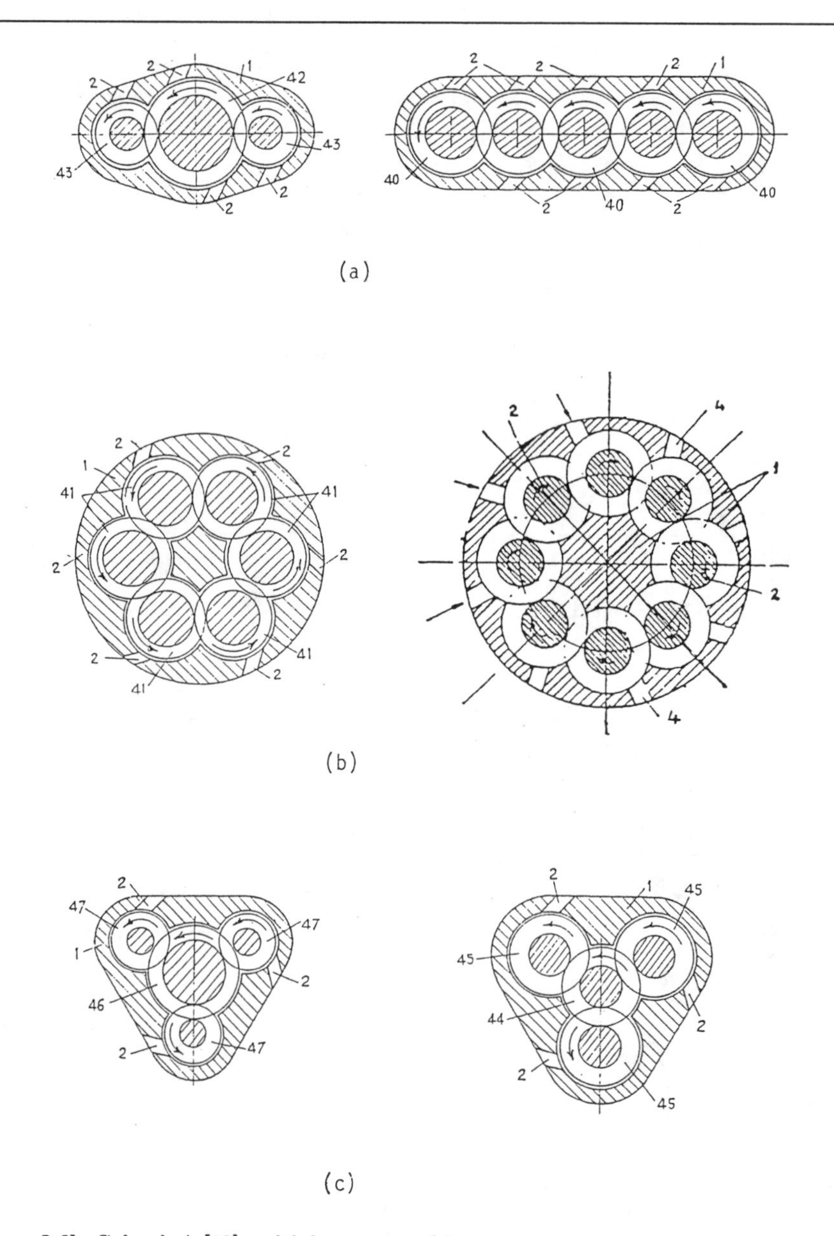

Figure 3.6b Colombo's [13] multiple screw machines.

Leistritz they sought to develop kneading action by varying pitch and flight/channel dimensions in the direction of flow. At the I.G. Farbenindustrie Wolfen Works in the 1940s, Erdmenger with Walter Meskat and A. Geberg working under the direction of K. Riess turned their attention to twin screw mixing and processing machines. They chose, however, to work instead with intermeshing co-rotating rather than counter-rotating machines. This was described in two 1944 patent applications by Meskat and Erdmenger [27,28] (Fig. 3.7).

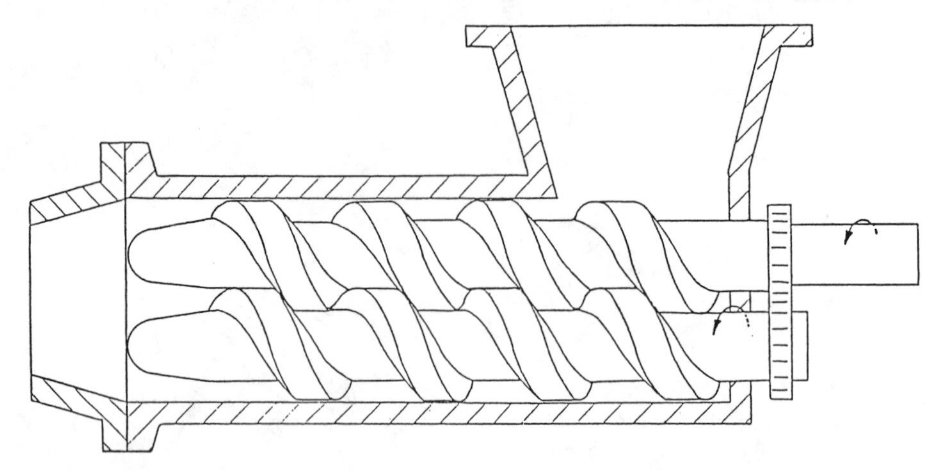

Figure 3.7 Meskat-Erdmenger [27] intermeshing co-rotating twin screw extruder.

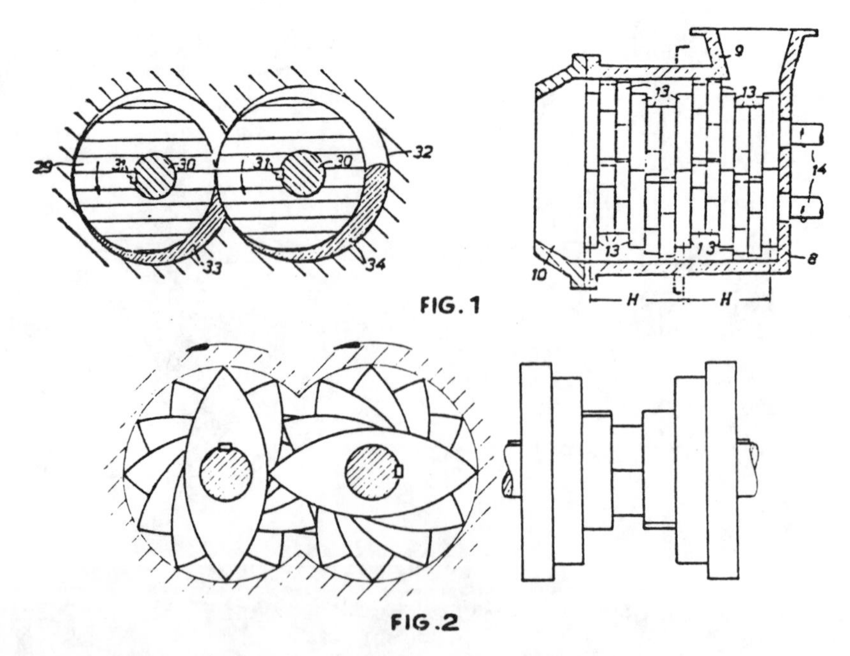

Figure 3.8 Erdmenger's [30] single- and double-tipped kneading disc mixer.

Following the end of World War II, the Americans occupied the area near Bitterfeld. When the area was to be turned over to the Russians, the Americans offered the Wolfen Works scientists and engineers the opportunity to move into Allied-occupied Western Germany. Riess was able to obtain a high executive position with Bayer in Leverkusen. Through Riess, Erdmenger was able to obtain a position with Bayer in Leverkusen and Meskat, a position with Bayer in Dormagen.

Erdmenger devoted himself to twin screw machines in Leverkusen. In September 1949, Erdmenger filed for patents on two new co-rotating twin screw mixing machines. One device was the intermeshing co-rotating labyrinth screw machine [29]. The second was a co-rotating kneading disc machine [30]. Discs with one and two lobes were used. This is shown in Figure 3.8. Machines of this type were manufactured for Bayer by Leistritz [31]. The kneading discs were keyed onto the shafts in a modular manner. A patent on the kneading disc mixer was applied for in the United States. However, only the cam-type single lobe design was allowed [32] apparently because of the Nelson 1932 patent [10]. The kneading disc machine was applied by Erdmenger to various applications. A 1953 patent application of Telle and Erdmenger [33] uses it for producing powders from agglomerates.

In Dormagen, Meskat had been joined by J. Pawlowski to work on extrusion technology. In December 1950, Meskat and Pawlowski [34] filed for a patent on a modular intermeshing co-rotating twin screw extruder with both right-handed and left-handed screw elements. The right-handed screw elements were forward pumping and developed pressure. The left-handed screw elements were backward pumping and reduced pressure. This is shown in Figure 3.9. The patent issued only in 1956.

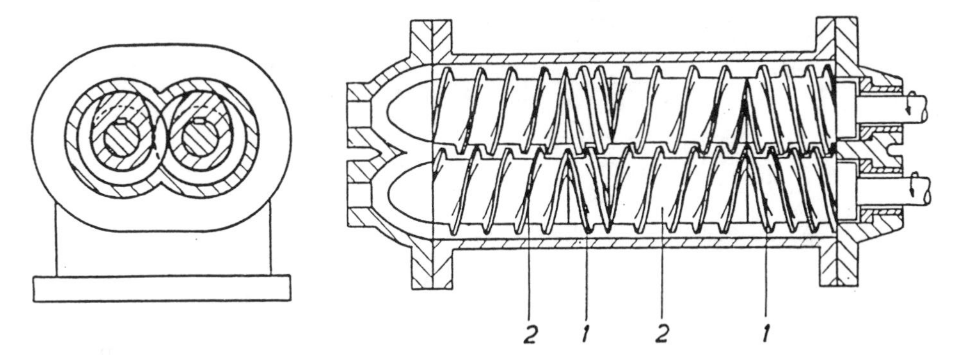

Figure 3.9 Meskat-Pawlowski [34] modular intermeshing co-rotating twin screw extruder with right-handed and left-handed screw elements.

A devolatilizing twin screw arrangement involving two transversely arranged intermeshing co-rotating twin screw extruder was described in a December 1951 patent application by Winkelmuller et al. [35] of the Bayer group. This device and its principle are shown in Figure 3.10.

In 1950-51 the Bayer team published several papers in the open (German) literature. A 1950 paper by Meskat [36] mentions recent developments including counter-rotating twin screw mixers and the Buss KoKneader but not co-rotating twin screw machines. A 1951 paper by Riess and Meskat [37] in the same journal on screw machines for the process technology of viscous and plastic fluids describes the Meskat-Erdmenger [27] patent application and its use to produce green carbon electrodes from coke and pitch. They also describe the

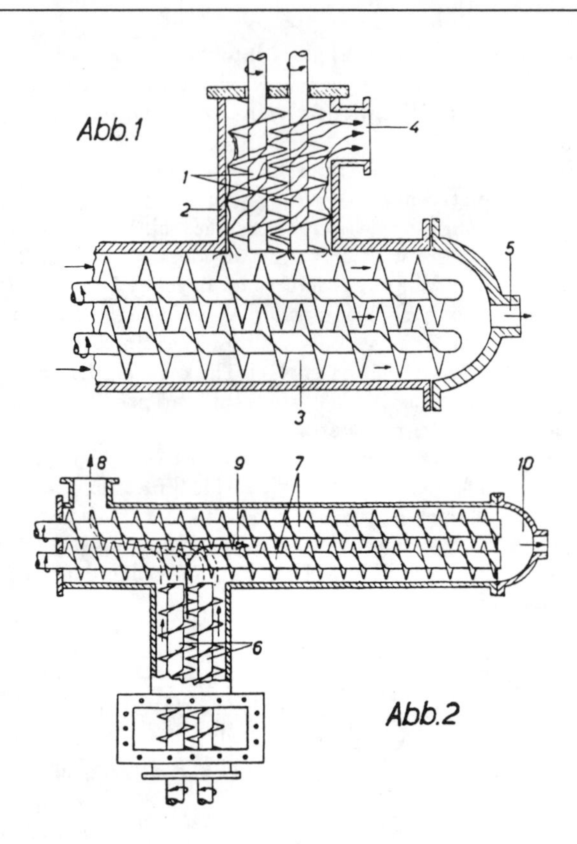

Figure 3.10 Winkelmuller et al. [35] devolatilization machine.

Meskat-Pawlowski [34] modular twin screw machine with forward and backward pumping screw elements. The same year, a paper by Riess and Erdmenger [31] on continuous kneading processes describes the Meskat-Erdmenger [27] twin screw machine, and Erdmenger's [30] kneading disc mixer. Riess [38] returned to the subject in a 1955 paper which discusses the use of the Meskat-Pawlowski device to prepare spinning solutions and to esterify cellulose. This is probably the first paper in the open literature to discuss reactive twin screw extrusion. We show a remarkable figure from this paper [38] showing the complexity of combinations of twin screw machines being used by Bayer in this period (Fig. 3.11).

In 1953, Erdmenger applied for patents for a kneading disc mixer with triple-tipped kneading discs (Fig. 3.12). He received a German patent in 1956 [39] and an American patent in 1957 [40].

The year 1953 also saw the joining of Bayer with Werner and Pfleiderer of Stuttgart to manufacture a commercial modular twin screw machine [1,41]. Werner and Pfleiderer appear to have had an exclusive license of the Bayer co-rotating twin screw technology. The Werner and Pfleiderer team was led by G. Fahr and H. Ocker and included R. Fritsch and H. Herrmann. The first prototype was built in 1955. 1958-59 saw Erdmenger filing patents in Germany and the USA for a modular twin screw machine with both kneading discs and screw

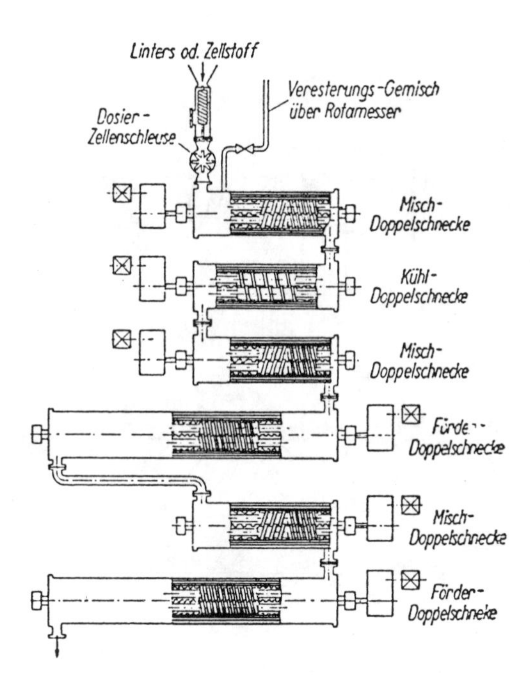

Figure 3.11 Bayer process for continuous esterification of cellulose [38].

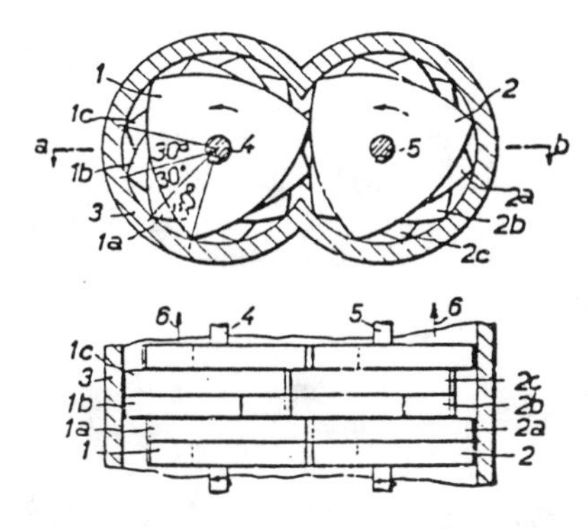

Figure 3.12 Erdmenger's triple-tipped kneading disc mixer.

elements (Fig. 3.13). The patent emphasizes the modularity of the machine and suggests the occurrence and control of starvation in the machine. A U.S. patent was received in 1964 [42].

In 1959, Fritsch and Fahr of Werner and Pfleiderer [43] announced the commercialization of a modular intermeshing co-rotating twin screw extruder with screw and triple-tipped kneading disc elements. The machine was called ZSK (Zwei Schnecken

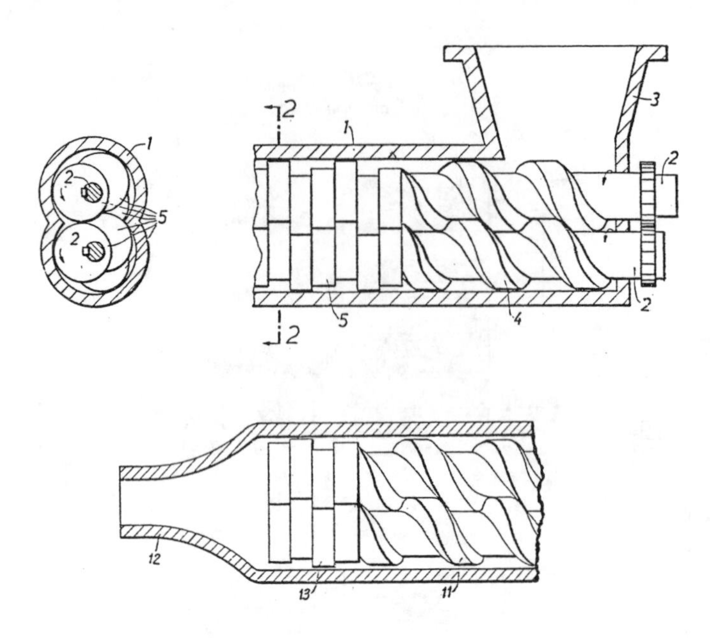

Figure 3.13 Erdmenger's [41] modular intermeshing co-rotating twin screw extruder.

Kunststoffe) System Erdmenger. Application to kneading, devolatilization mixing and granulating was described as was compounding glass fibers into thermoplastics.

During the 1960s, the modular intermeshing co-rotating twin screw extruder was the subject of papers by Erdmenger [44,45] and by Herrmann [46]. The first Erdmenger paper [44] describes the development of twin screw devolatilization machines. The second paper [45] involves fundamental studies of flow mechanisms in modular intermeshing corotating twin screw machines. Erdmenger's individual efforts at Bayer continued. In the mid 1960s, he developed a new self-wiping intermeshing co-rotating screw profile [47].

3.2.4 Ellermann's Continuous Mixing Machines

A most interesting development of twin screw machines is associated with Wilhelm Ellermann. Like Erdmenger, Ellermann began his career working with counter-rotating twin screw machines and later moved on to the development of co-rotating twin screw mixers. During the early 1940s (if not earlier), Ellermann invented an intermeshing counter-rotating twin screw machine consisting of machined shafts with a sequence of a screw, kneading rotor and a second screw section. A patent was filed [48] in January 1941 but did not issue until January 1945. The name of the inventor and the assignee were purposely omitted on the patent. The machine was however manufactured by Krupp in the Grusonwerk and marketed in the 1940s as the Knetwolf. Herrmann, in both his monograph [1] and later review paper [41], publishes photographs taken by Ellermann of the machined screw shafts of the Knetwolf. This machine was to be applied to kneading synthetic rubber.

In the early 1950s, we find Ellermann associated with the firm Josef Eck und Sohne of Dusseldorf. A patent filed in 1951 by Ellermann [49] and assigned to Eck involves two machined counter-rotating screw shafts consisting of machined screws, with screws nearest the feed being intermeshing, followed by an intermeshing kneading rotor and subsequently by a pair of tangential screw kneading to the output. This was marketed by Eck as the "Eck-Mixtruder."

A subsequent patent by Kulgen et al. [50], which was filed in 1953 and assigned to Eck, separates the final pair of screw elements into separate housings, removes the screw thread from the shafts and places them on the housings. The housing is removable and can be changed when different materials are processed.

The machines described above were all counter-rotating. In 1953, Ellermann [51] filed a patent assigned to Eck for a co-rotating twin screw machine consisting of machined shafts containing intermeshing co-rotating twin screw elements followed by intermeshing kneading elements and an output section where the shafts enter separate housings which have screw threads cut into them. This is shown in Figure 3.14.

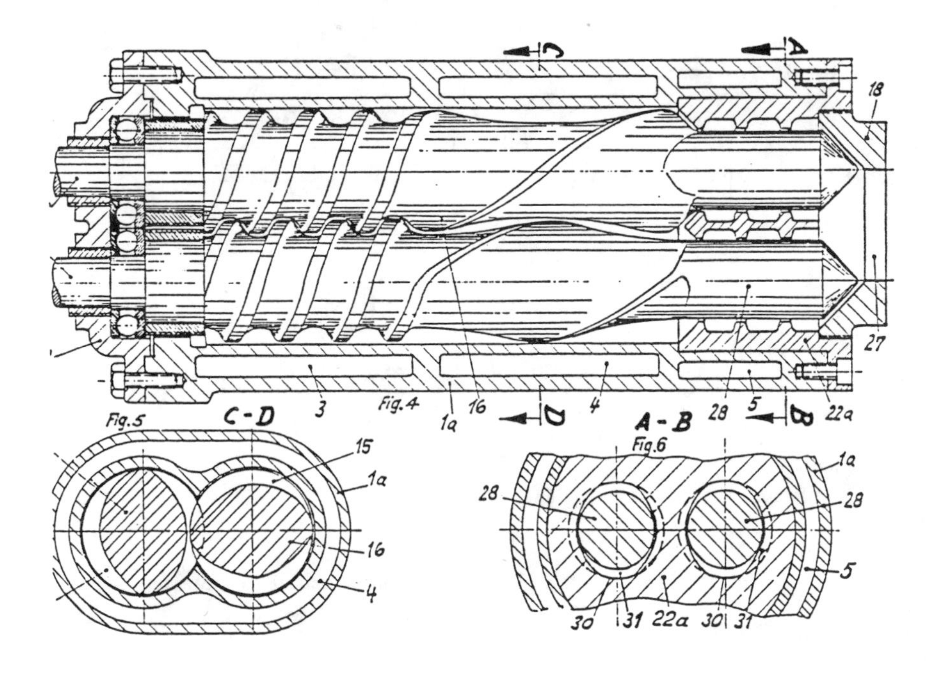

Figure 3.14 Ellermann's [51] intermeshing co-rotating twin screw machine patent.

However, Ellermann's co-rotating machine was not to be manufactured by Josef Eck und Sohne. The patent was licensed to Krauss-Maffei which is based in Munich. The machine was marketed as the Doppelschneckenmischer (DSM) [52]. The machine as described by Proksch [52] differs from that described in the Ellermann patent [51] in having the shafts threaded rather than the machine housing in the final output section.

3.2.5 Erdmenger-Bayer Four-Screw Machine

The activities of Erdmenger and his co-workers at Bayer in Leverkusen on intermeshing co-rotating twin screw machines continued. The next major development was a four-screw machine consisting of two pairs of intermeshing co-rotating twin screw combination posed at each other in a tangential counter-rotating manner. This is described in a 1960 application for a German Auslegeschrift issued in 1961 to Erdmenger and Oetke [53]. The intention of the machine was for devolatilization of high viscosity liquids. The machine is shown in Figure 3.15. It was called a Vierwellenschnecken Verdampfer. In several publications in succeeding years, Erdmenger [44,45,54] was to describe this machine. As with the modular intermeshing co-rotating twin screw extruder, Bayer licensed the technology to Werner and Pfleiderer of Stuttgart who manufactured the machine as the Vierwellige Schneckenverdampfer VDS-V. This is described by Herrmann [1].

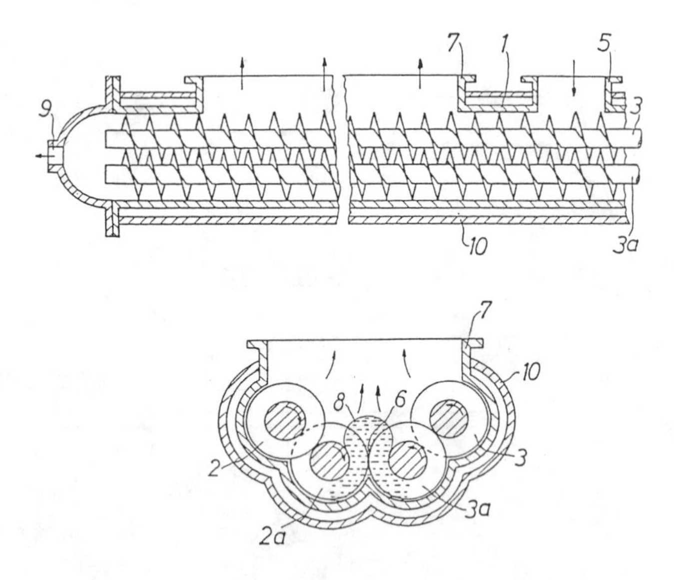

Figure 3.15 Erdmenger and Oetke's [53] modular intermeshing co-rotating twin screw extruder.

3.2.6 Readco-Baker Perkins Development

An independent but later development of a modular intermeshing co-rotating twin screw machine took place at Readco (formerly the Read Machinery Company of York, Pennsylvania) in the USA. The machine is described in patents applied for in 1962 by Bernard A. Loomans and Ambrose K. Brennan, Jr. [55,56] (Fig. 3.16). The patents and technology were transferred to Baker Perkins of Saginaw, Michigan, prior to the issuing of the patents. Several employees, including Loomans, joined Baker Perkins in Saginaw. The patents, when issued, were assigned to Baker Perkins. The machine as described in these patents consists of

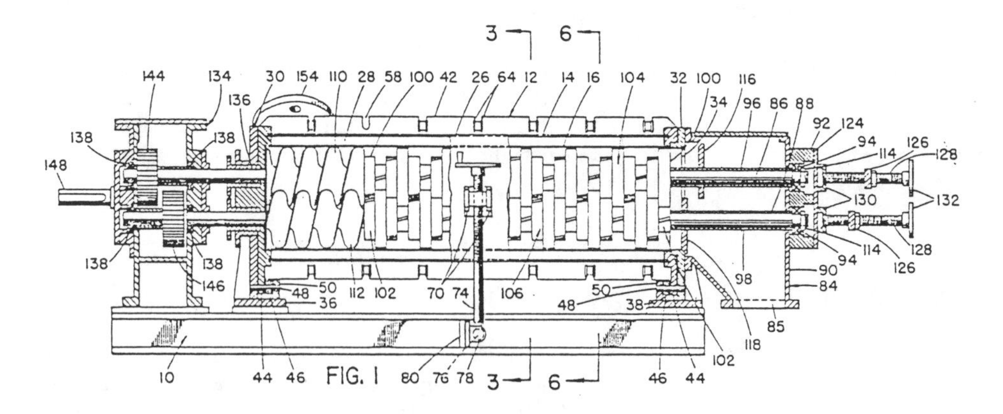

Figure 3.16 Loomans-Brennan [55] modular intermeshing co-rotating twin screw extruder.

two parallel co-rotating shafts having modular pairs of keyed selfwiping screw and two-tipped paddle elements. The paddles resemble the "paddles" of Nelson's [10] 1932 patent and the kneading discs of Erdmenger's later patents [30,32,39,40,42]. However, it is to be remembered that the Nelson patent, which was long since expired in 1962, was held valid for the double-tipped paddle and Erdmenger's U.S. patents only cited singleand triple-tipped designs. The Loomans-Brennan patents also describe a barrel which may be vertically split along two mating barrel sections.

As noted, Loomans joined Baker Perkins in Saginaw, Michigan, and continued his development of the modular intermeshing co-rotating twin screw machine on patents filed from 1967 on [57-60]. Most interesting of these patents is Loomans' development of a new self-wiping screw profile (Fig. 3.17) which Baker Perkins has incorporated as a new element into their modular twin screw machine.

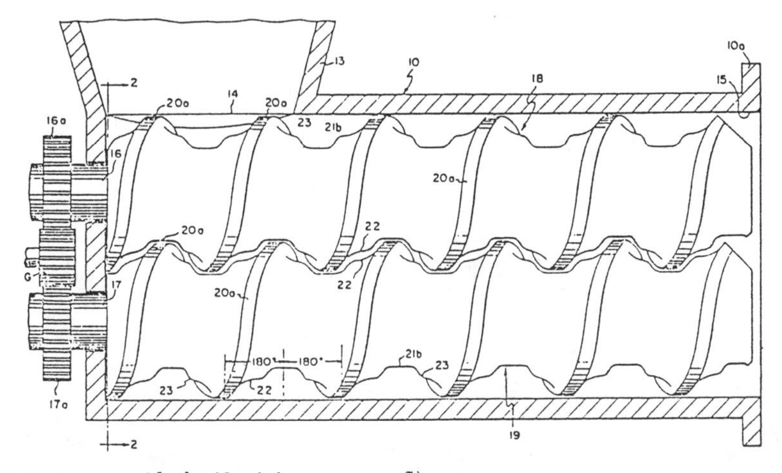

Figure 3.17 Loomans' [60] self-wiping screw profile.

In 1969, Ambrose K. Brennan, Jr. [61], still in the Readco laboratories but now with Teledyne which had absorbed Readco, filed a patent on a modular intermeshing co-rotating twin screw extruder. The key part of this machine is the design of the paddle region. The paddles which are double-tipped are disposed to 90° out-of-phase on the two different screw shafts. The assembly of paddles is arranged in a helix. This is argued by Brennan to give better shear and mixing. Subsequently, Teledyne-Readco began the manufacture of modular intermeshing co-rotating twin screw extruders.

3.2.7 Throttling Devices

The placement of Venturi-type devices in single screw extruders as throttling devices for reducing pressure and removing volatiles has a long history [62]. Such a device was first introduced into intermeshing co-rotating twin screw extruders by Roberto Colombo [63] in a 1961 patent application which was assigned to LMP. Colombo describes in this patent a modular twin screw machine which contains only right-handed screw elements and Venturi throttles. He describes the application of removing bubbles in pipe extrusion.

A 1977 patent application by Todd [64] discusses an adjustable flow controlling saddle element for a modular intermeshing co-rotating twin screw extruder which would operate like a throttle.

3.2.8 Recent Bayer Developments

Rathjen and Ullrich [65,66] of Bayer AG, in 1980, developed new screw elements containing slices in their flights. These elements were developed to aid in the continuous crystallization of low molecular weight compounds (Fig. 3.18). The slices seem to help in breaking up the crystals and thus aid in producing a continuous desirable product.

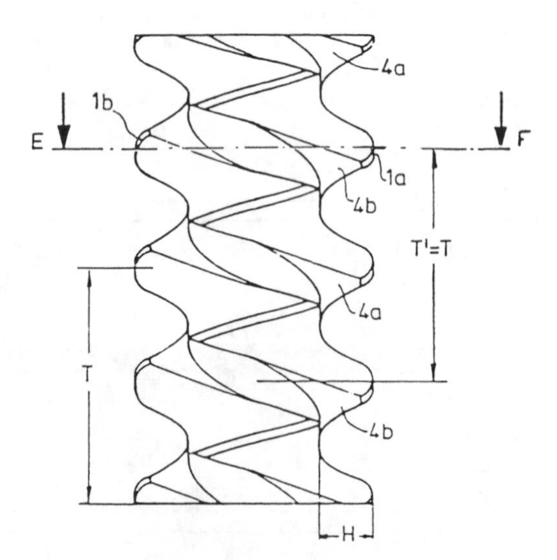

Figure 3.18 Rathjen and Ullrich's [65,66] modified screw element.

3.2.9 Hermann Berstorff Developments

Hermann Berstorff Gmbh of Hannover, Germany [67-69] introduced a modular intermeshing co-rotating machine in 1981. They proposed machines divided into sections where (i) the screw has a relatively large diameter and the barrel has a relatively small diameter for the purpose of high torque, plasticizing, and homogenizing, (ii) the screw has a small diameter and the barrel has a large diameter for degassing.

Berstorff has also introduced "mixing gears" as well as kneading discs for the purpose of mixing. Berstorff's unpatented kneading gears give improved distributive mixing without the severe shearing action of kneading discs. Another innovation of Berstorff are screw elements which are incompletely self-wiping. The leading screw flight is not self-wiping, but the neutral flight is.

3.2.10 Recent Werner and Pfleiderer Developments

Werner and Pfleiderer [70] has introduced screw elements with slices in their flights into their machines. These are called "screw mixing elements" and seem to be intended for giving gentler mixing than kneading block elements. They provide distributive mixing without high intensity shear zones. Werner and Pfleiderer also have been making a transition from three-lobed to two-lobed screws and kneading discs. There has also been a movement to systems with greater diameter ratios of the maximum to the minimum cross-section. The purpose is to offer greater free volume. A second thrust has been better gear boxes, designs, and materials of construction to allow higher torques and throughputs.

3.2.11 Newer Machine Manufacturers

Additional machinery manufacturers have entered the market place to manufacture intermeshing co-rotating twin screw machines.

One such manufacturer of modular intermeshing co-rotating twin screw extruders is Maschinenfabrik Paul Leistritz of Nuremburg who developed the intermeshing counter-rotating machine. Leistritz has also pioneered a modular laboratory machine which may be operated in a co-rotating or counter-rotating manner with suitable elements.

Japan Steel Works, Mistsubishi Heavy Industries, Toshiba Machine, and Kobe Steel have developed modular intermeshing co-rotating twin screw extruders. They also manufacture a laboratory machine which may be either co-rotating or counter-rotating.

Betol Machinery Ltd, Luton, Bedfordshire, England, has introduced a modular intermeshing co-rotating twin screw machine [71].

Older Colombo licensees such as Ikegai of Japan and Clextral (formely C.A.F.L.) of France now make modular Erdmenger machines.

3.3 APPLICATIONS

A large patent technology has been developed on applications of intermeshing co-rotating twin screw extruders. Much of the early patent literature is associated with Werner and Pfleiderer [72-79] in patents filed from 1964 on. These patents variously treat application technologies such as blending of different polymers and their homogenization [73,76], devolatilization [74], polymerization [75], the continuous manufacture of chocolate [77], preparation of thermoset compounds [78], and the continuous manufacture of soap [79]. The patents outline the roles of the screw and kneading disc elements in the operation of the machine. Ocker [74], for instance, notes the role in left-handed elements reducing pressure and preceding devolatilization regions. Illing [75] discusses how screw and kneading disc elements interact in carrying out polymerization reactions.

Reactive extrusion in modular intermeshing co-rotating twin screw extruders has gained increasing attention. We have already cited the 1943 I.G. Farbenindustrie application [14] and the Bayer AG 1950s efforts [38] on reactive extrusion in co-rotating twin screw machines. The 1964 Illing patent application [75] opens up a new period. We find similar patents in succeeding years from other sources. In 1969, Wheeler, Irving and Todd [80] of Baker Perkins patented continuous polycondensation of pre-polymers of polyester and polyamide in their machine and subsequently Upjohn [81] obtained patents on continuous polycondensation of polyurethanes. The Bayer AG team of Erdmenger et al. [82] became openly involved in reactive extrusion developing processes for crosslinkable powdered lacquers. Ullrich and his coworkers [83] subsequently patented a process for producing polyurethanes. Since this period in the early 1970s, the patent literature on reactive extrusion has grown enormously. The use of intermeshing co-rotating twin screw extruders as continuous reactors has been expanded upon by Illing [84] as well as by Werner and Pfleiderer engineers [85,86].

3.4 FLOW SIMULATION

3.4.1 General

The intermeshing co-rotating twin screw extrusion machines as marketed today are complex in character generally being modular and containing several different types of machine elements. We must thus discuss the simulation of flow in each type of element individually. Further, we must consider the flow in the composite system formed from the individual modules.

It will be our purpose in this section to critically summarize efforts at simulation of flow in different categories of machine elements in modular intermeshing co-rotating twin screw extruders. We begin in the next section with right-handed screw elements. We then move on successively to left-handed screw elements and kneading discs.

3.4.2 Right-Handed Screw Elements

The greatest efforts made to analyze flow in elements of intermeshing co-rotating twin screw extruders have been for right-handed screw modules [8,45,87-97]. The earliest efforts aimed at seeking flow mechanisms were certainly by the inventors themselves. This is clear in the U.S. patent of Easton [8]. Erdmenger [45] carried out basic experimental studies of the material motions and described the figure "8" flow around the screws and the role of barrel drag. Armstroff and Zettler [87] of BASF, recognizing as did Easton and Erdmenger, that the screw channel is open all along the length of the screw applied the well established one-dimensional drag and pressure flow in single screw extruder equations [98] to represent the flow. i.e. they write the flow rate Q as:

$$Q = \frac{1}{2}\pi DN\cos\theta HW F_D - \frac{H^3W}{12\eta}\frac{\Delta p}{L}F_P \tag{1}$$

where N is screw speed, H, channel depth, W, channel width, θ, helix angle, η, viscosity, p, pressure, L, channel length, and F_D and F_P shape factors.

This view was expanded by Herrmann and Burkhardt [88,89] of Werner and Pfleiderer in 1978. They described both the velocity and shear stress fields across the cross-section of intermeshing co-rotating twin screw as compared to full intermeshing co-rotating twin screw machines. We envisage a coordinate system with the direction along the screw channel being "1," the direction normal to the screw being "2," and that in the transverse being "3" (Fig. 3.19). If we only consider shearing between the screw and barrel, the Navier-Stokes equations for a lubricated shear dominated flow become:

$$0 = -\frac{\partial p}{\partial x_1} + \frac{\partial \sigma_{12}}{\partial x_2} = -\frac{\partial p}{\partial x_1} + \eta\frac{\partial^2 v_1}{\partial x_2^2} \tag{2}$$

This has the solution:

$$v_1(x_2) = \pi DN\cos\theta\left[\frac{x_2}{H}\right] - \frac{H^2}{2\eta}\frac{\partial p}{\partial x_1}\left[\left[\frac{x_2}{H}\right] - \left[\frac{x_2}{H}\right]^2\right] \tag{3}$$

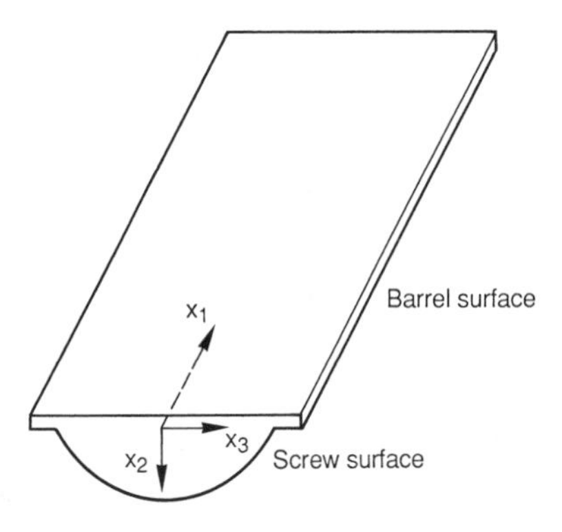

Figure 3.19 Flow in screw channel in intermeshing co-rotating twin screw extruder.

In 1976, Hans Werner of Werner and Pfleiderer presented a Dr. Ing. Dissertation [90] at the University of Munich that presented rather profound studies of flow in both screw and kneading disc elements. This was later published in part by Werner and Eise [91]. Werner argues that the flow in screw elements is starved and there is no pressure gradient.

$$\frac{\partial p}{\partial x_1} = 0 \tag{4}$$

Werner notes that the equations of motion reduce to:

$$0 = \frac{\partial^2 v_1}{\partial x_2^2} + \frac{\partial^2 v_1}{\partial x_3^2} \tag{5}$$

i.e. Laplaces equation. According to Werner's results:

$$Q = \alpha \phi N \tag{6}$$

where ϕ is fill factor.

In 1980, Denson and Hwang [92] presented a simulation of flow in a fully-filled self-wiping co-rotating twin screw channel. They solved:

$$0 = -\frac{\partial p}{\partial x_1} + \eta \left[\frac{\partial^2 v_1}{\partial x_2^2} + \frac{\partial^2 v_1}{\partial x_3^2} \right] \tag{7}$$

The flow problem is formulated again with the coordinate system embedded in the screw and the barrel is considered to move over it. The screw boundary is presumed to follow the shape of the self-wiping screw profile, i.e. it has the form:

$$v_1(x_1, H, x_3) = U \tag{8a}$$

$$v_1[x_1, h(x_3), x_3] = 0 \tag{8b}$$

where

$$x_2 = h(x_3) \tag{9}$$

defines the screw root profile. The profile shape $h(x_3)$ used by Denson and Hwang was in a form determined by Booy [93]. Velocity fields and pressure profiles were computed using finite element methods.

Denson and Hwang calculated quantitative screw characteristic curves, i.e.:

$$Q = F[\Delta p, N] \tag{10}$$

were computed. Their results were given in dimensionless form.

This was done as follows: Let us first explore the representations of Eq (10). We may rewrite this as:

$$\frac{Q}{2\pi R^3 N \cos\theta} = \frac{1}{2}\left[\frac{HW}{R^2}\right] F_D - \left[\frac{1}{12}\left[\frac{H^3 W}{R^4}\right] F_p\right] \frac{R}{2\pi N \cos\theta} \frac{\Delta p}{L} \tag{11}$$

This suggests the plot:

$$\frac{Q}{2\pi R^3 N \cos\theta} \text{ vs. } \frac{R}{2\pi R N \cos\theta} \frac{\Delta p}{L} \tag{12}$$

Denson and Hwang [92] represent their computations in this manner. It is presented in Figure 3.20.

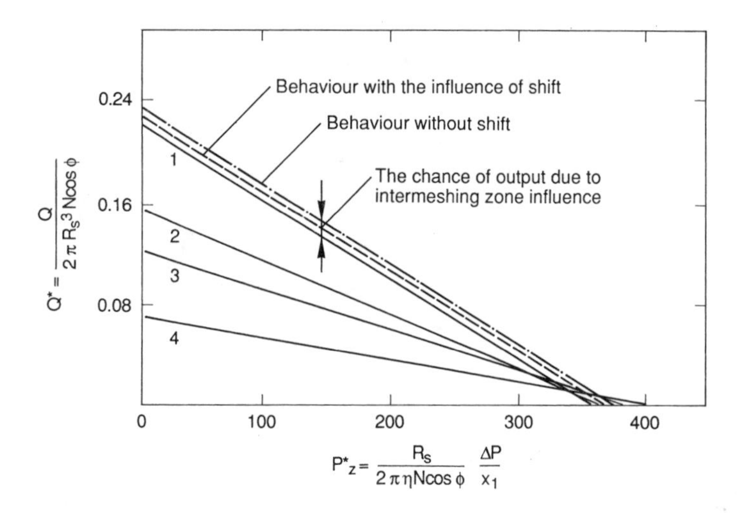

Figure 3.20 Comparison of screw characteristic curves for calculations of Denson and Hwang [92] and Szydlowski and White [95] for a Newtonian fluid.

Some years later, the same problem was taken up by Szydlowski and White [95], but using a different procedure. These authors considered transverse as well as longitudinal flows across the screw channel. They consider the intermeshing region between screws. This is done by considering the differing positions of the screw flights on the two different screws relative to the channel. The melt must skew around the screw flight thickness as suggested by Figure 3.21. Transverse shearing by the screw flights was neglected. In essence, these authors begin with:

$$0 = -\frac{\partial p}{\partial x_1} + \frac{\partial \sigma_{12}}{\partial x_2} = -\frac{\partial p}{\partial x_1} + \eta \frac{\partial^2 v_1}{\partial x_2^2} \tag{13a}$$

$$0 = -\frac{\partial p}{\partial x_3} + \frac{\partial \sigma_{32}}{\partial x_2} = -\frac{\partial p}{\partial x_3} + \eta \frac{\partial^2 v_3}{\partial x_2^2} \tag{13b}$$

subject to the boundary conditions:

$$v_1(x_1, h_b, x_3) = U_1 = U \cos \theta \tag{14a}$$

$$v_3(x_1, h_b, x_3) = U_3 = -U \sin \theta \tag{14b}$$

$$v_1(x_1, h_s(x_1, x_3), x_3) = 0 \tag{14c}$$

$$v_3(x_1, h_s(x_1, x_3), x_3) = 0 \tag{14d}$$

The screw barrel distance is:

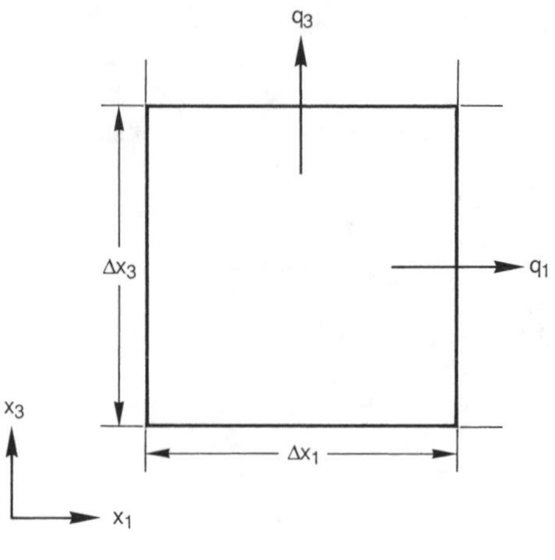

Figure 3.21 Szydlowski and White [95] screw flow analysis.

$$H = H(x_1, x_3) \tag{15}$$

which includes the intermeshing region where the screws meet. The calculations were carried out using a modification of the FAN technique of Tadmor et al. [100]. They proceeded by integrating Eqs (13a) and (13b) to give:

$$q_1 = \bar{v}_1 H = \frac{1}{2} U_1 H - \frac{H^3}{12\eta} \frac{\partial p}{\partial x_1} \tag{16a}$$

$$q_3 = \bar{v}_3 H = \frac{1}{2} U_3 H - \frac{H^3}{12\eta} \frac{\partial p}{\partial x_3} \tag{16b}$$

and making a balance of q_1 and q_3 fluxes (Fig. 3.22), i.e.

$$q_1(x_1 + \Delta x_1, x_3)\Delta x_3 + q_3(x_1, x_3 + \Delta x_3)\Delta x_1 - q_1(x_1, x_3)\Delta x_3 - q_3(x_1, x_3)\Delta x_1 = 0 \tag{17}$$

A specified machine flux was generally used as a boundary condition for Eq (17). It is also possible to proceed by specifying inlet and outlet pressures. Screw characteristic curves of the form of Eq (10) were computed for different screw element geometries. These are presented in Figure 3.20, where they are contrasted with the results of Denson and Hwang [92]. For equivalent screw geometries the results are very similar.

It was also possible to consider the effects of flight thickness in the intermeshing region on screw performance. This is shown in Figure 3.22 where we show the effect of flight thickness on the extra pressure drop in the intermeshing region. Only when the screw flight takes up a significant fraction of the surface of the channel is there a large effect.

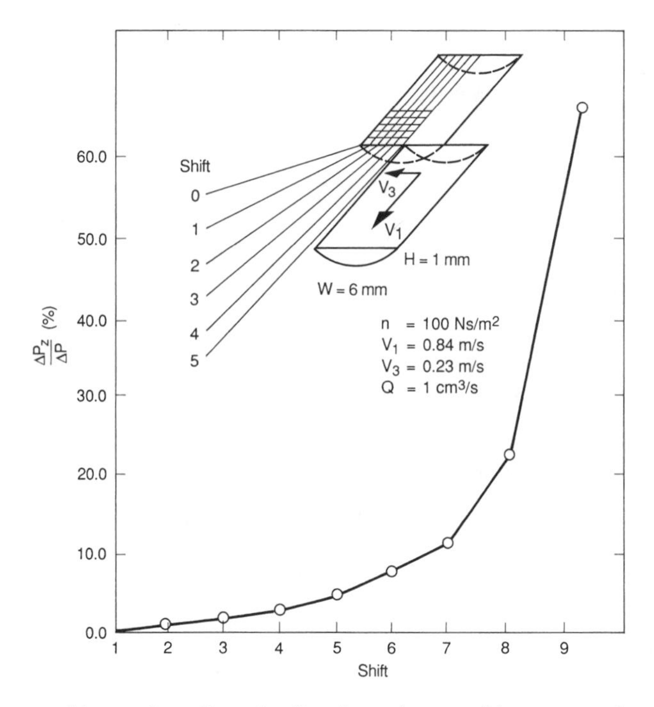

Figure 3.22 Intermeshing region effects for flow in an intermeshing co-rotating twin screw extruder [95].

Polymer melts are non-Newtonian and viscoelastic. Non-Newtonian flow in the screw section of an intermeshing co-rotating twin screw extruder has been modeled by Wang and White [97]. They approach this problem both using the method of Szydlowski and White [95] described above and by a finite element procedure. First, we consider the former method. The shear stresses σ_{12} and σ_{32} of Eq (13) are again given by:

$$\sigma_{12} = \eta \frac{\partial v_1}{\partial x_2} \qquad \sigma_{32} = \eta \frac{\partial v_3}{\partial x_2} \tag{18}$$

but shear viscosity, η, depends velocity gradients. This is considered to be through the second invariant of the rate of deformation tensor [98-103]. This leads to:

$$\eta = \eta \left[\left[\frac{\partial v_1}{\partial x_2} \right]^2 + \left[\frac{\partial v_3}{\partial x_2} \right]^2 \right] \tag{19a}$$

and if we presume power law fluid behavior:

$$\eta = K\, II_d^{\frac{n-1}{2}} = K\left[\left(\frac{\partial v_1}{\partial x_2}\right)^2 + \left(\frac{\partial v_3}{\partial x_2}\right)^2\right]^{\frac{n-1}{2}} \tag{19b}$$

We then have in place of Eq (13):

$$0 = -\frac{\partial p}{\partial x_1} + \frac{\partial \sigma_{12}}{\partial x_2} = -\frac{\partial p}{\partial x_1} + \frac{\partial}{\partial x_2} K\left[\left(\frac{\partial v_1}{\partial x_2}\right)^2 + \left(\frac{\partial v_3}{\partial x_2}\right)^2\right]^{\frac{n-1}{2}} \frac{\partial v_1}{\partial x_2}$$

(20)

$$0 = -\frac{\partial p}{\partial x_3} + \frac{\partial \sigma_{32}}{\partial x_2} = -\frac{\partial p}{\partial x_3} + \frac{\partial}{\partial x_2} K\left[\left(\frac{\partial v_1}{\partial x_2}\right)^2 + \left(\frac{\partial v_3}{\partial x_2}\right)^2\right]^{\frac{n-1}{2}} \frac{\partial v_3}{\partial x_2}$$

Eqs (20) may be integrated:

$$\frac{\partial v_1}{\partial x_2} = \frac{1}{\eta}\frac{\partial p}{\partial x_1}[x_2 - C'] \tag{21a}$$

$$\frac{\partial v_3}{\partial x_2} = \frac{1}{\eta}\frac{\partial p}{\partial x_3}[x_2 - C'''] \tag{21b}$$

suggesting the following form of Eq (19b) (compare Griffith [102]):

$$\eta = K\left[\frac{1}{\eta^2}\left[\frac{\partial p}{\partial x_1}\right]^2 (x_2 - C')^2 + \frac{1}{\eta^2}\left[\frac{\partial p}{\partial x_3}\right]^2 (x_2 - C''')^2\right]^{\frac{n-1}{2}} \tag{22a}$$

$$\eta = K\left[\left[\frac{\partial p}{\partial x_1}\right]^2 (x_2 - C')^2 + \left[\frac{\partial p}{\partial x_3}\right]^2 (x_2 - C''')^2\right]^{\frac{n-1}{2}} \tag{22b}$$

Integrating Eq (21) to obtain v_1 and $\bar{v}_1$, as well as v_2 and $\bar{v}_2$ yields:

$$q_1 = \bar{v}_1 H = \int_0^H \int_0^{x_2} \left[\left[x_2 \frac{\partial p}{\partial x_1} + C'\right]^2 + \left[x_2 \frac{\partial p}{\partial x_3} + C'''\right]^2\right]^{\frac{1-n}{2n}} \left[x_2 \frac{\partial p}{\partial x_1} + C'\right] dx_2 dx_2$$

(23a, b)

$$\Sigma_3 = \bar{v}_3 H = \int_0^H \int_0^{x_2} \left[\left[x_2 \frac{\partial p}{\partial x_1} + C'\right]^2 + \left[x_2 \frac{\partial p}{\partial x_3} + C'''\right]^2\right]^{\frac{1-n}{2n}} \left[x_2 \frac{\partial p}{\partial x_3} + C'\right] dx_2 dx_2$$

Screw characteristic curves for power law fluids calculated on this basis are presented in Figure 3.23. The geometry chosen for analysis was a two-tip screw element with screw radius of 15.3 mm. The screw pitch is 42 mm, the center line distance is 26 mm and the tip clearance is 0.075 mm. This screw geometry is based upon the design of an element of a Werner Pfleiderer ZSK-30 twin screw extruder which is available in our laboratory. As with single screw extruders [98,102], decreasing power law exponent is found to deteriorate screw pumping capabilities.

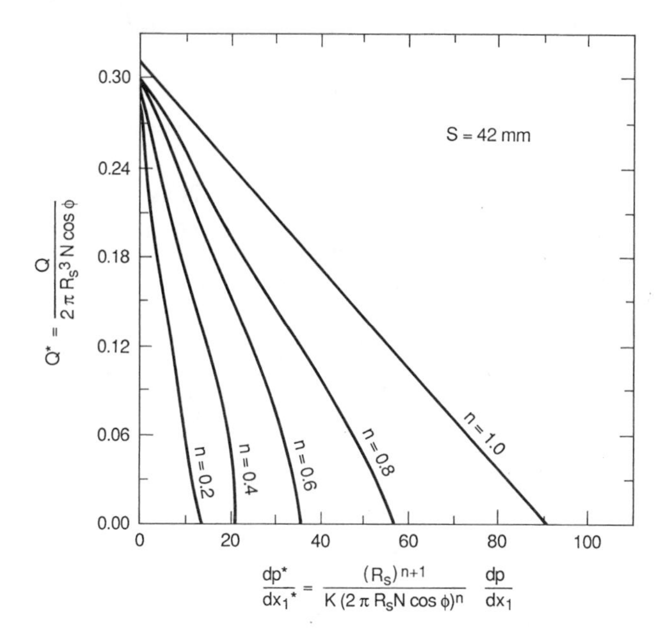

Figure 3.23 Influence of power law exponent on screw characteristic curves after Wang and White [97].

A more general view of this problem has also been taken by Wang and White [97]. They seek to simulate flow in a screw channel for a non-Newtonian fluid with velocity field:

$$\underset{\sim}{v} = v_1(x_2, x_3)\underline{e}_1 + v_2(x_2, x_3)\underline{e}_2 + v_3(x_2, x_3)\underline{e}_3 \tag{24}$$

They use as equations of motion:

$$0 = -\frac{\partial p}{\partial x_1} + \frac{\partial}{\partial x_2}\left(\eta\frac{\partial v_1}{\partial x_2}\right) + \frac{\partial}{\partial x_3}\left(\eta\frac{\partial v_1}{\partial x_3}\right)$$

$$0 = -\frac{\partial p}{\partial x_2} + \frac{\partial}{\partial x_2}\left(2\eta\frac{\partial v_2}{\partial x_2}\right) + \frac{\partial}{\partial x_3}\left(\eta\frac{\partial v_2}{\partial x_3} + \eta\frac{\partial v_3}{\partial x_2}\right) \tag{25}$$

$$0 = -\frac{\partial p}{\partial x_3} + \frac{\partial}{\partial x_2}\left(\eta\frac{\partial v_2}{\partial x_3} + \eta\frac{\partial v_3}{\partial x_2}\right) + \frac{\partial}{\partial x_3}\left(2\eta\frac{\partial v_3}{\partial x_3}\right)$$

with the boundary conditions of Eq (14) plus the restriction of v_2 being zero on all of the surfaces. The shear viscosity now becomes a function of all velocity gradients:

$$\eta = K\left[\left[\frac{\partial v_1}{\partial x_2}\right]^2 + \left[\frac{\partial v_1}{\partial x_3}\right]^2 + \left[\frac{\partial v_3}{\partial x_2} + \frac{\partial v_2}{\partial x_3}\right]^2 + 2\left[\frac{\partial v_2}{\partial x_2}\right]^2 + 2\left[\frac{\partial v_3}{\partial x_3}\right]^2\right]^{\frac{n-1}{2}} \tag{26}$$

The problem was solved by a finite element method using the variational principle:

$$J = \int \frac{II_d^{\frac{n-1}{2}}}{n+1} ds - \int p\left[\frac{\partial v_2}{\partial x_2} + \frac{\partial v_3}{\partial x_3}\right] ds \tag{27}$$

This was solved using an iterative procedure described elsewhere [97]. The down-channel (axial) velocity profiles as described above along the transverse direction (x_3-direction) are shown in Figures 3.24a and 3.24b at open discharge and closed discharge respectively. While there are only positive components in the open-discharge case, some negative components

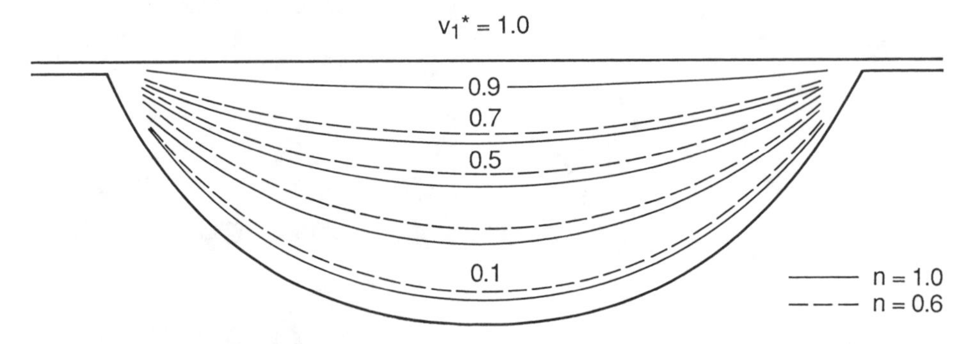

Figure 3.24a Isovels for open discharge in a self-wiping screw channel for power law fluid [97].

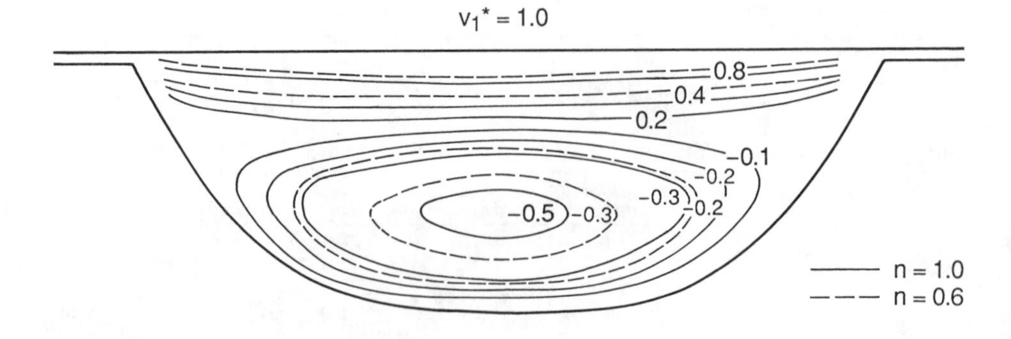

Figure 3.24b Isovels for closed discharge in a self-wiping screw channel for closed discharge [97].

exist in the closed-discharge case. As expected, the magnitudes of the velocity components of the power-law case are always smaller than that of the Newtonian case due to the shear-thinning effect. Furthermore, the difference between the power-law case and the Newtonian case is larger for closed discharge when we compare Figures 3.24a and 3.24b. It is because the flow of the fluids are subject to higher shear at closed discharge.

We applied the above results to calculate the pumping behavior. The difference between the finite element method and the modified FAN technique is no more than 10%.

3.4.3 Left-Handed Screw Elements

The characteristics of left-handed intermeshing co-rotating screw elements must have been realized by Meskat and Pawlowski [34] of Bayer, who have first used them in their 1950 patent application. Those aware of the extrusion literature of the 1950s and 60s know that both Meskat and Pawlowski were very perceptive individuals who made major contributions to the basic understanding of extrusion. The first explicit simulation of flow in left-handed elements was given much more recently by White and Szydlowski [104]. Essentially, the simulation method of the previous section remains valid but the boundary conditions differ. Left-handed elements drag the fluid towards the hopper. In place of Eq (3), we have:

$$v_1(x_2) = -U\cos\theta\left[\frac{x_2}{H}\right] - \frac{H^2}{2\eta}\frac{\partial p}{\partial x_1}\left[\left[\frac{x_2}{H}\right] - \left[\frac{x_2}{H}\right]^2\right] \tag{28}$$

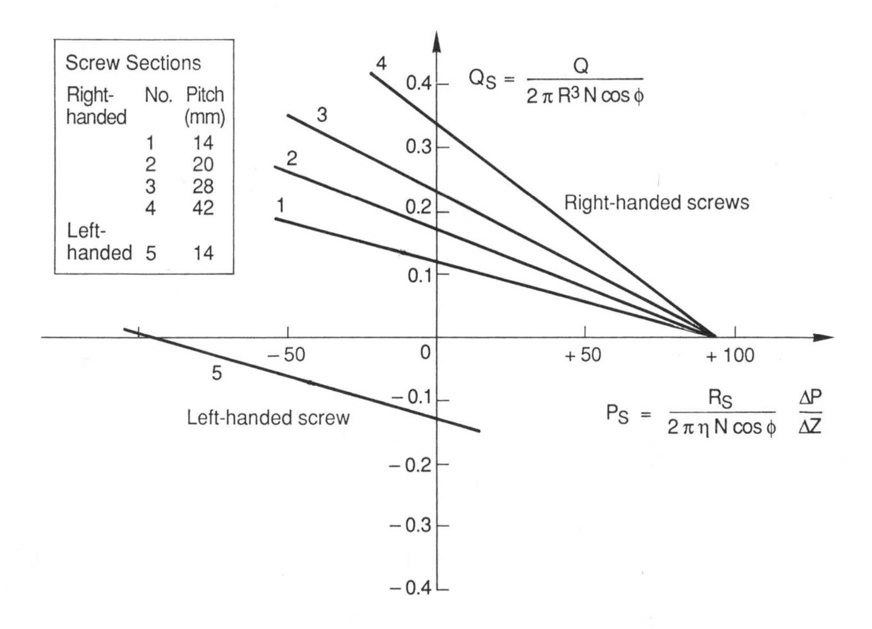

Figure 3.25 Screw characteristic curves for right-handed and left-handed screw elements.

and in place of Eq (1):

$$Q = -\frac{1}{2} U \cos\theta HW F_D - \frac{H^3 W}{12\eta} \frac{\Delta p}{L} F_p \tag{29}$$

Figure 3.25 presents screw characteristic curves for left-handed as opposed to right-handed screws. It can be seen that if Q is positive, the Δp must be negative. The pressure must decrease along the flow directly. Thus left-handed screw elements are suitable for inducing devolatilization.

Figure 3.26 shows calculations for a power law fluid in a left-handed screw element.

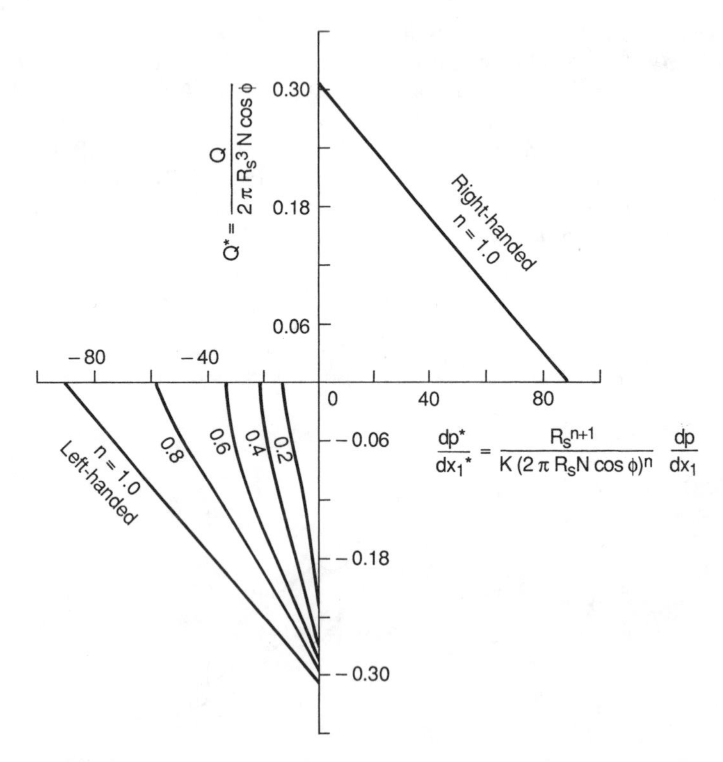

Figure 3.26 Screw characteristic curves for left-handed screw element for power law fluid.

3.4.4 Kneading Disc Elements

Paddles or kneading disc elements were introduced by Nelson [9] and Erdmenger [25,29,31,32,39,40]. It is not clear from their patents to understand to what extent they understood the types of fluid motions induced by these elements. The first efforts to simulate these motions is due to Werner [90,91]. A more quantitative modeling was developed by Szydlowski et al. [105], and later Szydlowski and White [106] for Newtonian fluids, and Szydlowski and White [107], and Wang and White [108] for non-Newtonian fluids.

As in screw elements, one proceeds by flattening out the kneading disc/barrel system. A coordinate system was embedded in the kneading discs and the barrel is considered to move past it. The coordinate system is placed so that "1" is along the screw axis, "3" is in the circumferential direction, and "2" is perpendicular to the kneading disc axis (see Fig. 3.27). The formulation is based on the hydrodynamic lubrication theory of Reynolds [109]. This involves neglect of normal stresses which in Newtonian and viscous non-Newtonian fluids are associated with elongational flows.

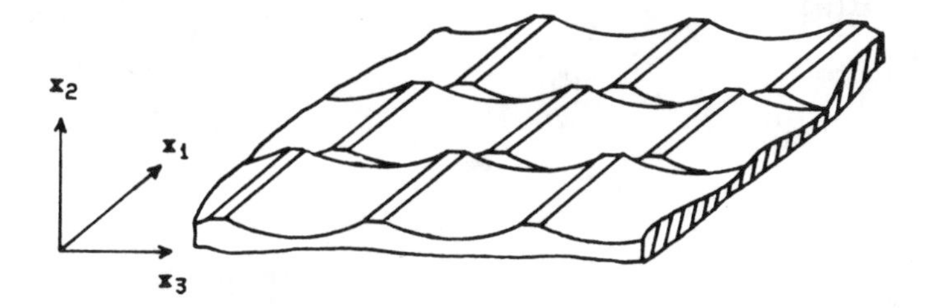

Figure 3.27 Flattened out kneading disc region.

The arguments of the above paragraph lead us to Eqs (13) with boundary conditions:

$$v_1(x_1, h_b, x_3) = 0 \tag{30a}$$

$$v_3(x_1, h_b, x_3) = U$$

$$v_1(x_1, h_s(x_1, x_3), x_3) = 0 \tag{30b}$$

$$v_3(x_1, h_s(x_1, x_3), x_3) = 0$$

Szydlowski et al. [105] have solved this problem using the same procedure as Szydlowski and White [95] applied to screw elements. This involves a balance of q_1 and q_3. Fluxes have the same functional form as in Eq (16). These have the form:

$$q_1 = -\frac{H^3}{12\eta}\frac{\partial p}{\partial x_1} \tag{31a}$$

$$q_3 = \frac{1}{2}U_3 H - \frac{H^3}{12\eta}\frac{\partial p}{\partial x_3} \tag{31b}$$

as there is only pressure flow in the "1" direction and "3" is the circumferential direction of the screw shaft. This method is equivalent to numerically solving Reynolds equation for the pressure in hydrodynamic lubrication theory. It yields a pressure field which must be differentiated to give a mean velocity, or q, flux field.

Typical computed pressure fields in the kneading disc regions are shown in Figure 3.28. These are based on the work of Szydlowski et al. [105]. When one computes fluid fluxes, it is found that the kneading discs exert a pumping characteristic. The helical distribution of kneading disc tips produces a roughly screwlike motion of the fluid. It is possible to calculate

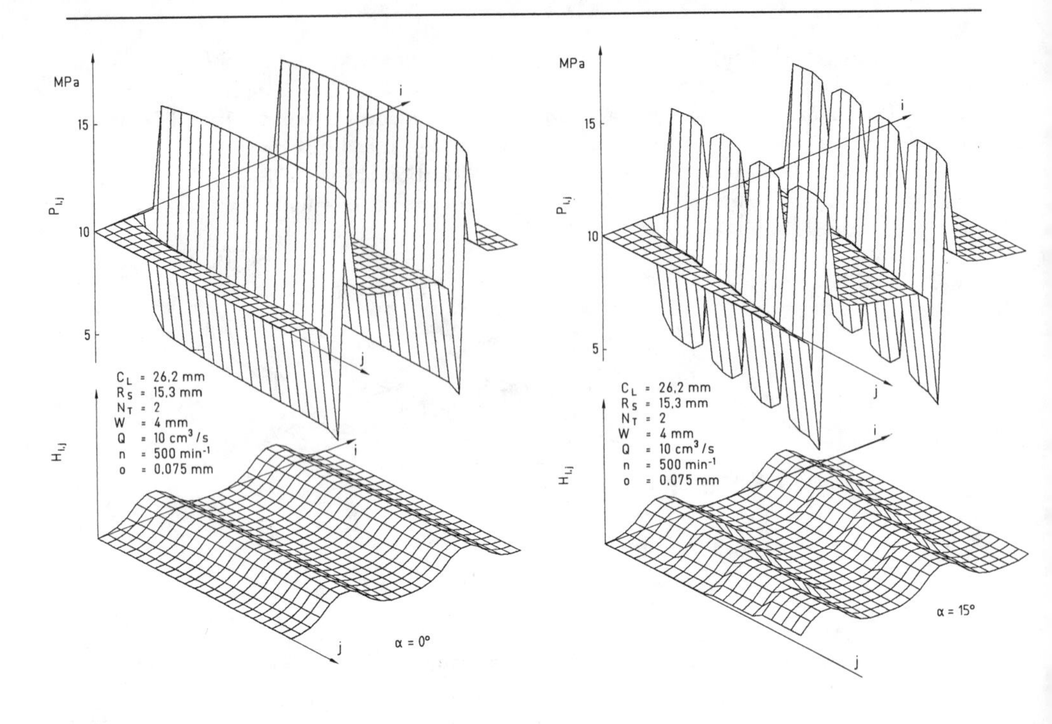

Figure 3.28 Pressure fields in kneading disc region based on calculations of Szydlowski et al. [105].

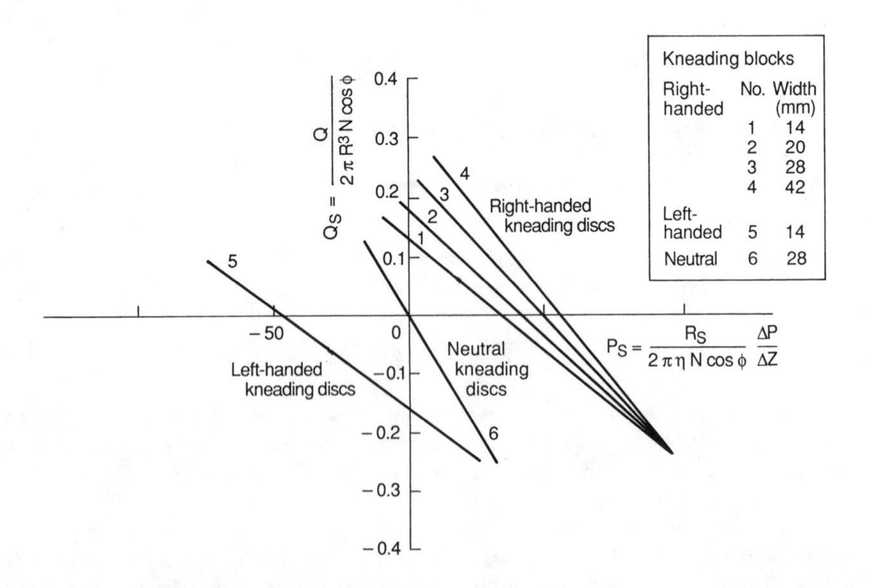

Figure 3.29 Screw characteristic curves of kneading discs after White and Szydlowski [104].

equivalent screw characteristic curves. These are summarized in Figure 3.29 based on the work of White and Szydlowski [104]. Depending on the arrangement of the discs, it is possible to obtain either positive or negative drag flow, i.e. there are right-handed and left-handed kneading disc arrangements. Secondly, the slopes of the "screw characteristic" curves for kneading discs are larger than for true screws. This means they are less able to pump against high pressure than screws.

The inability of helically-arranged kneading discs to pump as well as screws is due to pressure induced backward flow, Q_b, between the tips. This is shown in Figure 3.30. This has been computed based on the above modeling by Szydlowski et al. [105]. To quantify their results they define fractions:

$$f_L = \frac{Q_L}{Q_L + Q_b + Q_c} \tag{32a}$$

$$f_b = \frac{Q_b}{Q_L + Q_b + Q_c} \tag{32b}$$

$$f_c = \frac{Q_c}{Q_L + Q_b + Q_c} \tag{32c}$$

where Q_L is the forward longitudinal flow and Q_c is the circumferential flow. f_L, f_b, and f_c are the fractions of flow in the longitudinal, backward, and circumferential directions.

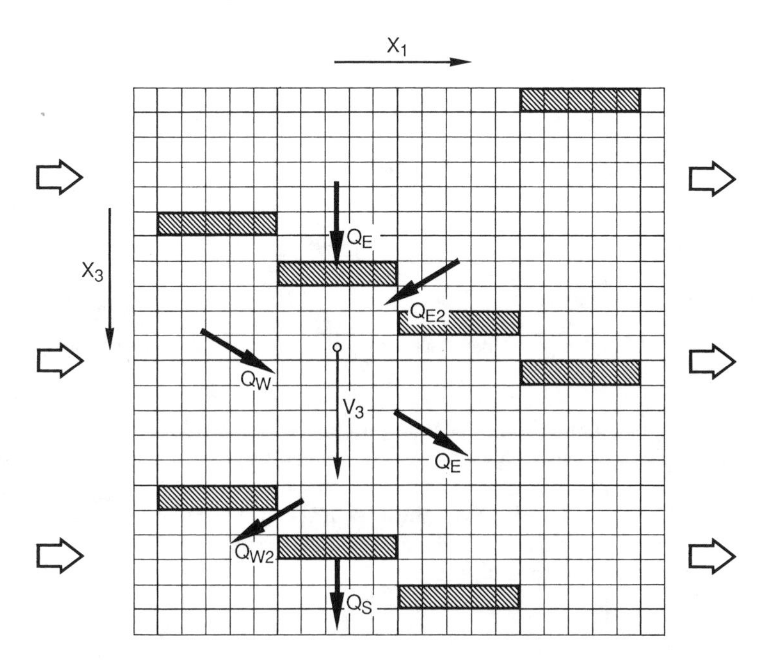

Figure 3.30 Fluxes and mechanisms of backflow in kneading discs after Szydlowski et al. [105].

The influence of non-Newtonian fluid flow characteristics in the kneading disc region has been analyzed by Szydlowski and White [107] and by Wang et al. [108]. They describe various methods of incorporating non-Newtonian flow effects. The first authors suggest 3 methods. Method 1 is based on replacing Eq (31) with:

$$q_1 = -\frac{H^3}{12K(U_3/H)^{n-1}} \frac{\partial p}{\partial x_1} \tag{33a}$$

$$q_3 = \frac{1}{2} U_3 H - \frac{H^3}{12K(U_3/H)^{n-1}} \frac{\partial p}{\partial x_1} \tag{33b}$$

Method 2 is based upon:

$$q_1 = -\frac{H^3}{12\bar{\eta}} \frac{\partial p}{\partial x_1} \tag{33c}$$

$$q_3 = \frac{1}{2} U_3 H - \frac{H^3}{12\bar{\eta}} \frac{\partial p}{\partial x_3} = \frac{1}{2} U_3 H \, F\left[\frac{H}{6K(U_3/H)^n} \left[\frac{\partial p}{\partial x_3}\right]^n\right] \tag{33d}$$

Here $\bar{\eta}$ is an effective viscosity calculated from Eq (33d) which represents the solution of the problem:

$$0 = -\frac{\partial p}{\partial x_3} + K \left[\left[\frac{\partial v_3}{\partial x_2}\right]^2\right]^{\frac{n-1}{2}} \frac{\partial v_3}{\partial x_2} \tag{34}$$

i.e. one which considers transverse shear flow only. Method 3 uses:

$$q_1 = -\frac{H^3}{12\bar{\eta}} \frac{\partial p}{\partial x_1} \tag{35a}$$

$$q_3 = \frac{1}{2} U_3 H - \frac{H^3}{12\bar{\eta}} \frac{\partial p}{\partial x_3} \tag{35b}$$

where $\bar{\eta}$ is determined from the shear rates of the exact Newtonian solution which are introduced into Eq (19b).

Wang et al. [108] have modeled flow in this region using Eqs (19) to (23). We believe this to be the best of the four methods. We show pumping characteristic curves for power fluids calculated on this basis in Figure 3.31. It can be seen that decreasing power law exponent decreases pumping ability as the curves fall off more sharply.

We display in Figure 3.32, based on the work of Wang et al. [108] the variation of f_L, f_b, and f_c with processing conditions using apparatus dimensions from Werner and Pfleiderer ZSK-30 elements. These indicate that generally:

$$f_L > f_b >> f_c \tag{36}$$

The backward flux between the kneading disc ties is very significant.

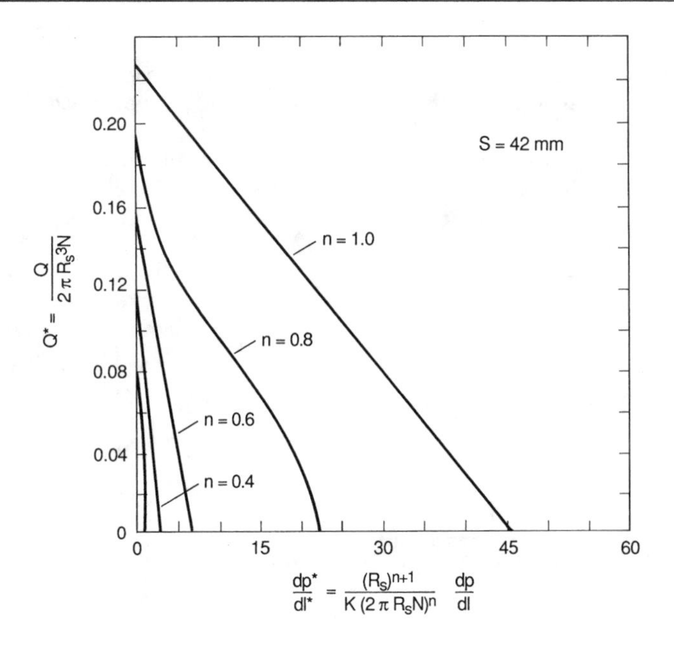

Figure 3.31 Screw characteristics of kneading discs for power law fluids after Wang et al. [108].

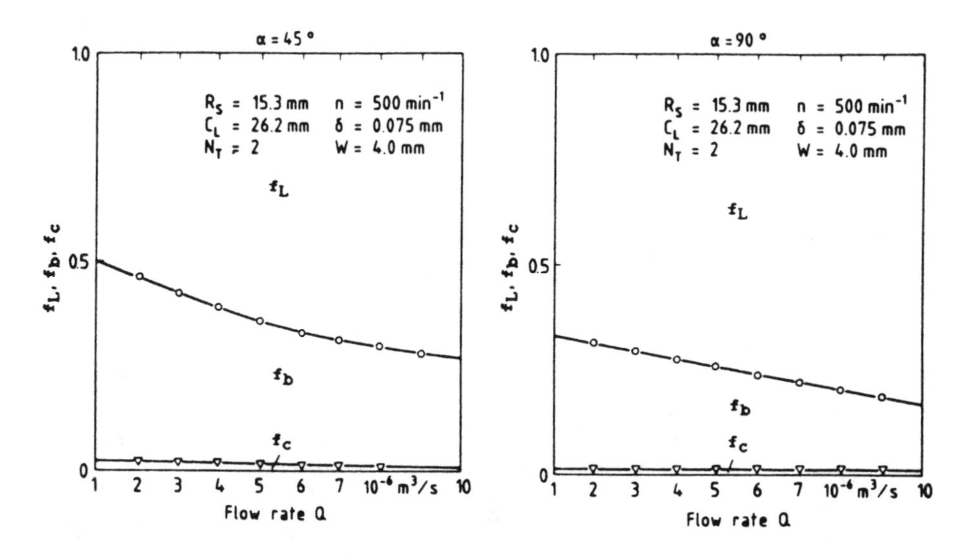

Figure 3.32 Factors f_L, f_b, and f_c and their variation with process conditions [108].

Szydlowski and White [106] have taken a more sophisticated look at this problem. They have noted that the motion of the kneading discs causes a peristaltic pumping effect because of changing local volume in the intermeshing region. They analyzed this making a balance on q_1 and q_3 of Eqs (17) coupled with an accumulation term. Szydlowski and White [105] predicted that right-handed kneading discs have an additional periodic forward pumping motion and left-handed kneading discs have a periodic backward pumping capability. Typical effects are shown in Figure 3.33.

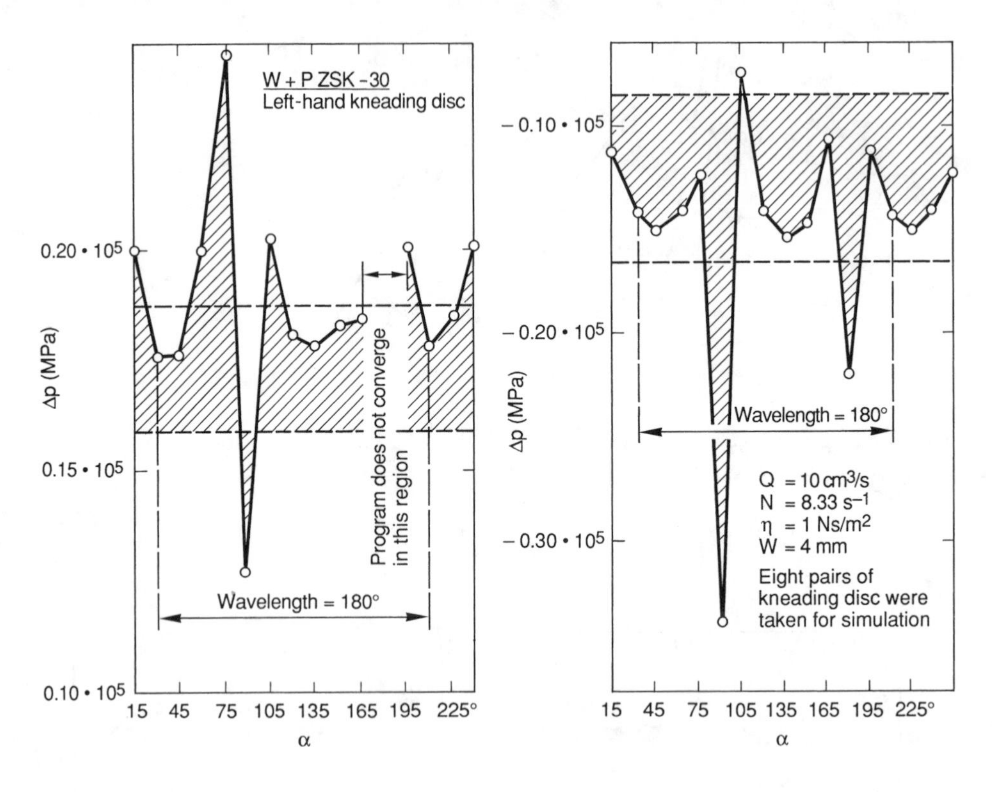

Figure 3.33 Peristaltic pumping effects in intermeshing regions.

3.4.5 Modular Machine

The characteristics of modular intermeshing co-rotating twin screw extruders were first modeled by White and Szydlowski [104] for Newtonian fluids. They noted that if the extruder is fully filled, we may represent the flow of a Newtonian fluid in an element by:

$$Q = A_j N - \frac{B_j}{\eta} \frac{\Delta p_j}{L_j} \tag{37}$$

The total pressure along the length of the screw is:

$$\Delta p = \Sigma\, \Delta p_j \tag{38}$$

Eqs (37) and (38) may be combined to give:

$$Q = \left[\frac{\Sigma \frac{A_j \; L_j}{B_j}}{\Sigma \; \frac{L_j}{B_j}} \right] N - \frac{1}{\eta} \left[\frac{1}{\Sigma \; \frac{L_j}{B_j}} \right] \Delta p \tag{39}$$

The formulation of Eqs (38) to (39) may obviously be extended to non-Newtonian fluids. This has been carried out by Wang et al. [108].

In practice, however, modular intermeshing twin screw extruders usually operate under starved conditions. One may show this by back calculating from the die using Eq (37). When the pressure falls to zero, starvation develops. With a left-handed screw or kneading disc element, filling again arises and pressure increases in the direction of the hopper. With right-handed screws and kneading discs, the pressure again falls off as one moves backward. This is shown in Figure 3.34.

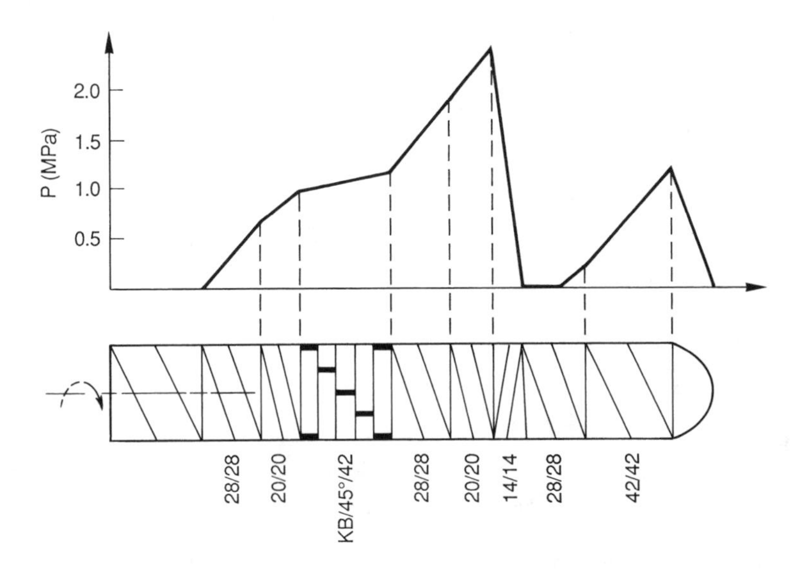

Figure 3.34 Typical pressure fields and location of starved regions in a modular intermeshing co-rotating twin screw extruder.

Wang et al. [108] indicated that quantitative predictive calculations of locations of starvation and fully-filled regions in a modular machine can be made for a power law fluid model using the procedure described in the above paragraph.

ACKNOWLEDGMENT

The authors research on intermeshing co-rotating twin screw extruders has been supported by the Monsanto Chemical Company.

NOMENCLATURE

A_j: Coefficient of Eq (37)
B_j: Coefficients of Eq (37)
$\underline{e}_j$: A unit vector
Fp: Drag and pressure shape factors
f_L f_b f_c: Fractions of longitudinal backward and circumferential flow
H_p: Screw channel depth:
$h(x_3)$: Self-wiping channel profile
L: Screw channel length
N: Screw rotation rate
n: Power law exponent
p: Pressure
Q: Flow rate
q_j: Flux defined by $\bar{v}_j H = \int_0^H v_j dx_2$
U: Barrel velocity
U_1: Barrel velocity component c screw channel
U_3: Barrel velocity component normal to screw channel
v_j: Mean velocity
W: Channel width
x_j: Distance coordinate
η: Shear viscosity (33)
$\bar{\eta}$: Mean viscosity of Eq (33)
θ: Angle of screw flight with normal place to screw axis
ϕ: Fill factor of screw
σ_{12}: Stress component

REFERENCES

1. Herrmann, H., "Schneckenmaschinen in der Verfahrenstechnik," Springer, Berlin, 1972.
2. White, J.L., Szydlowski, W., Min, K., and Kim, M.H., *Adv. Polym. Tech.*, *7*, 295 (1987).
3. Schutz, F.C., *SPE J.*, *18*, 1147 (1962).
4. Prause, J.J., *Plas. Tech.*, (Nov.) 41 (1967), (Feb.) 29 (1968), (March) 52 (1968).
5. Coignet, F., U.S. pat. (filed Dec. 21, 1869) 93,035 (1869).
6. Wünsche, A., German pat. (filed Sept. 12, 1901) 131,392 (1901).
7. Easton, R.W., British pat. (filed Sept. 25, 1916) 109,663 (1917).
8. Easton, R.W., U.S. pat. (filed June 2, 1920) 1,468,379 (1923).
9. La Casse, H., U.S. pat. (filed Aug. 25, 1919) 1,356,296 (1919).
10. Nelson, W.K., U.S. pat. (filed June 4, 1931) 1,868,671 (1932).

11. Pease, F.F., U.S. pat. (filed Aug. 17, 1933) 2,048,286 (1936).
12. "LMP and Polytal, the Extrusion Specialists," LMP, Turin (1986), LMP Impianti, LMP, Turin (1987).
13. Colombo, R., Italian pat. (filed Feb. 6, 1939) 370,578 (1939), Swiss pat. (filed Jan. 3, 1941) 220,550 (1942).
14. Anonymous, German pat. (filed July 24, 1943) 895,058 (1953).
15. Lavorazione Materie Plastisch, British pat. (filed May 2, 1946) 629,109 (1949).
16. Colombo, R., U.S. pat. (filed Aug. 7, 1947) 2,563,396 (1951).
17. Colombo, R., Canadian pat. (filed Dec. 27, 1949) 517,911 (1955).
18. Greenwood, S.H., *Rubber World*, *129*, 73 (1953).
19. Greenwood, S.H., *SPE J.*, (March) 20 (1953).
20. Stoeckhert, K., *Kunststoffe*, *40*, 195 (1950).
21. Schaerer, A.J., *Kunststoffe*, *44*, 105 (1954).
22. Beyer, S., *Kunststoffe*, *48*, 157 (1958).
23. Baigent, K.H., *Trans. Plastics Inst.*, 135 (1956).
24. Baigent, K.H., U.S. pat. (filed July 23, 1957) 2,916,769 (1959).
25. Erdmenger, R., "Schneckenmaschinen für die Hochviskosverfahrenstechnik," Bayer, Leverkusen (1978).
26. Burghauser, F., U.S. pat. (filed Nov. 24, 1936) 2,115,006 (1938); Leistritz, P., Burghauser, F., German pat. (filed Dec. 1, 1935) 682,787 (1939); Burghauser, F., Erb, K., German pat. (filed Feb. 5, 1938) 690,990 (1940); Burghauser, F., Erb, K., U.S. pat. (filed Feb. 1, 1939) 2,231,357 (1941).
27. Meskat, W., Erdmenger, R., German pat. (filed July 7, 1944) 862,668 (1953).
28. Meskat, W., Erdmenger, R., German pat. (filed July 28, 1944) 872,732 (1953).
29. Erdmenger, R., German pat. (filed Sept. 24, 1949) 815,641 (1951).
30. Erdmenger, R., German pat. (filed Sept. 29, 1949) 813,154 (1951).
31. Riess, K., Erdmenger, R., *VDI Zeit*, 93, p. 633 (1951).
32. Erdmenger, R., U.S. pat. (filed Sep. 20, 1950) 2,670,188 (1954).
33. Telle, O., Erdmenger, R., German pat. (filed July 22, 1953) 945,086 (1953).
34. Meskat, W., Pawlowski, J., German pat. (filed Dec. 10, 1950) 949,162 (1956).
35. Winkelmuller, W., Erdmenger, R., Neidhardt, S., Hirschberg, G., Fortuna, B., German pat. (filed Dec. 30, 1951) 915,689 (1954).
36. Meskat, W., *Chem. Ing. Tech.*, *22*, 516 (1950).
37. Riess, K., Meskat, W., *Chem. Ing. Tech.*, *23*, 205 (1951).
38. Riess, K., *Chem. Ing. Tech.*, *27*, 457 (1955).
39. Erdmenger, R., German pat. (filed July 28, 1953) 940,109 (1956).
40. Erdmenger, R., U.S. pat. (filed July 27, 1954) 2,814,472 (1957).
41. Herrmann, H., in "Kunststoffe - ein Werkstoff Macht Karriere," W. Glenz, Ed., Hanser Publishers, Munich (1985).
42. Erdmenger, R., U.S. pat. (filed Aug. 17, 1959) 3,122,356 (1964).
43. Fritsch, R., Fahr, G., *Kunststoffe*, *49*, 543 (1959).
44. Erdmenger, R., *Chem. Ing. Tech.*, *34*, 751 (1962).
45. Erdmenger, R., *Chem. Ing. Tech.*, *36*, 175 (1964).
46. Herrmann, H., *Kunststoff Gummi*, *3*, 217 (1964), *Chem. Ing. Tech.*, *38*, 25 (1966).
47. Erdmenger, R., U.S. pat. (filed Mar. 26, 1963) 3,254,367 (1967).
48. Anonymous (Ellermann?) German pat. (filed Jan. 31, 1941) 750,509 (1945).
49. Ellermann, W., German pat. (filed July 4, 1951) 879,164 (1953).
50. Kungen, J., Hilger, F., Hilger, H., Hilger, P., British pat. (filed Oct. 8, 1953) 738,461 (1955).

51. Ellermann, W., German pat. (filed July 31, 1953) 935,634 (1955); U.S. pat. (filed Oct. 12, 1953) 2,693,348 (1954).
52. Proksch, W., *Kunststoff Gummi*, *3*, 426 (1964).
53. Erdmenger, R., Oetke, W., German Ausleschrift (filed Mar. 16, 1960) 1,111,154 (1961).
54. Erdmenger, R., Ullrich, M., *Chem. Ing. Tech.*, *42*, 1 (1970).
55. Loomans, B.A., Brennan, A.K., U.S. pat. (filed Mar. 21, 1962) 3,195,868 (1965).
56. Loomans, B.A., Brennan, A.K., U.S. pat. (filed Mar. 21, 1962) 3,198,491 (1965).
57. Loomans, B.A., U.S. pat. (filed Jan. 12, 1967) 3,423,074 (1969).
58. Loomans, B.A., Todd, D.B., U.S. pat. (filed Nov. 24, 1969) 3,613,160 (1970).
59. Loomans, B.A., U.S. pat. (filed Mar. 22, 1971) 3,719,350 (1973).
60. Loomans, B.A., U.S. pat. (filed Oct. 29, 1973) 3,900,187 (1975).
61. Brennan, A.K., U.S. pat. (filed Nov. 14, 1969) 3,618,902 (1971).
62. Price, R.B., U.S. pat. (filed Mar. 26, 1913) 1,156,096 (1915).
63. Colombo, R., U.S. pat. (filed Dec. 27, 1961) 3,114,171 (1963).
64. Todd, D.B., U.S. pat. (filed Jul 27, 1977) 4,136,968 (1979).
65. Rathjen, C., Ullrich, M., European pat. appl. (filed Oct. 11, 1980) 0 049 835 (1982).
66. Ullrich, M., Rathjen, C., *Chem. Ing. Tech.*, *58*, 590 (1986).
67. Anders, D., U.S. pat. (filed July 7, 1981) 4,423,960 (1984).
68. Anders, D., *Kunststoffe*, *74*, 367 (1984).
69. Anders, D., *SPE Tech. Papers*, *41*, 15 (1984).
70 Kapfer, K., *SPE Tech. Papers*, *46*, 96 (1988).
71. Ess, J.L., Hornsby, P.R., *Plast. Rubber Proc. Appl.*, *8*, 147 (1987).
72. Boden, H., Ocker, H., Pfaff, G., Worz, W., U.S. pat. (filed Nov. 13, 1964) 3,305,894 (1967).
73. Fritsch, R., Kuhner, H.H.O., U.S. pat. (filed Feb. 28, 1967) 3,392,962 (1968).
74. Ocker, H., U.S. pat. (filed Nov. 15, 1968) 3,525,124 (1970).
75. Illing, G., U.S. pat. (filed Oct. 23, 1964) 3,536,680 (1970).
76. Koch, H., U.S. pat. (filed Nov. 20, 1968) 3,608,868 (1971).
77. Ocker, H., U.S. pat. (filed Aug. 8, 1972) 3,682,086 (1972).
78. Herrmann, H., Ocker, H., U.S. pat. (filed Jul 31, 1973) 3,749,375 (1973).
79. Ocker, H., U.S. pat. (filed Sep. 7, 1971) 3,764,114 (1973).
80. Wheeler, D.A., Irving, H.F., Todd, D.B., U.S. pat. (filed Oct. 30, 1969) 3,630,689 (1971).
81. Rausch, W.K., McClellan, G.R., U.S. pat. (filed Dec. 5, 1969) 3,642,964 (1972).
82. Erdmenger, R., Ullrich, M., Germedonk, R., Pedain, J., Quiring, B., Wingler, F., U.S. pat. (filed Jan. 21, 1971) 3,725,340 (1973).
83. Ullrich, M., Meisert, E., Eitel, A., U.S. pat. (filed Jul 17, 1974) 3,963,679 (1976).
84. Illing, G., *Mod. Plast.*, (August) 70 (1969).
85. Mack, W.A., Herter, R., Chem. Eng. Prog., 72 (1), 64 (1976).
86. Eise, K., *Plast. Compound.*, (Jan./Feb.), 44 (1986).
87. Armstroff, O. and Zettler, H.D., *Kunststofftechnik*, *12*, 240 (1973).
88. Herrmann, H., Burkhardt, K., in the 5th Leobener Kunststoffkolloquium Doppelschneckenextruder, p. 11, Lorenz Verlag (1978).
89. Burkhardt, K., Herrmann, H., Jakopin, S., *Plast. Compound.*, (Nov./Dec.) 73 (1978).
90. Werner, H., Dr.-Ing. Dissertation, University of Munich (1976).
91. Werner, H., Eise, K., *SPE Tech. Papers*, *37*, 181 (1979).
92. Denson, C.D., Hwang, B.K., *Polym. Eng. Sci.*, *20*, 965 (1980).
93. Booy, M.L., *Polym. Eng. Sci.*, *20*, 220 (1980).

94. Eise, K., Curry, J., Nangeroni, J.F., *Polym. Eng. Sci.*, *23*, 642 (1983).
95. Szydlowski, W., White, J.L., *Adv. Polym. Technol.*, *7*, 177 (1987).
96. Kalyon, D., Gotsis, A., Gogos, C., Tsenoglou, C., *SPE Tech. Papers*, *46*, 64 (1988).
97. Wang, Y., White, J.L., *J. Non-Newt. Fluid Mech.*, *32*, 19 (1989).
98. Rauwendaal, C., "Polymer Extrusion," Hanser Publishers, Munich (1985).
99. Criminale, W.O., Ericksen, J.L., Filbey, G.L., *Arch. Rat. Mech. Anal.*, *1*, 410 (1958).
100. Tadmor, Z., Broyer, E., Gutfinger, C., *Polym. Eng. Sci.*, *14*, 660 (1974).
101. Bird, R.B., Stewart, W.E., Lightfoot, E.N., "Transport Phenomena," John Wiley & Sons, New York (1960).
102. Griffith, R.M., *IEC Fund.*, *1*, 180 (1962).
103. Bird, R.B., Armstrong, R.C., Hassager, O., "Dynamics of Viscoelastic Fluids, Vol. 1," John Wiley & Sons, New York (1977).
104. White, J.L., Szydlowski, W., *Adv. Polym. Technol.*, *7*, 419 (1987).
105. Szydlowski, W., Brzoskowski, R., White, J.L., *Int. Polym. Proc.*, *1*, 207 (1987).
106. Szydlowski, W., White, J.L., *Int. Polym. Proc.*, *2*, 142 (1988).
107. Szydlowski, W., White, J.L., *J. Non-Newt. Fluid Mech.*, *28*, 29 (1988).
108. Wang, Y., White, J.L., Szydlowski, W., *Int. Polym. Proc.*, *4*, 262 (1989).
109. Reynolds, O., *Phil. Trans. Roy. Soc.*, *177*, 157 (1886).

CHAPTER 4

NUMERICAL SIMULATION OF THE TRANSPORT PROCESSES IN A TWIN-SCREW POLYMER EXTRUDER

by T.H. Kwon, Y. Jaluria, M.V. Karwe and T. Sastrohartono

Department of Mechanical and Aerospace Engineering
Rutgers University, New Brunswick, NJ 08903
U.S.A.

ABSTRACT
4.1 INTRODUCTION
4.2 LITERATURE REVIEW
4.3 THE PRESENT PROBLEM
4.4 ANALYSIS
4.4.1 Mathematical Model for the Flow and Heat Transfer in a Single-Screw Extruder
4.4.2 Mathematical Model for the Flow in the Intermeshing Zone of a Twin-Screw Extruder
4.5 NUMERICAL SOLUTION
4.6 NUMERICAL RESULTS AND DISCUSSION
4.6.1 Results on the Translation Zone
4.6.2 Choice of Parameter Values
4.6.3 Results for Screw Moving Case
4.6.4 Results for the Intermeshing Zone Based on the FEM Model
4.7 MODELING OF THE TRANSPORT IN A TWIN-SCREW EXTRUDER
4.8 CONCLUSIONS
ACKNOWLEDGEMENTS
NOMENCLATURE
REFERENCES

ABSTRACT

A new simplified approach has been proposed toward the numerical simulation of the transport phenomena in co-rotating and counter-rotating twin-screw extrusion processes. For the sake of analysis the flow domain inside a twin-screw extruder may be divided into two regions: (i) the translation region, which represents a flow similar to that in a single-screw channel and (ii) the mixing or intermeshing region, which is located between the two screws. To simplify the problem, the flow region in a twin-screw extruder is divided into sequences of these two domains. A finite-difference method is employed for the developing flow and temperature fields in the translation region, while a finite element method is employed for determining the mixing and the transport in the intermeshing region.

Numerical results are first obtained for the translation zone, on the basis of a single-screw extruder model. These indicate that the temperature rise in the down channel direction has a small effect on the corresponding velocity field, which is determined mainly by the total volumetric rate. I is also found that heat is transferred from the barrel to the flowing material in the beginning portion of the extruder, whereas heat may be transferred from the material to the barrel further downstream, under certain conditions. Among the many results obtained, the most notable is the one regarding the residence time distribution obtained from this model. The results obtained agree with those reported in the literature.

Numerical calculations are also carried out, using a finite element method, to study the flow behavior in the mixing region of a twin-screw extruder. In order to indicate the extent of mixing, a mixing parameter, defined as the ratio of "the volume flow rate of the material which crosses over to the other channel" to "the total volume flow rate in a given screw channel," was introduced. It was found that the mixing parameter increases as the power-law index is decreased, for a given throughput or volumetric flow rate. It increases with an increase in the throughput, for a given power-law index. It was also found to increase as the depth of the screw channel is decreased, for a given material.

The models for the translation and the mixing zones are combined to obtain the transport in the entire twin-screw extruder. The basic procedure to do so is outlined.

4.1 INTRODUCTION

Screw extrusion is one of the most important manufacturing methods in industries such as those related to plastics, metals, pharmaceuticals, and food. A lot of effort has been directed at understanding the transport phenomena underlying the extrusion process and at building a scientific basis for the design of extruders. As a result, scientific understanding has been greatly enhanced by many experimental, analytical, and numerical studies, especially for single-screw extrusion. However, much less work has been done, analytically or numerically, on twin-screw extrusion processes, mainly because of the complexity involved. The present study is aimed at a more detailed understanding of the flow and the associated heat transfer in twin-screw extruders.

4.2 LITERATURE REVIEW

There is a considerable amount of literature on analytical, experimental, and numerical studies on single-screw extruders and twin-screw extruders. Because of the long history of usage of single-screw extruders in plastics industry, most of the effort has been directed at single-screw plastic extrusion. Therefore, scientific knowledge has been obtained mostly for single-screw extruders for polymeric materials whose rheological behavior has been characterized in terms of a non-Newtonian power-law fluid. The literature on twin-screw extrusion is relatively scarce, mainly because of its complicated geometry and flow behavior.

In one of the pioneering studies on the flow in a single-screw extruder, Griffith [1] solved for the flow of an incompressible fluid through a screw extruder with the velocity and the temperature profiles essentially the same as those in a channel of infinite width and length. The effects of curvature and leakage, across the flights, were ignored. Zamodits and Pearson [2] obtained numerical solutions for a fully developed, two-dimensional, isothermal and non-Newtonian flow of polymer melts in infinitely wide rectangular screw channels. Rauwendaal [3] and Lappe and Potente [4] have discussed the throughput-pressure relationship for single-screw extruders.

The solid conveying zone has been mathematically modeled by Klein [5,6] and recently by Derezinski [7] and Peng et al. [8]. Extensive work has been done on the melting mechanisms for plastic extrusion by Donovan [9], Lindt [10] as well as by Chung and Chung [11]. Lidor and Tadmor [12] and Bigg and Middleman [13] have carried out a theoretical analysis to determine the residence time distribution function and the strain distribution in plasticating screw extruders. Tadmor and Gogos [14] and Fenner [15-17] have solved the flow of a polymer in the feed, compression, and metering sections of an extruder. Fenner [16] also solved the case of the temperature profile developing along the length of the screw channel. Agur and Vlachopolous [18] have studied the flow of polymeric materials, which included a model for flow of solids in the feed hopper, a model for solid conveying zone and a model for melt conveying zone. Fukase et al. [19] presented a plasticating model for a single-screw extruder including the melting of solid beads of a polymer. Elbirli and Lindt [20,21] developed a model of the melting process for two types of barrier screws. Mokhtarian and Erwin [22,23] developed a mathematical model for mixing in a single-screw extruder.

Few investigators [24,25] have attempted to solve the flow of food materials in a single-screw extruder. Generally, food materials are assumed to have rheological behavior similar to polymers [10,26]. Bruin et al. [27] have discussed the flow of biopolymers in an extruder. Recently, Mohamed and Morgan [28] presented the results for average heat transfer coefficients in a single-screw extruder for the flow of non-Newtonian food materials. They have discussed the effect of viscous dissipation and convective cooling on the transport process.

There is relatively little published literature on the study of the twin-screw extruder, mainly because of the complexity involved. White et al. [29] published a useful compiled record of the development of the technology and the analysis of flow mechanisms in twin screw extruders. This covered the intermeshing counter-rotating twin screw extruders, the tangential counter-rotating twin screw extruders, and the intermeshing co-rotating twin screw extruders. Burkhardt et al. [30] has investigated the difference in the mixing mechanisms between co-rotating and counter-rotating twin-screw extrusion process for a Newtonian fluid. Nichols [31,32] and, Nichols and Yao [33] have investigated the pumping characteristics of the counter-rotating tangential twin-screw extruders. Howland and Erwin [34] experimentally studied the mixing characteristics in counter-rotating tangential twin-screw extruders. Regarding the modeling of twin-screw extruders, Janssen [35] has extensively discussed the flow characteristics in fully intermeshing counter-rotating twin-screw extruders with a C-shaped chamber. Secor [36] developed a mass transfer model for a co-rotating twin-screw extruder. Wyman [37] has analyzed the flow of Newtonian fluids in shallow channels of twin-screw extruder using a simple theoretical model in which the interest was on the down-channel flow only, so that it is applicable to both co- and counter-rotating twin-screw extruders. Yacu [38] has simulated the energy input, pressure and velocity distributions in co-rotating twin-screw extruder, based on global energy and force balance using a uni-directional analysis.

Some reports have discussed the use of twin-screw extruders for materials processing. Hold [39] discussed the technological development of continuous mixers with counter-rotating non-intermeshing rotors. Illing [40] reported that the twin screw extruder meets the requirements of the direct extrusion process of nylon products from lactams. Jakopin [41] has reported the advantage of using twin-screw extruders for compounding of fillers. The twin-screw extruder offers the ability to perform several processing functions in a single operation during the compounding process, particularly the intermeshing co-rotating twin screw compounder with interchangeable screw configuration which offers system versatility.

The residence time distribution (RTD) is one of the important characteristics in extruders [42,43]. Todd [44], and Kao and Allison [45] have performed experimental studies on the residence time distribution in fully intermeshing co-rotating twin-screw extruders. Wolf et al. [46] have also reported the experimental results of RTD for a fully intermeshing counter-rotating twin-screw extruder. Herrmann and Eise [47] have experimentally studied the RTD in fully intermeshing counter-rotating twin-screw extruders, and in both co- and counter-rotating intermeshing twin-screw extruders. Janssen [35] and Walk [43] have also extensively discussed the same topic. Most of the experimental results have been compared with theoretical model predictions.

The geometry of a fully wiped co-rotating twin screws has been investigated by Booy [48]. Later, Booy [49] also proposed a mathematical model for the isothermal flow of a Newtonian liquid through co-rotating twin-screw extruders. The two cases investigated were: (i) the fully filled channel case, and (ii) the partly filled channel case. Herrmann and Jacopin [50] have categorized twin-screw extruders into several kinds, such as intermeshing, non-intermeshing, partially intermeshing, lengthwise, crosswise, open, or closed. Eise et al. [51] have done experiments on various types of intermeshing co-rotating and counter-rotating twin screw extruders and compared the operating principles and processing characteristics of each type.

Denson and Hwang [52,53] have studied the performance of self-wiping, co-rotating twin-screw extruders under effects of parameters, such as the leakage and cross-channel flow, and the axial pressure gradient. Hwang [54] developed the numerical simulation for both the cross-channel and down-channel flows of Newtonian fluids and discussed the intermeshing region of a fully-wiped co-rotating twin-screw extruder. Maheshri and Wyman [55] have also investigated numerically the combined cross- and down channel-flow in an idealized, leak proof, intermeshing twin-screw extruder.

4.3 THE PRESENT PROBLEM

The primary goal of this study is to understand the transport phenomena during the polymer extrusion process through a co-rotating or counter-rotating twin-screw extruder. The detailed understanding of the transport characteristics requires the determination and prediction of the following: velocity and pressure fields, residence time distribution (RTD), shear strain and stress distributions, temperature distribution, thermal stresses, etc.

The flow pattern within a twin-screw extruder is extremely complex to deal with as a whole at once. In the present study, a simplified approach has been introduced to enhance the scientific understanding of the basic phenomena of the flow and heat transfer. There are various types of twin-screw extruders which are classified according to the screw geometry, the degree of meshing, and the sense of rotation. Amongst them, as a first attempt towards the goal described above, a certain type of twin-screw extruder configuration has been focused on, namely cross-wise fully intermeshing, lentgth-wise open, or tangential type. Figure 4.1 shows a schematic diagram of the cross-sectional geometry of such a twin-screw extruder. The flow region in such a twin-screw extruder may be considered to consist of translation regions, similar to the those in single-screw extruders, and mixing regions (or intermeshing zones), located at the central area between the two screws. The complete twin-screw extruder can then be modeled as two series of translation and mixing regions in an alternating order, as depicted in Figure 4.2. The flow behavior in a translation region is very similar to that in the channel of a single-screw extruder. In partially intermeshed twin-screw extruders, a portion of the material entering the mixing region from one screw channel may stay with the same screw, while the other part crosses over to the other screw channel. The material leaving the mixing region then enters the two adjacent translation regions of the two screw channels. The complete numerical solution of twin-screw extruders involves analysis of the flow in these two basic regions. With this simplified approach, two numerical models have been developed which are: (i) a numerical model for a single-screw extruder to represent the translation regions, using a finite difference method, and (ii) a finite element model (FEM) to study the flow and mixing characteristics in the mixing zones.

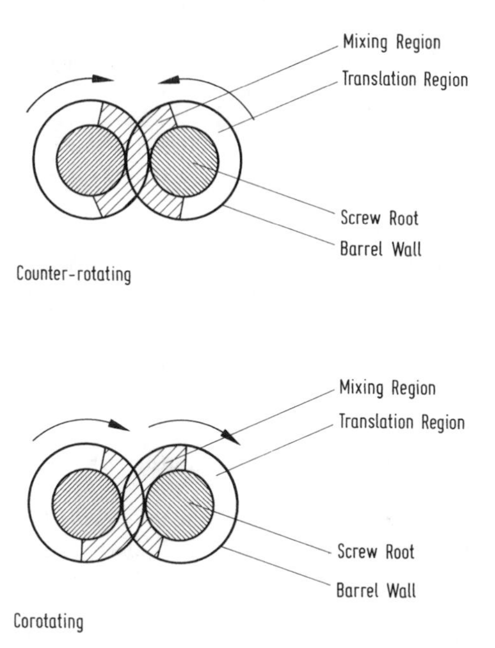

Figure 4.1 Schematic diagram of the cross section of a twin-screw extruder.

The numerical model for a single-screw extruder can handle both fully developed and developing flow as well as the temperature field along the down channel direction of a screw.

Finite difference computations are carried out over a wide range of governing dimensionless parameters such as Griffith Number, Peclet Number, dimensionless volume flow rate, and the power law index for non-Newtonian power-law fluids.

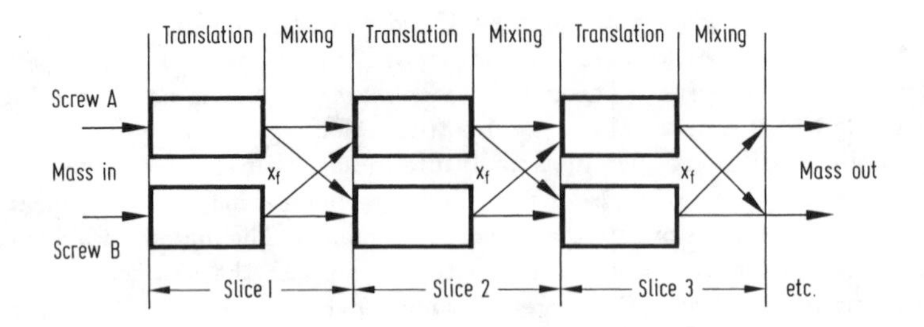

Figure 4.2 Flow in a twin-screw extruder represented by a series of translation and mixing zones.

The transport phenomena in twin-screw extruders is significantly affected by the mixing mechanisms in the intermeshing zone. The effort here includes a study of the characteristic of this flow and the effect of various parameters, such as the geometry of the extruder (diameter of the screw root, diameter of the barrel, distance between the screw axes), the processing parameters (such as the speed and direction of rotation of the screw), and the material properties on the mixing.

Finally, the numerical simulation in the two regions are to be combined for the flow analysis in the entire twin-screw extruder. The fundamental considerations and methodology for coupling the two simulations are discussed. Thus, a mathematical model for the simulation of a full twin-screw extruder is obtained.

4.4 ANALYSIS

4.4.1 Mathematical Model for the Flow and Heat Transfer in a Single-Screw Extruder

The simplified geometry of a single-screw extruder and the cross section of a screw channel are shown in Figure 4.3. For ease of visualization and analysis, the coordinate system is fixed to the screw root and the barrel is moved in a direction opposite to the screw rotation. For steady, developing, two-dimensional flow of a homogeneous fluid in a single-screw extruder with shallow channels, i.e., for $H \ll W$ in Figure 4.3, the conservation of mass gives

$$\frac{\partial u}{\partial x}+\frac{\partial w}{\partial z}=0 \qquad (1)$$

The channel width W is assumed to be infinite and the flow velocities are assumed to be fully developed. Therefore, both the terms in Eq (1) vanish. After applying the lubrication approximation [56] in the x and z directions, the equations for the conservation of momentum become

$$\frac{\partial p}{\partial x} = \frac{\partial(\tau_{yx})}{\partial y}, \tag{2}$$

$$\frac{\partial p}{\partial y} = 0, \tag{3}$$

$$\frac{\partial p}{\partial z} = \frac{\partial(\tau_{yz})}{\partial y} \tag{4}$$

where u and w are the velocity components in the x and z directions, respectively, p is the hydrostatic pressure and τ is the shear stress. The velocity, v, in the y direction is taken as zero in this model. This assumption is valid for screw channels having a very large W/H ratio. The effects of curvature are assumed to be negligible. Also, the clearance between the screw flights and the barrel is assumed to be small enough to neglect the leakage across the flights from one screw channel to the neighboring channel. Even though this model is a highly simplified one, it is sufficient to bring out the overall nature of the transport phenomena.

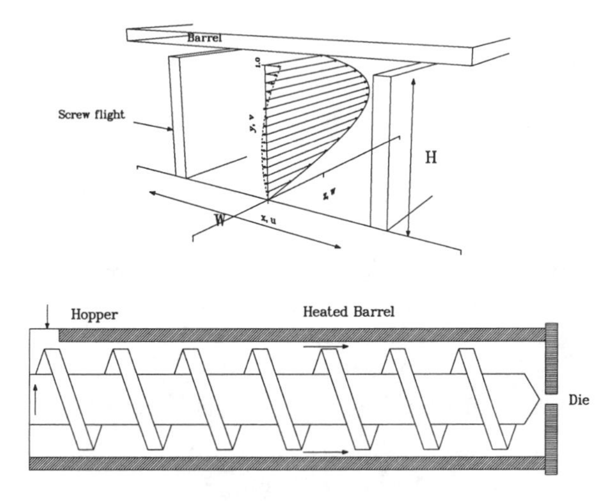

Figure 4.3 Simplified geometry of a single-screw extruder, screw geometry, and velocity profiles.

Within a screw channel, the temperature as well as the velocity fields may change along the length of the channel. If the barrel is maintained at a fixed temperature, the velocity and temperature profiles will approach a fully developed situation, .i.e., they do not change in form along the channel length [57]. The energy equation governing the temperature in such cases is obtained by neglecting the axial convection terms. The energy equation then becomes

$$0=\frac{\partial(k\frac{\partial T}{\partial y})}{\partial y}+\tau_{yx}\frac{\partial u}{\partial y}+\tau_{yz}\frac{\partial w}{\partial z} \tag{5}$$

In the presence of strong viscous dissipation effects and/or heat addition from the barrel, the thermal convection along the z-direction (along the channel length) is significant. Therefore, the temperature field develops along the z-direction. The velocities may also change with the downstream position as a result of this change in temperature [17]. However, it is assumed that this change is negligible. In such cases, Eq (5) is modified as

$$\rho C w\frac{\partial T}{\partial z}=\frac{\partial(k\frac{\partial T}{\partial y})}{\partial y}+\tau_{yx}\frac{\partial u}{\partial y}+\tau_{yz}\frac{\partial w}{\partial z} \tag{6}$$

where T is the temperature, ρ the density, C the specific heat and k the thermal conductivity of the fluid.

The shear stresses τ_{xy} and τ_{yz} are given by

$$\tau_{yx}=\mu\frac{\partial u}{\partial y},\quad \tau_{yz}=\mu\frac{\partial w}{\partial y} \tag{7}$$

where μ is the molecular viscosity. Since most of the practical materials undergoing extrusion process behave as non-Newtonian fluids, the following constitutive equation may be used to represent the dependence of μ on T and on the strain rate $\dot{\gamma}$ [17,29]:

$$\mu=\mu_0(\frac{\dot{\gamma}}{\dot{\gamma}_0})^{n-1}\exp\{-b(T-T_0)\} \tag{8}$$

where $\dot{\gamma}=[(\partial u/\partial y)^2+(\partial w/\partial y)^2]^{1/2}$, b is the temperature coefficient of viscosity, n is the power law index and the subscript o denotes the reference conditions.

The boundary conditions are prescribed in terms of Figure 4.4 as:

for all $z \neq 0$: at $y=0$: $u=0,\ w=0,\ \frac{\partial T}{\partial y}=0$ (adiabatic screw)

at $y=H$: $u=V_{bx},\ w=V_{bz},\ T=T_b$

for $z=0$: $T=T_i,\ u=u_{dev},\ w=w_{dev}$

where V_{bz} and V_{bx} are the z and x components, respectively, of the barrel velocity V_b and the subscript "dev" denotes the solution for the developed case at the inlet temperature T_i. The screw has been taken as isothermal, at T_b, in most studies reported in literature. However, a more practical circumstance is represented by the adiabatic condition at the screw surface. The proper approach is to take into account the conduction within the screw, in which case a conjugate heat transfer problem is involved [58]. Such conjugate effects are not considered here, though they are planned for consideration in the future.

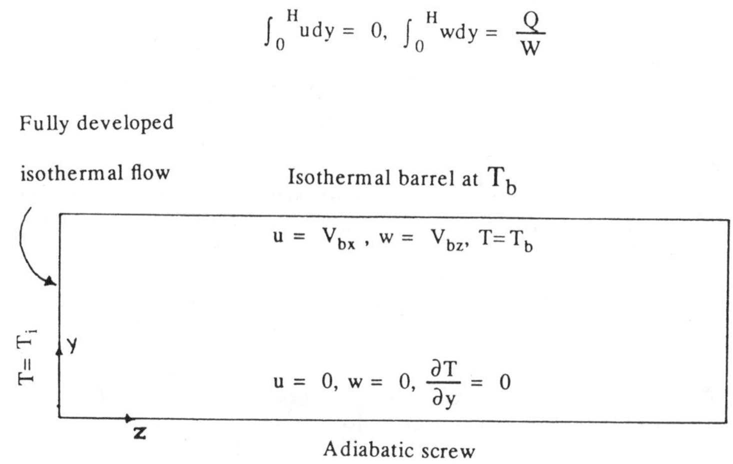

$$\int_0^H u\,dy = 0, \quad \int_0^H w\,dy = \frac{Q}{W}$$

Fully developed

isothermal flow

Isothermal barrel at T_b

$u = V_{bx}$, $w = V_{bz}$, $T = T_b$

$T = T_i$

y

$u = 0$, $w = 0$, $\frac{\partial T}{\partial y} = 0$

z

Adiabatic screw

Figure 4.4 Boundary conditions in the single-screw model.

Two additional constraints arise due to the flow considerations. The net flow in a direction normal to the screw flights (i.e., x axis in Fig. 4.3) must be zero and the net flow in the down channel direction (i.e., z axis in Fig. 4.3) must be equal to the total volumetric flow rate, Q. These constraints are given by the equations:

$$\int_0^H u\,dy = 0, \quad \int_0^H w\,dy = \frac{Q}{W} \tag{9}$$

The above equations are non-dimensionalized in terms of the following dimensionless variables

$$x^* = x/H, \; y^* = y/H, \; z^* = z/H$$

$$u^* = u/V_{bz}, \; w^* = w/V_{bz}$$

$$\theta^* = \frac{T - T_i}{T_b - T_i}, \; p^* = p/\bar{p}, \; \bar{\mu} = \frac{\mu_0}{\dot{\gamma}_0^{\,n-1}}\left[\frac{V_{bz}}{H}\right]^{n-1} e^{-b(T_i - T_0)}$$

$$\bar{p} = \bar{\mu} V_{bz}/H, \; \dot{\gamma}^* = \dot{\gamma}/(V_{bz}/H)$$

Then, the dimensionless equations for momentum conservation become

$$\frac{\partial p^*}{\partial x^*} = \frac{\partial}{\partial y^*}\left(\frac{\partial u^*}{\partial y^*}[\dot{\gamma}^*]^{(n-1)/2}\exp(-\beta\theta^*)\right) \tag{10}$$

$$\frac{\partial p^*}{\partial z^*} = \frac{\partial}{\partial y^*}\left(\frac{\partial w^*}{\partial y^*}[\dot{\gamma}^*]^{(n-1)/2}\exp(-\beta\theta^*)\right) \tag{11}$$

where w^* and u^* are the dimensionless velocity components and p^* is the dimensionless pressure.

By choosing an appropriate non-dimensionalization, the effects of shear [59] and of the dependence of viscosity on temperature, are separated in the energy equation. This is desirable since the two effects may be distinct in practice. However, in literature, the two are often considered together [15-17]. The energy equation, in dimensionless form, is thus obtained as:

$$\mathrm{Pe}\, w^* \frac{\partial \theta^*}{\partial z^*} = \frac{\partial^2 \theta^*}{\partial y^{*2}} + G\,[\dot{\gamma}^*]^{(n+1)/2}\exp(-\beta\theta^*) \tag{12}$$

where Pe is termed as the Peclet number in the literature and G the Griffith number. These are defined in the nomenclature and are well known parameters from the heat transfer literature. The corresponding dimensionless boundary conditions are:

at the screw, i.e., at $y^* = 0$: $u^* = 0,\ w^* = 0,\ \frac{\partial \theta^*}{\partial y^*} = 0$ (adiabatic screw)

at the barrel, i.e., at $y^* = 1$: $u^* = \tan(\phi),\ w^* = 1,\ \theta^* = 1.$

at the inlet, i.e., at $z^* = 0$: $\theta^* = 0,\ u^* = u^*_{dev},\ w^* = w^*_{dev}$

The constraints on the flow are obtained, in dimensionless form, as

$$\int_0^1 u^* dy^* = 0 \tag{13}$$

$$\int_0^1 w^* dy^* = q_v = \frac{Q/W}{HV_{bz}} \tag{14}$$

Here, the parameter q_v represents the dimensionless throughput, or the volumetric flow rate, emerging from the extruder.

Thus, the parameters that govern the numerical solution are:

1. Peclet Number $\mathrm{Pe} = V_{bz}H/\alpha$. This parameter indicates the relative importance of thermal convection as compared to conduction in the fluid.
2. Griffith Number $G = \bar{\mu}V^2_{bz}/k(T_b - T_i)$. This parameter indicates the relative importance of viscous dissipation as compared to thermal conduction in the fluid.
3. Dimensionless volume flow rate, or the throughput, denoted by q_v.
4. $\beta = b(T_b - T_i)$. This relates to the effect of temperature on viscosity.
5. n : power law index, where $n = 1$ applies for Newtonian fluids.
6. ϕ : screw helix angle.

4.4.2 Mathematical Model for the Flow in the Intermeshing Zone of a Twin-Screw Extruder

The flow inside the twin-screw extruder is very complicated in nature and is very much dependent on the design and operating conditions of the extruder. For the purposes of simulating the flow inside the intermeshing zone, a relatively simple, special kind of twin-screw extruder was adopted in the simulation, namely the co- and counter-rotating tangential twin-screw extruder. Furthermore, it was assumed that the pitch of the screw is much larger than the depth of the screw, so that a cross-section of the extruder can be made without interference of the screw flights. This cross-section is shown in Figure 4.1, lying on the x-y plane of the chosen coordinate system. The z-axis is pointed out of the paper along the extruder axis. The computational domain of interest is shown in Figure 4.1 as the area between the two screws identified as "mixing region." The problem formulation is simplified by assuming that the flow in the extruder is incompressible and creeping, i.e., inertia effects are negligible, with the viscosity model of a Newtonian or a power-law non-Newtonian fluid. Based on those assumptions, the velocity gradient in the z-direction can be reasonably neglected, compared with the velocity gradients in the x- and y-directions. Neglecting terms involving $\partial/\partial z$, the conservation equations for mass and momentum may be written, for a two-dimensional and steady flow, as follows:

$$\frac{\partial u}{\partial x}+\frac{\partial v}{\partial y}=0 \tag{15}$$

$$\frac{\partial p}{\partial x}=\frac{\partial \tau_{xx}}{\partial x}+\frac{\partial \tau_{xy}}{\partial y} \tag{16}$$

$$\frac{\partial p}{\partial y}=\frac{\partial \tau_{xy}}{\partial x}+\frac{\partial \tau_{yy}}{\partial y} \tag{17}$$

with

$$\tau_{xx}=2\mu\frac{\partial u}{\partial x},\quad \tau_{yy}=2\mu\frac{\partial v}{\partial y}\quad \text{and}\quad \tau_{xy}=\mu\left(\frac{\partial u}{\partial y}+\frac{\partial v}{\partial x}\right) \tag{18}$$

where u and v represent velocity components in the x and y directions, respectively, p is the pressure, and τ represents the stress tensor components in the Cartesian coordinate system.

In the current study, the viscoelastic effect is ignored. The flow in the mixing region is considered in terms of a pure isothermal viscous material. The viscosity $\mu(\dot{\gamma},T)$ is of power-law type, as described in Eq (8). The generalized shear rate can be represented as follows:

$$\dot{\gamma}=[2D_{ij}D_{ij}]^{1/2}=\left[2\left(\frac{\partial u}{\partial x}\right)^2+\left(\frac{\partial u}{\partial y}+\frac{\partial v}{\partial y}\right)^2+2\left(\frac{\partial v}{\partial y}\right)^2\right]^{1/2} \tag{19}$$

The boundary conditions are:

$$(\tau_{xx}-p)n_x+\tau_{xy}n_y=\bar{t}_x \quad\text{and}\quad \tau_{xy}n_x+(\tau_{yy}-p)n_y=\bar{t}_y \quad \text{on } S_t \tag{20}$$

$$u = \bar{u}, v = \bar{v} \text{ on } S_v \tag{21}$$

where $\bar{t}_x$, n_x, $\bar{u}$ and $\bar{t}_y$, n_y, $\bar{v}$ stand for the prescribed traction (force per unit area), outward normal unit vector and prescribed velocity components in the x and y directions, respectively. The boundary of the computational domain is divided into the velocity boundary S_v and the traction boundary S_t, where the velocity components and the traction components of the boundary conditions are applied, respectively.

It may be noted that, in the above formulation, the effect of the screw flight is neglected. Later on, however, the effect of the screw flight has also been addressed. In order to incorporate the flight effect, a velocity component in the z-direction, i.e. axial direction resulting from the screw rotation has been taken into account. Therefore, an additional equation was employed to represent the flow in the z-direction. Based on the assumption described above, the momentum equation along the z-direction can be represented as follows:

$$\frac{\partial p}{\partial z} = \frac{\partial \tau_{xz}}{\partial x} + \frac{\partial \tau_{yz}}{\partial y}, \tag{22}$$

with

$$\tau_{xz} = \mu \frac{\partial w}{\partial x} \text{ and } \tau_{yz} = \mu \frac{\partial w}{\partial y} \tag{23}$$

The corresponding boundary conditions are as follows:

$$w = \bar{w} \text{ on } S_w; \quad \frac{\partial w}{\partial x} n_x + \frac{\partial w}{\partial y} n_y = \bar{w}_n \text{ on } S_q \tag{24}$$

Then the generalized shear rate is represented, instead of Eq (19), as follows:

$$\dot{\gamma} = [2D_{ij}D_{ij}]^{1/2}$$

$$= [2(\frac{\partial u}{\partial x})^2 + (\frac{\partial u}{\partial y} + \frac{\partial v}{\partial x})^2 + 2(\frac{\partial v}{\partial y})^2 + (\frac{\partial w}{\partial x})^2 + (\frac{\partial w}{\partial y})^2]^{1/2} \tag{25}$$

The momentum equations in the x- and y- directions, Eqs (16) and (17), remain the same, but are effected by the flight effect through the viscosity function with the shear rate in Eq (25).

The effect of temperature change along the extruder can be also addressed. In such a non-isothermal case, the energy equation should be included with the proper incorporation of the temperature dependent viscosity. However, this case has not been addressed in the scope of the current work. Because of the complex geometry of the intermeshing zone of a twin-screw extruder, the Finite Element Method has been utilized for analyzing the flow in the intermeshing zone.

4.5 NUMERICAL SOLUTION

For the case of a single-screw extruder, the governing equations are Eqs (10)-(14). These are solved by means of finite difference techniques [17,35,60]. The computational domain and the finite-difference grid are shown in Figure 4.5. The computations were carried out over a 101 x 61 grid with $\Delta y^* = 0.0166$ and $\Delta z^* = 2.0$. With the boundary conditions in terms of u*, w* and θ^* given at any upstream z location, the energy equation, Eq (12), is solved to obtain the temperature distribution at the next downstream z location. With the temperature distribution thus obtained, the momentum equations, Eqs (10) and (11), are solved to obtain the velocity distribution there. The solution of the momentum equations employs the iterative Newton-Raphson method [60]. Thus, at a given z location, Eqs (10) and (11) are iteratively solved, using an implicit finite difference scheme, untill the constraints given by Eqs (13) and (14) are satisfied to within a specified convergence criterion. Since the energy equation, Eq (12), is parabolic in z, boundary conditions are necessary only at $z^* = 0$ to allow marching along the z direction and, thus, obtain the solution in the entire domain. To avoid instability in the numerical computations, the upper bound on the step size Δz^* is given by [17]:

$$\Delta z^* \lesssim 4(Pe)\,\Delta y^{*2} \tag{26}$$

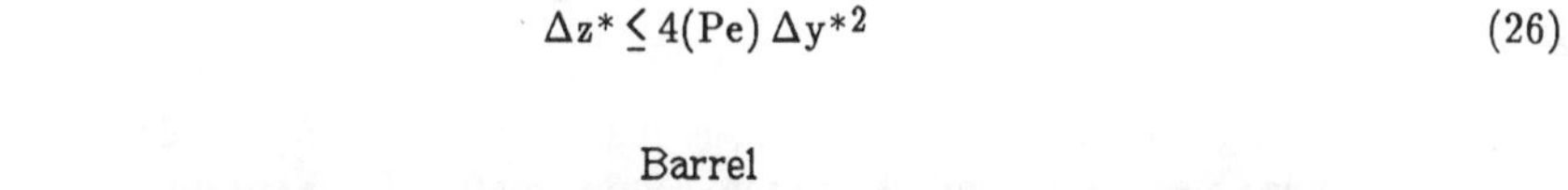

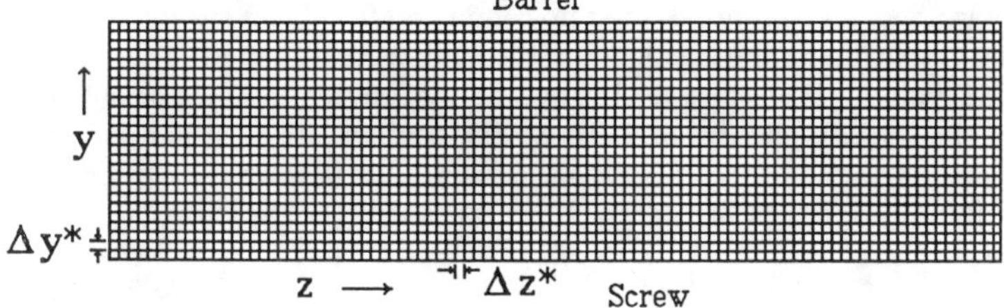

Figure 4.5 Computational domain and the finite-difference grid.

Thus, if the properties of the polymer material under consideration and the operating conditions of an extruder are given, the actual temperature and velocity profiles and other relevant quantities, such as mixing, residence time, etc., can be computed for an extrusion process.

The calculations for the intermeshing zone were carried out using a Finite Element model (FEM). The Finite Element package that has been developed consists of three major parts, i.e., the pre-processor, the main processor (main FEM program), and the post-processor.

The pre-processor is an automatic mesh generator with a special care for the velocity boundary conditions along the screw root surface. Considering the complicated geometry representation of a twin-screw extruder, a program is developed to prepare the input data for the mesh generation program. A special feature has been added to the program to distinguish the boundary nodes from the internal nodes, and to take care of the boundary conditions on the boundary in terms of the rotational speed of the screw. It is obvious that the pre-processor saves a great deal of time in preparing the data for the main-processor.

The main-processor is the finite element analysis program developed for a plane or axisymmetric creeping flow of Newtonian or generalized Newtonian fluids, such as a power-law model, Carreau model and Cross model [14] fluid. This finite element analysis program consists of two programs, the first one is the program for solving the momentum equations in x and y plane, while the second one is for solving the z-momentum equation.

The FEM formulation for the momentum equations in x- and y-directions is based on the principle of virtual power with the velocity and pressure being the primitive variables which are the unknowns in the system equations. A Lagrangian multiplier is introduced to satisfy the incompressibility condition. The resulted equations can be assembled to obtain the global matrix equation in the form [61]:

$$\begin{bmatrix} [K] & [G]^T \\ [G] & [0] \end{bmatrix} \begin{bmatrix} (u) \\ (-p) \end{bmatrix} = \begin{bmatrix} (F) \\ (0) \end{bmatrix} \tag{27}$$

where [K] is the global stiffness matrix, [G] is the global matrix for the incompressibility constraint, [0] is a zero matrix and (0) is a zero column matrix, (F) is a column matrix for the work equivalent force due to the traction and the body force, (u) and (-p) are the unknown velocity column matrix and pressure column matrix respectively. It should be noted that [K] is dependent on the velocity gradient and also on the temperature for a non-isothermal flow through the viscosity $\mu(\dot{\gamma},T)$. Therefore, iteration is needed to solve the nonlinear equation, Eq (27). The [K] matrix is updated at each iteration and the Newton-Raphson iteration scheme is used.

The finite element formulation for the z-momentum equation is based on the Galerkin method to result in:

$$[H]\,(w) = (f) \tag{28}$$

where [H] is the stiffness matrix, (w) is the column matrix of the unknown w velocity, and (f) is the column matrix due to the $\partial p/\partial z$ and boundary condition. The [H] matrix is also updated through iteration procedure.

When the z-momentum equation is used for solving the w velocity component, a more complicated iteration scheme is needed. As can be seen from the momentum equations, Eqs (16), (17) and (22), the three equations are coupled through the viscosity term. Therefore, those three equations have to be solved simultaneously by using an iteration scheme. In the present study, Eq (27) for x-y directions and Eq (28) for z-direction is alternatingly solved until a convergent solution is obtained.

The current finite element analysis program uses quadratic triangular elements with quadratic interpolation for the velocity field and linear interpolation for the pressure field. The pressure nodal points are located at the vertices of the triangular element. The solution is then displayed graphically by the post-processor. It displays the computed velocity field and the pressure contours. Thus, the post-processor helps the user in interpreting the characteristic of the flow and the pressure field very quickly. In addition, this processor calculates the total volume flow rate in a channel and the flow rate of the fluid that crosses over to the other screw channel. This determines the mixing parameter, as discussed earlier.

4.6 NUMERICAL RESULTS AND DISCUSSION

4.6.1 Results on the Translation Zone

The results presented here are based on the single-screw model and are in the moving barrel formulation. Upon obtaining the numerical solution for the velocity and the temperature fields, various quantities of interest such as heat input from the barrel, bulk temperature, shear stress, and pressure at various downstream locations, are calculated. Also, the residence time distribution (RTD) is calculated by numerically simulating an experimental procedure for estimating the RTD as explained later.

For the fully developed case when the temperature and the velocity fields are assumed not to vary downstream, the results are obtained by solving Eqs (2)-(4), and Eq (5), in which the convective transport of heat is ignored. Figure 4.6 shows typical w^* velocity profiles for different values of q_V. Figure 4.7 shows the characteristic curve in terms of the throughput, q_V, and the dimensionless pressure gradient, $\partial p^*/\partial z^*$, for different values of the Griffith number G and at $n = 0.5$. The situation corresponding to the drag flow, in which the pressure gradient is zero, is obtained for $q_V = 0.5$. In this case, the w^* velocity component varies linearly with y^*.

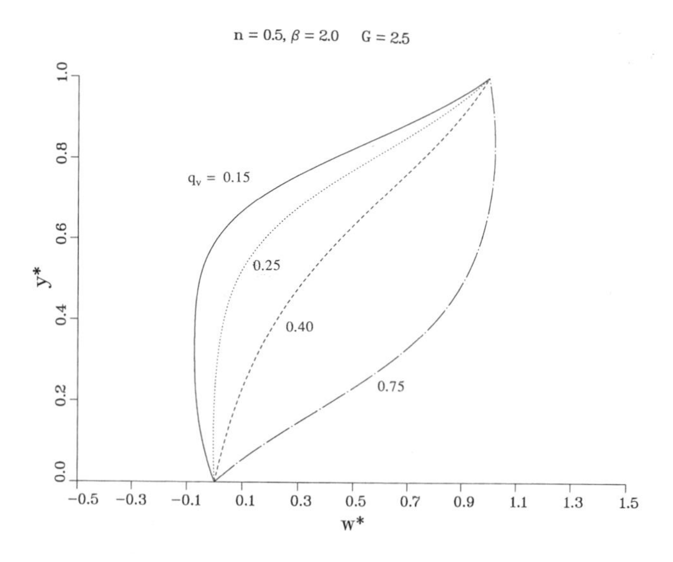

Figure 4.6 Profiles of the w^* velocity component at different values of q_V, for the developed case in the single-screw model.

For the developing case, when the temperature and velocity fields are changing downstream, the results are presented in terms of the calculated isotherms, isovelocity lines and other relevant quantities of interest, as discussed below.

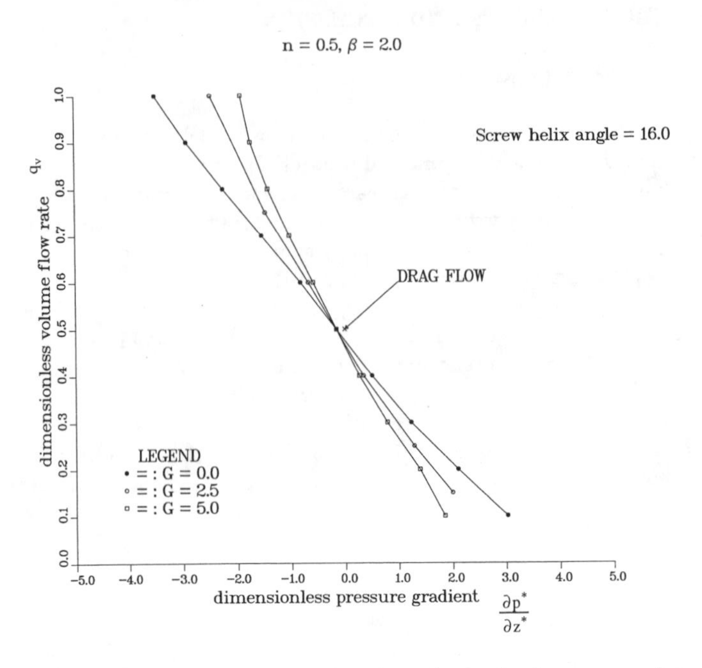

Figure 4.7 Characteristic curves for the developed case in the single-screw model.

Velocity and Temperature Fields

Figure 4.8 shows the general nature of the flow field in the y-z plane, obtained from a finite difference numerical solution for a Newtonian fluid with $n = 1$, $G = 0.008$, $Pe = 3427$, $\beta = 1.61$ and $q_V = 0.30$. These values are based on the conditions for a typical extrusion process. In Figure 4.8, the velocity is assumed to be fully developed at $z = 0$. The barrel temperature is then raised in a step change to a constant value of T_b and held at this value downstream. As seen from Figure 4.9, the temperature of the flow rises gradually with downstream distance because of the heat addition from the barrel. The heat conducted from the barrel is convected in the z direction by the flow. The effect of viscous dissipation is small for the chosen values of the parameters.

Figures 4.10 and 4.11 show the results for a non-Newtonian fluid with $n = 0.5$, $G = 10.0$, $Pe = 3427$, $\beta = 1.61$ and $q_V = 0.30$. Here, the temperature field is significantly different from the one shown in Figure 4.9. Due to increased viscous dissipation, the temperature of the flow is seen to increase above the barrel temperature. However, the corresponding changes in the velocity field are not significant. Beyond a certain distance along the z direction downstream, for $z^* > 100$, approximately in Figure 4.11, the fluid temperature near the barrel is higher than the barrel temperature and heat transfer occurs from the flow to the barrel. Several other operating conditions and fluid properties were considered and trends similar to these discussed above were obtained.

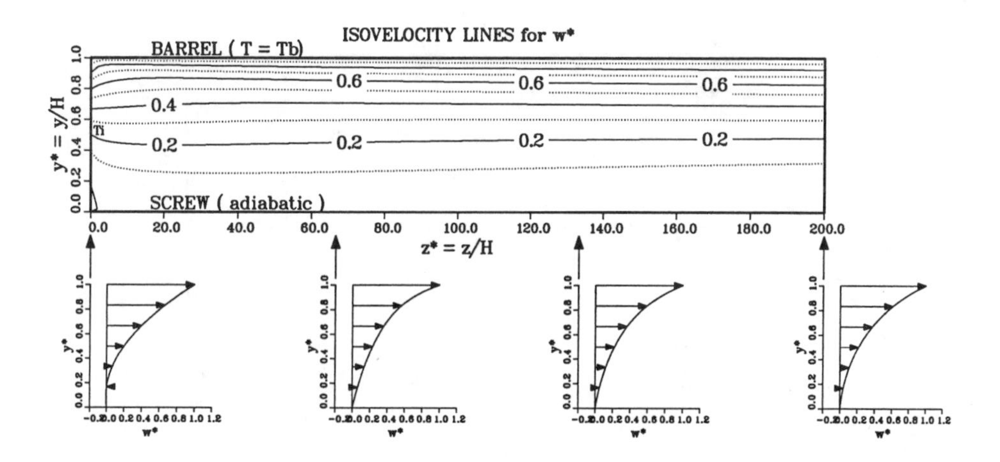

Figure 4.8 Lines of constant w*, and w* velocity profiles at four downstream locations for n = 1, G = 0.008, Pe = 3427, β = 1.61 and q_V = 0.30.

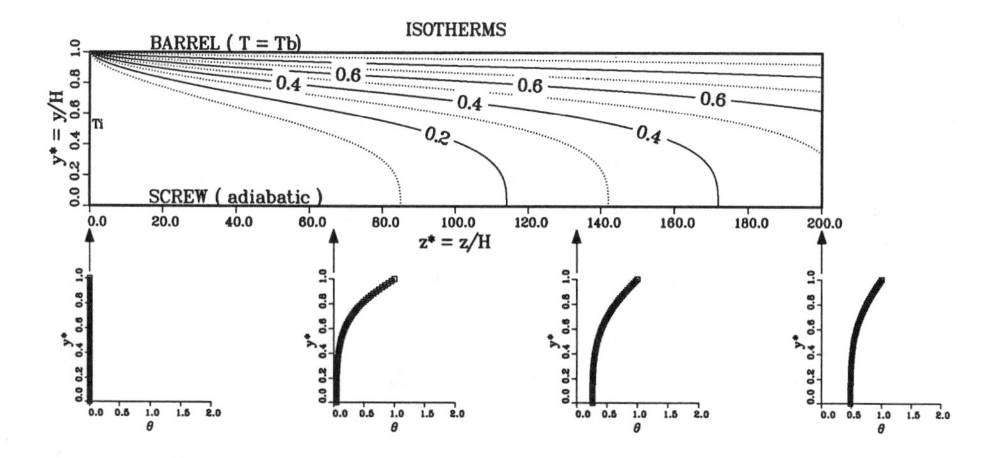

Figure 4.9 Isotherms and temperature profiles at four downstream locations for n = 1, G = 0.008, Pe = 3427, β = 1.61 and q_V = 0.30.

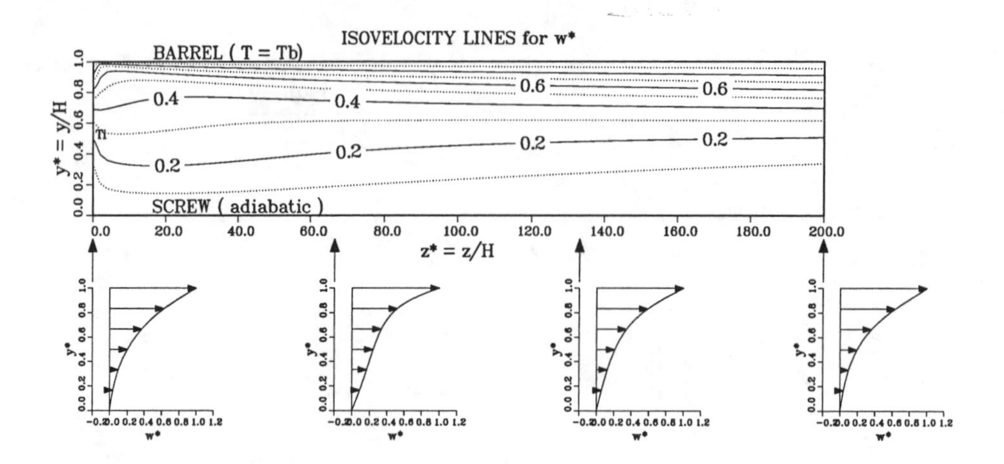

Figure 4.10 Lines of constant w*, and w* velocity profiles at four downstream locations for $n = 0.5$, $G = 10.0$, $Pe = 3427$, $\beta = 1.61$ and $q_V = 0.30$.

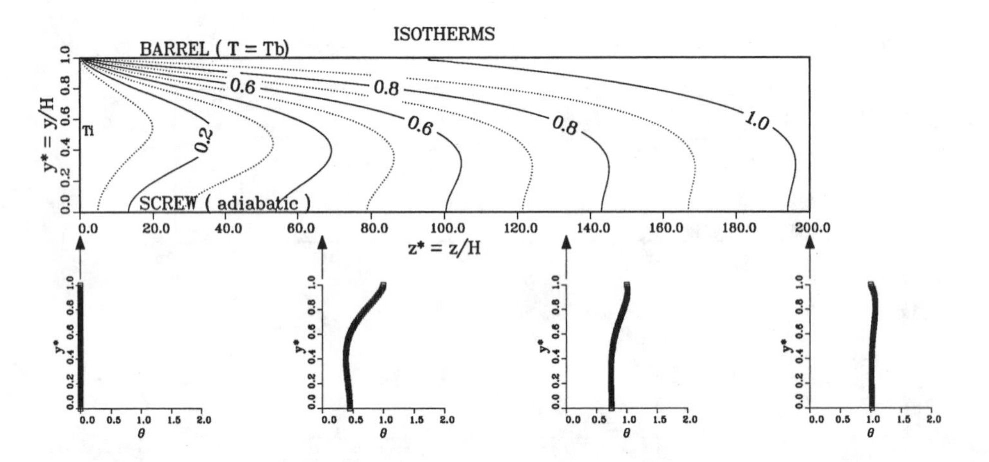

Figure 4.11 Isotherms and temperature profiles at four downstream locations for $n = 0.5$, $G = 10.0$, $Pe = 3427$, $\beta = 1.61$ and $q_V = 0.30$.

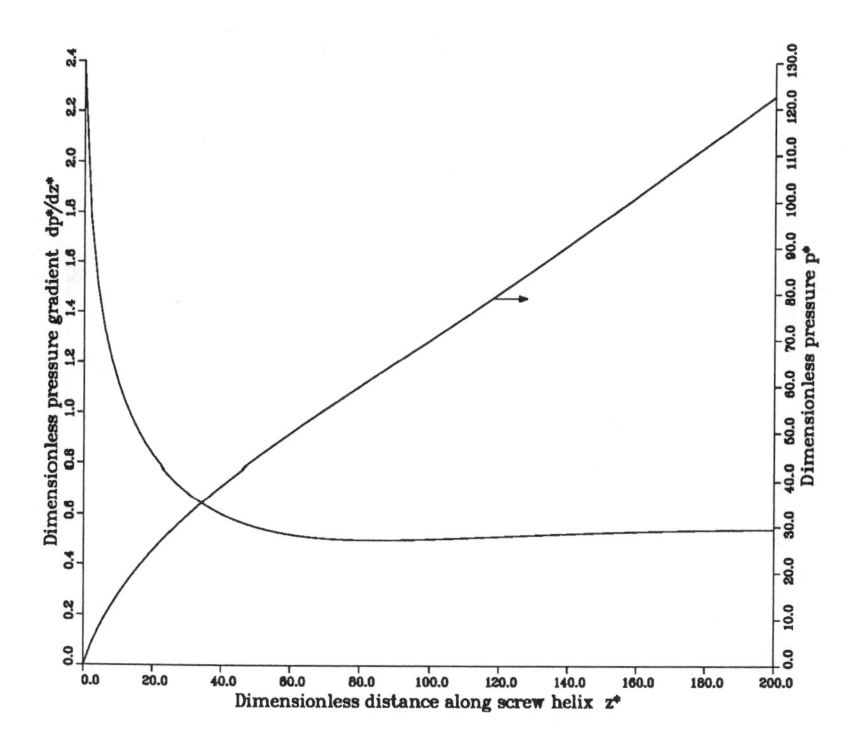

Figure 4.12 Variation of the dimensionless pressure gradient, $\partial p^*/\partial z^*$, and the dimensionless pressure p^* along z^*, for $n = 1$, $G = 0.008$, $Pe = 3427$, $\beta = 1.61$ and $q_V = 0.30$.

Pressure Gradients and Pressure

Figures 4.12 and 4.13 show the corresponding pressure gradient $\partial p^*/\partial z^*$ and the local pressure values obtained at various downstream locations for the two cases described above. As seen from Figures 4.12 and 4.13, higher pressure gradients are required for the case when $n = 1.0$, $G = 0.008$ than when $n = 0.5$, $G = 10.0$. The Griffith number, G, indicates the relative importance of viscous dissipation compared to conduction of heat. At higher values of G, more heat is generated due to the viscous dissipation than is conducted. This results in an increasing temperature in the flow downstream. In the case when $G = 10.0$, the temperature rise is substantial enough to cause a drop in viscosity according to the constitutive equation, Eq (8), and therefore, the pressure gradients are lower in this case for the same flow rate. However, in both cases, the value of q_V is 0.3, which is smaller than that for a simple drag flow ($q_V = 0.5$) with no pressure variation in the z direction. Therefore, as expected, an adverse pressure gradient exists for $q_V = 0.3$. This represents a reduced flow rate due to the presence of a die.

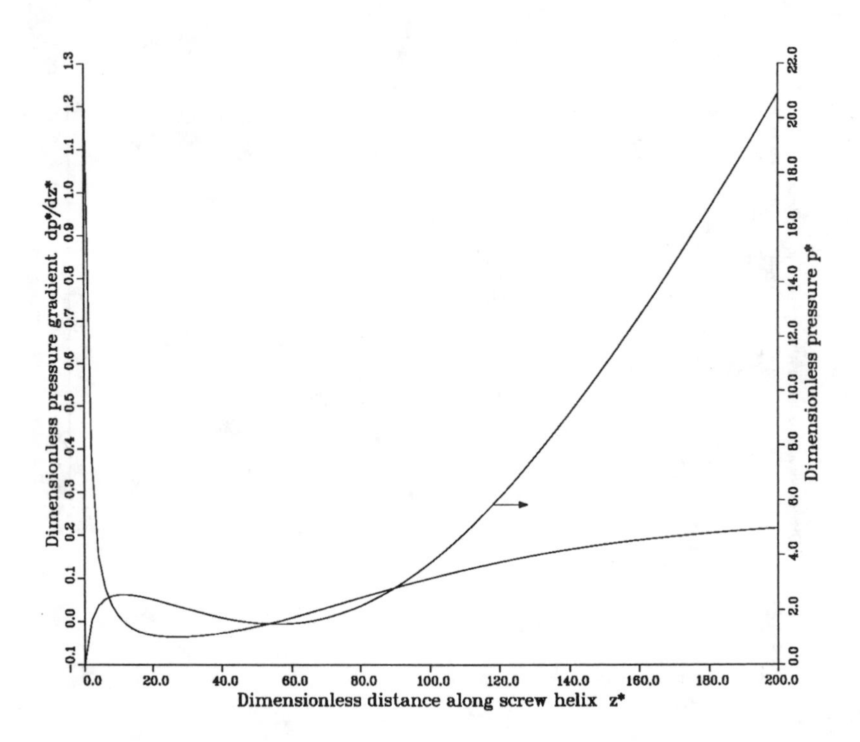

Figure 4.13 Variation of the dimensionless pressure gradient, $\partial p^*/\partial z^*$, and the dimensionless pressure p^* along z^*, for $n = 0.5$, $G = 10.0$, $Pe = 3427$, $\beta = 1.61$ and $q_v = 0.30$.

Bulk Temperature and Heat Input at the Barrel

Figures 4.14 and 4.15 show the variation of the bulk temperature in the flow and of the heat input from the barrel to the flow at various downstream locations. The bulk temperature, T_{bulk}, and the heat input from the barrel, q_{in}, are defined as follows:

$$T_{bulk} = \left[\int_0^H w\,T\,dy\right] / \left[\int_0^H w\,dy\right] \Rightarrow \theta^*_{bulk} = \frac{1}{q_v}\int_0^1 w^*\,\theta^*\,dy^* \tag{29}$$

$$q_{in} = k\left[\frac{\partial T}{\partial y}\right]_{barrel} \Rightarrow q^*_{in} = \left[\frac{\partial \theta^*}{\partial y^*}\right]_{barrel} \tag{30}$$

As seen from Figures 4.14 and 4.15, the bulk temperature rises monotonically. The heat flux at the barrel drops rapidly along the downstream direction. The heat conducted from the barrel is convected downstream by the flow. Therefore, a smaller amount of heat needs to

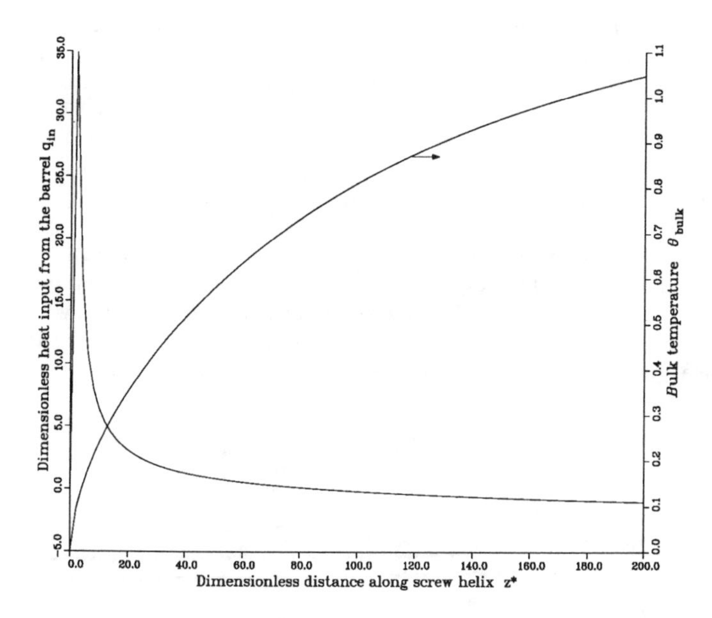

Figure 4.14 Variation of the dimensionless bulk temperature, θ^*_{bulk}, and the dimensionless heat input from the barrel, q^*_{in}, along z*, for n = 1, G = 0.008, Pe = 3427, $\beta = 1.61$ and $q_V = 0.30$.

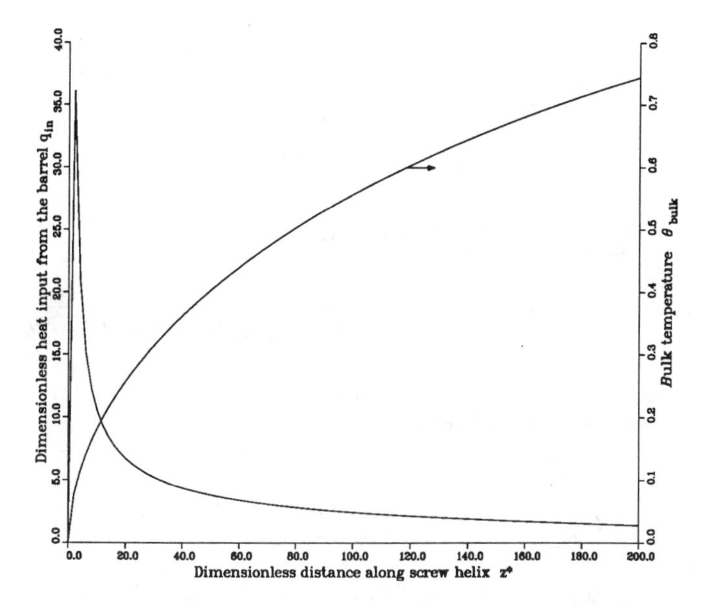

Figure 4.15 Variation of the dimensionless bulk temperature, θ^*_{bulk}, and the dimensionless heat input from the barrel, q^*_{in}, along z*, for n = 0.5, G = 10.0, Pe = 3427, $\beta = 1.61$ and $q_V = 0.30$.

be supplied to the barrel downstream in order to maintain it at a constant temperature T_b. When the heat generation due to viscous dissipation is substantial, so that the flow temperature rises above the barrel temperature, then heat must be removed from the barrel in order to maintain it at T_b. This was observed to be the case for G = 10.0 (Fig. 4.15). Such a circumstance is commonly encountered in plastic extrusion processes.

Shear Stresses

Figures 4.16 and 4.17 show the corresponding shear stresses τ_{xy}, τ_{zy}, and total shear stress τ, at the inner surface of the barrel and also at the screw root. Note that the shear stress at the barrel is not the same as that at the screw. In the absence of the pressure gradient, which applies for the drag flow situation, $q_v = 0.5$, the shear stress values at the barrel and at the screw root would be equal due to considerations of dynamic equilibrium [14].

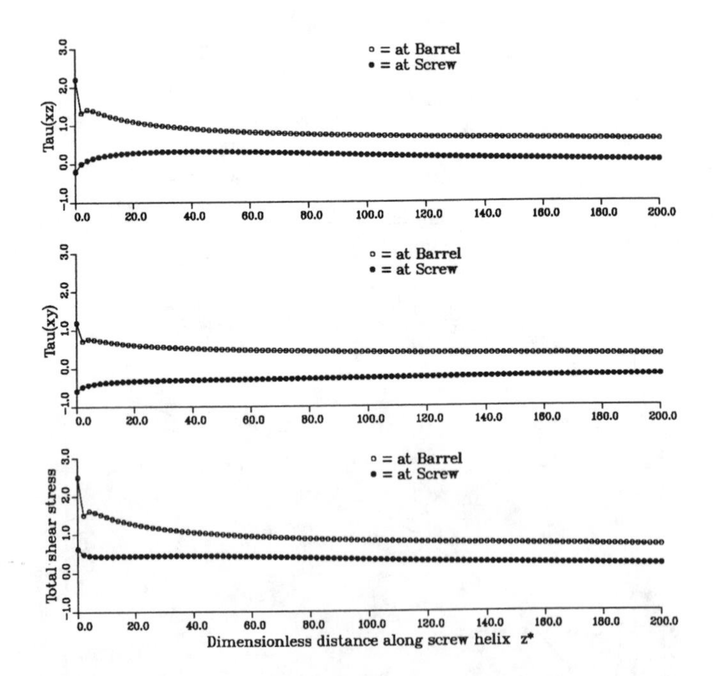

Figure 4.16 Dimensionless shear stress distribution at the barrel surface and the screw root for $n = 1$, $G = 0.008$, $Pe = 3427$, $\beta = 1.61$ and $q_v = 0.30$.

Residence Time Distribution

In order to obtain the residence time distribution (RTD), particle traces were obtained numerically by following the path of each particle as it moved from the inlet to the die. The amount of time required for a particle to cover the distance L (i.e., axial screw length), was calculated from the axial component of the velocity at that y location, for different z locations.

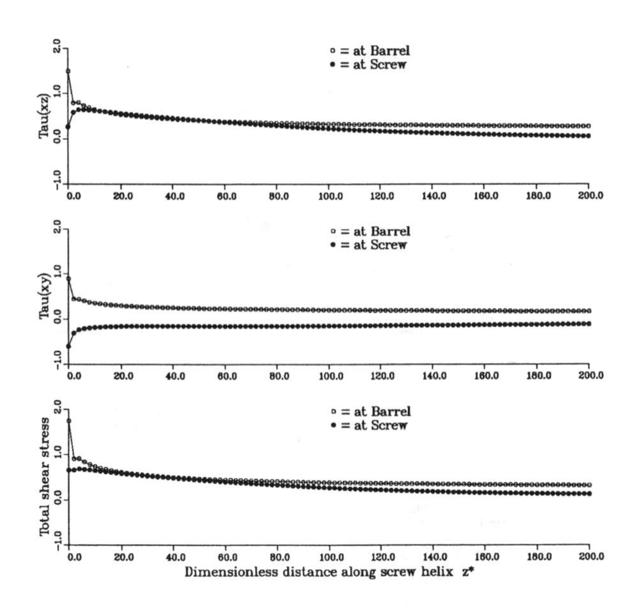

Figure 4.17 Dimensionless shear stress distribution at the barrel surface and the screw root for $n = 0.5$, $G = 10.0$, $Pe = 3427$, $\beta = 1.61$ and $q_V = 0.30$.

This is shown in Figure 4.18 for six particles at different y locations. It can be seen from Figure 4.18 that the particles near the barrel ($y^* = 1$) and the screw ($y^* = 0$), take a long time to come out, as compared to the particles in the middle portion of the screw channel. In fact, the particles at the screw and the barrel surfaces will never come out because of the no-slip conditions and also because the axial component of velocity is zero there. Also, the y position of a particle does not change with downstream distance. This is because the vertical component, v, of the velocity was neglected in the model. In practice, due to the finite width, W, the vertical component is not negligible, especially near the screw flights and therefore, the y position of a particle will change with the downstream distance.

In practice, the residence time distribution is obtained by releasing a fixed amount of dye near the inlet (hopper) and then measuring the flow rate of the dye material as it comes out of the die at the outlet. To simulate this, an idealized situation was considered, as shown in Figure 4.19. A slab of the dye material with a uniform thickness S, was introduced near the inlet. As the dye moved with the flow, it was traced numerically as described above. Since the velocity varied with y, the dye material at different y locations came out at different times. This is shown in Figure 4.20 in terms of a time versus y plot. The two curves correspond to the front and the back sides of the die slab. The length of the horizontal line drawn at each point on the plot, between the two curves, indicates the amount of time difference between the front and the back of the dye slab in coming out. This is approximately given by

$$\Delta t^*(y) = \frac{S}{V^*_{axial}(y)} \tag{31}$$

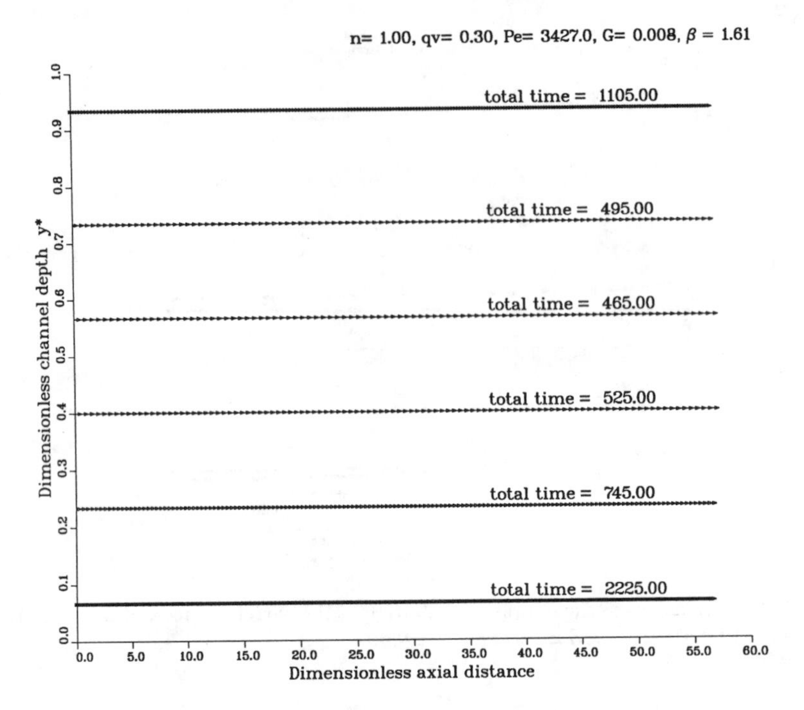

Figure 4.18 Particle traces for six particles at different y locations.

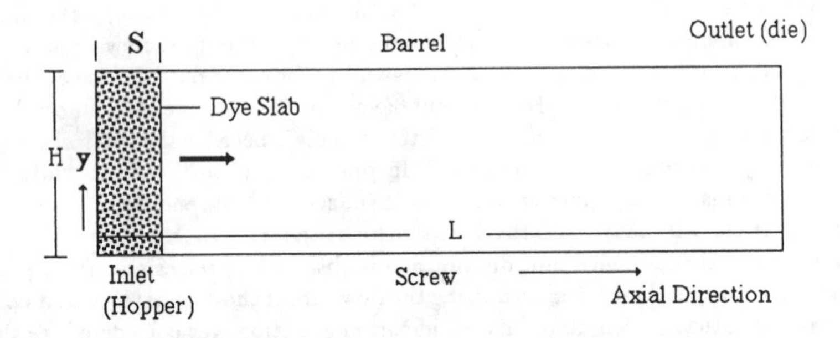

Figure 4.19 Schematic diagram showing the dye-slab and the computational domain for residence time distribution (RTD) calculations.

where the axial component of velocity V^*_{axial} is

$$V^*_{axial} = w^* \sin(\phi) - u^* \cos(\phi) \tag{32}$$

In the above calculations, it is assumed that at any y location, the change in thickness, S, of the slab as it moves from the inlet to the die is small. At the die, the flow rate of the dye, q_{dye}, is calculated by numerically integrating the quantity ($V^*_{axial} \cdot dy^*$) over H'^* within the shaded area shown in Figure 4.20. At any time t^*, the dye flow rate q_{dye} is given by:

$$q_{dye} = \int_{H'^*} V^*_{axial} \cdot dy^* \quad \text{where} \quad H'^* = H'/H \tag{33}$$

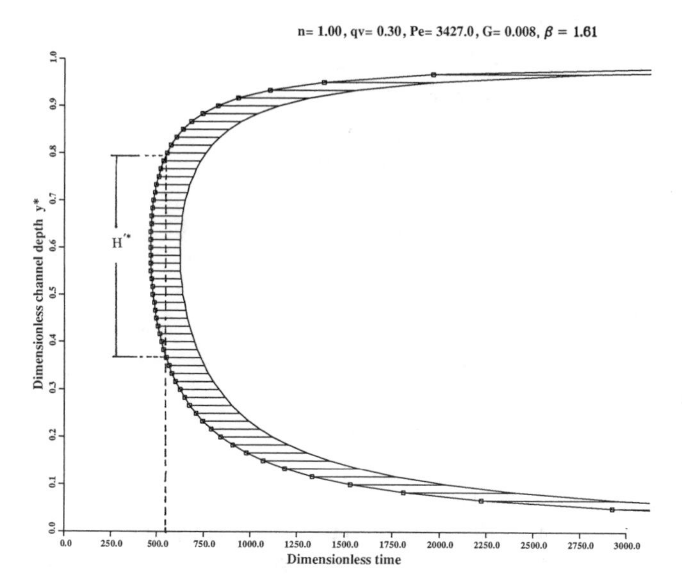

Figure 4.20 Residence time distribution, shown in terms of a y versus time plot for the dye slab.

The quantity q_{dye} (as seen by a stationary observer at the die) is plotted in Figure 4.21(a) as a function of time. The flow rate of the dye goes through a sharp peak. This indicates that at a certain point in time a large amount of dye material comes out. This happens mainly in the middle portion of the channel as can be inferred from Figure 4.20. The dye coming out near the barrel and the screw, contributes to the trailing portion of the curve in Figure 4.21(a). Near the barrel and the screw, the axial component of the velocity is relatively small and therefore, the material takes a much longer time in coming out of the extruder. The cumulative amount of die as a function of time is shown in Figure 4.21(b). This was obtained by numerical integration to obtain the area under the curve shown in Figure 4.21(a).

4.6.2 Choice of Parameter Values

The numerical computations discussed in the previous sections can, in general, be carried out for any arbitrarily chosen values of the various governing parameters. However, in actual practice, all the values can not be chosen independently of each other. For example, if the extruder geometry and extrudate material are fixed, we may choose the flow rate Q and the

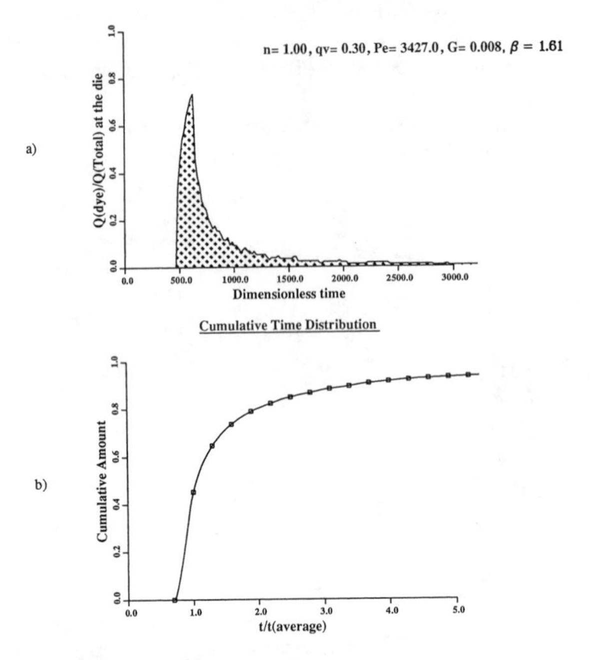

Figure 4.21 Residence time distribution
a) Dye flow rate as a fraction of the total flow rate at different times
b) Cumulative (normalized) amount of dye coming out at the die as a function of time.

rpm N. These fix the values of V_{bz}, q_v, the Peclet number Pe and the Griffith number G as follows:

$$V_{bz} = \pi DN/60, \quad q_v = Q/(V_{bz}WH), \quad Pe = V_{bz}H/\alpha,$$

$$G = \frac{\bar{\mu}\, V_{bz}^2}{k(T_b - T_i)} = \frac{\mu_o}{\dot{\gamma}_o^{\,n-1}} \left[\frac{V_{bz}}{H}\right]^{n-1} \frac{e^{-b(T_i - T_o)}\, \alpha^{1+n}}{H^{2n}\, k(T_b - T_i)} Pe^{1+n} \tag{34}$$

where D is the inner diameter of the barrel, H is the height of the screw channel and α is the thermal diffusivity of the fluid. The volume flow rate, Q, can be varied using different die openings, for a fixed rotational speed.

4.6.3 Results for Screw Moving Case

The results presented in the preceding sections are for the case in which the barrel is moved with respect to the screw. However, in practice, the barrel is fixed with respect to a stationary observer and the screw is rotating. The velocity profiles in such a frame of reference can be obtained by employing the following transformation:

$$w_s = w_b - V_{bz} \quad \text{and} \quad u_s = u_b - V_{bx}$$

which in dimensionless form become

$$w^*_s = w^*_b - 1.0 \quad \text{and} \quad u^*_s = u^*_b - \tan(\phi)$$

Here, the subscripts s and b correspond to the screw-moving and the barrel-moving cases, respectively. This transformation is required in order to couple the single-screw model with the model for the intermeshing zone discussed earlier. This allows us to obtain a complete model for the twin-screw extruder. The method for coupling is described later.

4.6.4 Results for the Intermeshing Zone Based on the FEM Model

In order to quantify the mixing characteristics, a mixing parameter is introduced in this study. This parameter is defined in terms of the following volume flow rates:

Q_b = volume flow rate of the material which crosses over from one screw to the other
Q_s = volume flow rate of the material that remains in the same screw
$Q_t = Q_b + Q_s$ = total volume flow rate.

The mixing parameter is then defined as:

$$x_f = \frac{Q_b}{Q_t} \tag{35}$$

An extensive study has been carried out on the mixing characteristics of twin-screw extruders, with two different sets of boundary conditions: (i) the case of linear velocity profiles at the inlets, and (ii) the case of other velocity profiles at the inlets. In the above cases, the screw flight effect is ignored. In the subsequent section, the effect of the screw flight will be discussed.

Simulation with Linear Velocity Profiles at the Inlets

The complete boundary conditions for the mixing region calculations are given as:

- At the inlets, a linear velocity profile was applied.
- At the outlets, a zero traction component in the x direction was applied together with a zero velocity in the y direction.
- At the barrel surface, a no-slip boundary condition was applied.
- At the screw root, the velocity components in x and y directions were prescribed for a given screw rotational speed.

The calculations were carried out using a finite element mesh with 80 elements, 201 velocity points, 61 pressure points, and 80 boundary points. The results of the numerical analysis on the mixing region of a co-rotating twin-screw extruder can be summarized as follows:

- The velocity field and the pressure contours were symmetrical about the vertical axis.
- Local maximum and minimum on the pressure field were found at the screw root.
- The mixing parameter was found to be essentially constant, regardless of the screw rotational speed.

Further study in this area includes the investigation of the effect of mesh refinement, geometry (such as radius of the barrel, radius of the screw root, distance between screw axes), and the material properties (such as the power-law index, temperature dependent coefficient for viscosity, etc.) on the mixing characteristics.

In the study that considered the effects of mesh refinement, the mesh was refined from 80 elements to 264 elements and then to 360 elements. In all cases, the mesh refinement gave reasonable and consistent results. Figure 4.22(a) shows the mesh discretization with 360 elements, 809 velocity points, 225 pressure points and 176 boundary points over the computational domain. Figure 4.22(b) shows a typical velocity field in the mixing region of a co-rotating twin-screw extruder at 60 rpm.

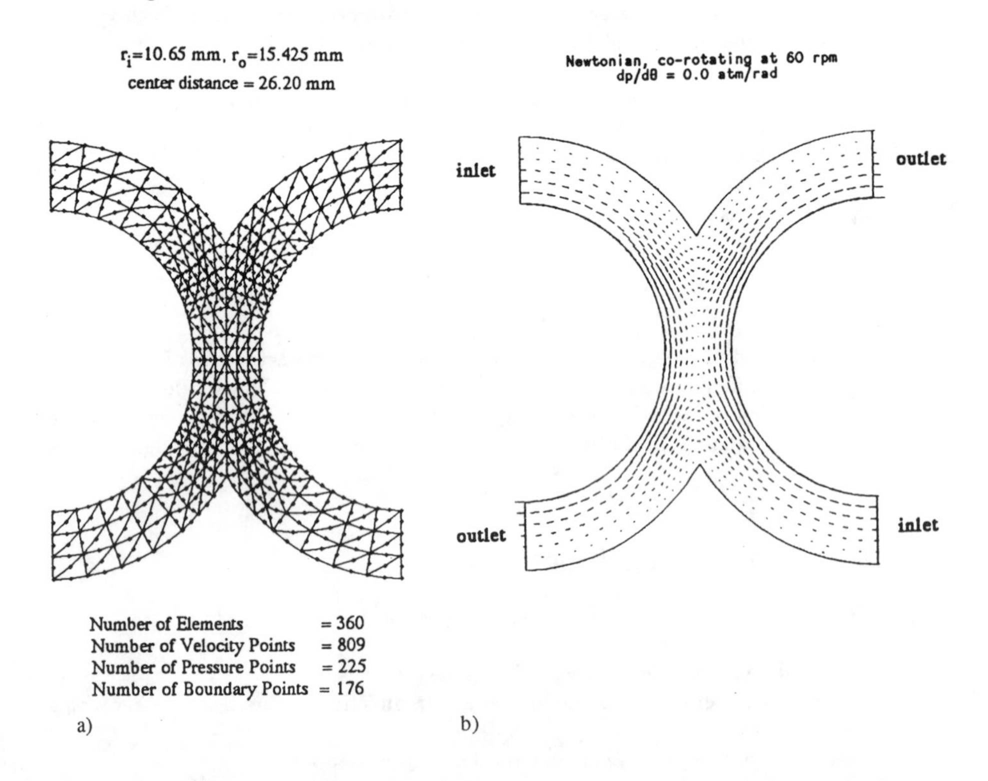

Figure 4.22 a) Mesh discretization, b) Velocity field of the mixing region of the twin-screw extruder (using 360 elements).

The power-law index n was varied to determine its effect on the mixing characteristics. Figure 4.23 shows variation of the mixing parameter with n. Figures 4.24(a) and 24(b) show the velocity profiles over half of the vertical and the horizontal midplanes, respectively. As can

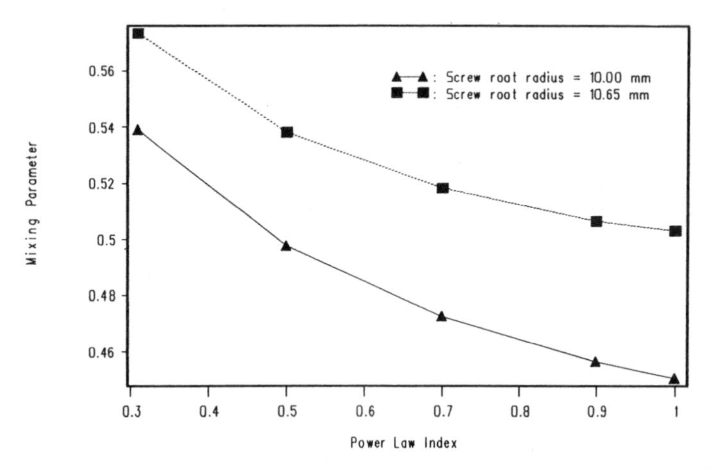

Figure 4.23 Effect of the power-law index on the mixing parameter using a linear velocity profile as the boundary condition at the inlet (using 360 elements).

be seen from Figure 4.23, the mixing parameter decreases as the power law index n increases. This fact can also be observed from Figures 4.24(a) and 4.24(b). In Figure 4.24(b), the velocity at the horizontal midplane is lower for a smaller power-law index. Since the area under this curve represents the volume flow rate of the material that remains in a given screw, it is clear that this flow rate is lower as the power-law index decreases. The velocity profile at the vertical mid-plane, Figure 4.24(a), shows the opposite trend. The area under this curve represents the volume flow rate of the material that crosses over to the other screw. Therefore, the volume flow rate of the material that goes to the other screw increases as the power-law index n decreases, while the flow rate of the material that remains in the screw also decreases. Therefore the resulting mixing parameter increases as the power-law index decreases.

In order to understand the effects of geometry on the mixing characteristics of the flow in the intermeshing region, one example was worked out by changing the screw root radius while keeping other dimensions and parameters constant. By using 360 elements and two different screw root radii, the result of the simulation is shown in Figure 4.23. The mixing parameter, shown in Figure 4.23, decreases as the screw root radius is decreased for all values of the power-law index n.

Simulation with Different Boundary Conditions at the Inlet

It should be emphasized again that the results discussed above were obtained by using linear velocity profiles at the inlets. However, this boundary condition is only one among many different possible boundary conditions. The proper boundary condition is probably the pressure or a pressure gradient at the inlets to the computational domain. Nonlinear velocity profiles may also be specified at the inlets.

For a Newtonian fluid, the tangential velocity profile in the flow between rotating concentric cylinders, with a given value of pressure gradient $dp/d\theta$, is:

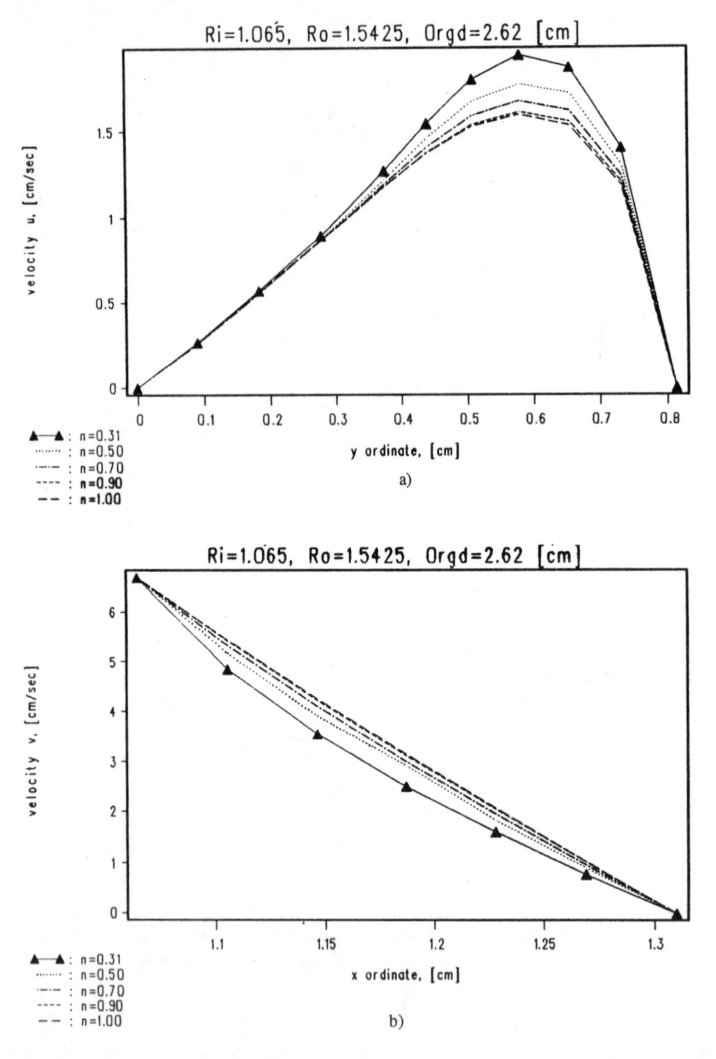

Figure 4.24 Velocity profile in the twin-screw extruder: a) at the vertical midplane, b) at the horizontal midplane (using 360 elements).

$$u_\theta = \frac{1}{\mu}\frac{dp}{d\theta}\frac{1}{4r}\frac{(r^2-r_o^2)}{(r_i^2-r_o^2)}\left[\frac{(r_i^2-r_o^2)}{(r^2-r_o^2)}(2r^2\ln r - 2r_o^2\ln r_o - r^2 + r_o^2) + (2r_o^2\ln r_o - 2r_i^2\ln r_i + r_i^2 - r_o^2)\right] + \frac{(r^2-r_o^2)}{r(r_i^2-r_o^2)}\omega_i r_i^2 \tag{36}$$

where r_i and r_o are the radii of the inner and outer cylinders, respectively, ω_i is the rotational speed of the inner cylinder, μ is the viscosity of the fluid, and r is the radial distance from the center of the cylinders. If the pressure gradient $dp/d\theta$ is known, one can find the tangential velocity profile for a Newtonian fluid at the inlets to the mixing region by using Eq (36). The complete boundary conditions for this case are then given as follows:

- Specify the tangential velocity profile at the inlets. This profile may be obtained from Eq (36) for a given value of $dp/d\theta$. At the outlets, apply a zero traction component in the x direction and a zero velocity in the y direction.
- At the barrel surface, apply the no-slip boundary condition.
- At the screw root, the velocity components in x and y directions are prescribed for a given screw rotational speed.
- Prescribe $p = 0.0$ at the bottom of the left outlet.

Three typical forms of the pressure gradient are of interest. The first one is the favorable pressure gradient, or a negative $dp/d\theta$, that produces a decreasing pressure along the direction of the flow. The second one is a zero pressure gradient and the third one is the adverse pressure gradient or a positive $dp/d\theta$, which produces an increasing pressure along the direction of the flow. The results of the simulation with these three types of pressure gradients, using a finite element mesh with 360 elements, 809 velocity points, 225 pressure points, and 176 boundary points, are presented in Figures 4.25 through 4.27. The pressure contours for three different values of $dp/d\theta$ are shown in Figure 4.25. An interesting fact is shown in Figure 4.26(a). The velocity profile at the horizontal midplane was the same for all values of $dp/d\theta$, while the velocity profile at the vertical midplane, shown in Figure 4.26(b) shows that the velocity at this midplane decreases as the value of $dp/d\theta$ increases. Figure 4.27 shows that the mixing parameter decreases as the value of $dp/d\theta$ increases. In other words, the favorable pressure gradient results in higher total flow rate and a higher mixing parameter, while the adverse pressure gradient results in a lower total flow rate and a lower mixing parameter. Therefore, the total volume flow rate greatly affects the mixing parameter.

It may be noted that the basic pattern of the pressure contours in the central region looks alike for the three difference pressure gradients, as can be seen in Figure 4.25. There exist local maximum and local minimum points of pressure at the screw root. It should be mentioned here that the comparison between the prescribed tangential velocity profile at the inlets and the computed velocity profile at the outlets shows an almost perfect agreement. This observation verifies that the tangential velocity profile described above is appropriately applied in this example. The above results were obtained for Newtonian fluid. The case of Non-Newtonian fluid requires a numerical integration for obtaining the tangential velocity profile.

Study of the Screw Flight Effect

It should be mentioned here that the results of the above simulation of the intermeshing zone were obtained as if there were no screw flight. In more recent work, the effect of the screw flight on the flow characteristic in the mixing region of the extruder was included. In this case, the three velocity components at the inlet side of the mixing region were incorporated as the boundary conditions. Those three velocity components can be obtained from the simulation of the translation region for a specified power-law index of the material. Equations (27) and (28) were solved iteratively, where the viscosity and the stiffness matrix were updated at each iteration step. The boundary conditions for solving the x-y momentum equations were the

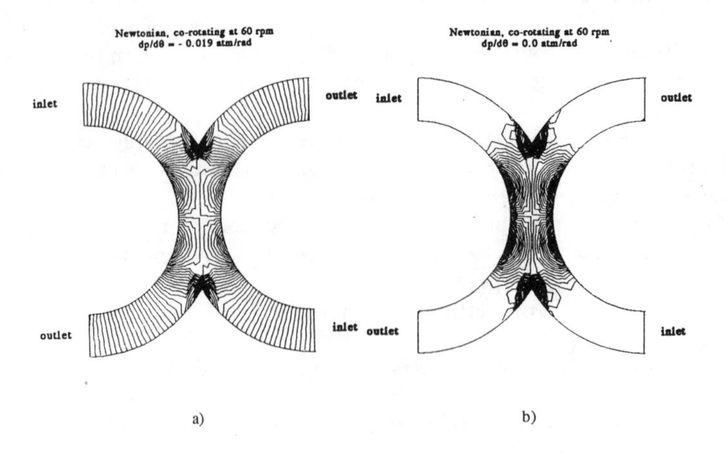

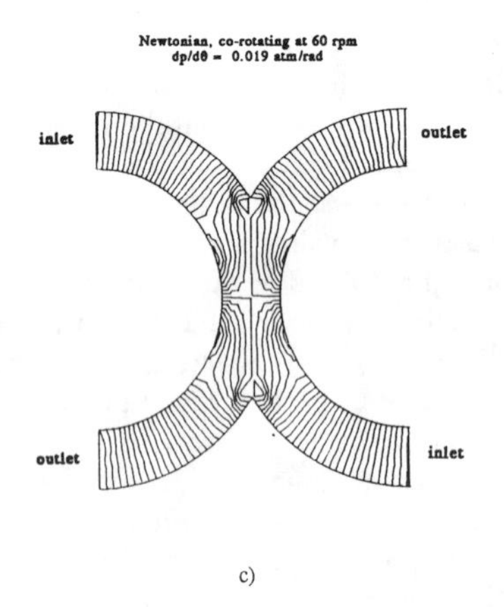

Figure 4.25 Pressure contours in the mixing region of the twin-screw extruder a) with a favorable pressure gradient, b) with zero pressure gradient, c) with an adverse pressure gradient (using 360 elements).

same as in the previous section, while a set of additional conditions was added for solving the z momentum equation. These boundary conditions were obtained from the simulation of the translation region. The boundary conditions for the z-momentum equation were as follows:

- Prescribe w velocity component at the inlet. Prescribe $dw/dn = 0$ at the outlet.
- Prescribe $w = 0$ at the barrel surface and the screw surface.
- Prescribe a certain dp/dz value obtained from the translation region simulation.

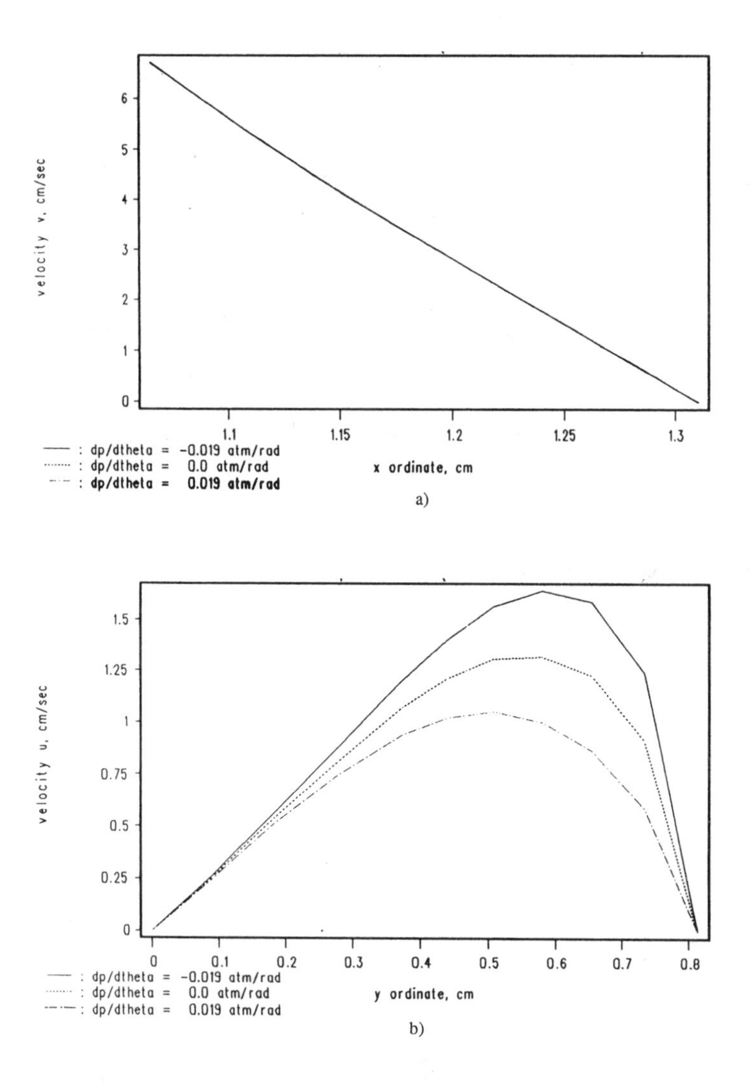

Figure 4.26 Velocity profile in the twin-screw extruder for various pressure gradients: a) at the horizontal midplane, b) at the vertical midplane (using 360 elements).

Using the above boundary conditions, Eqs (27) and (28) were solved for the three velocity components and pressure. In order to study the screw flight effect, the same velocity profiles in the x and y directions for each power-law index as before were also applied, but at this time solving only the x- and y-momentum equation. The result of this simulation then represents the flow in the intermeshing zone without considering the screw flight effect.

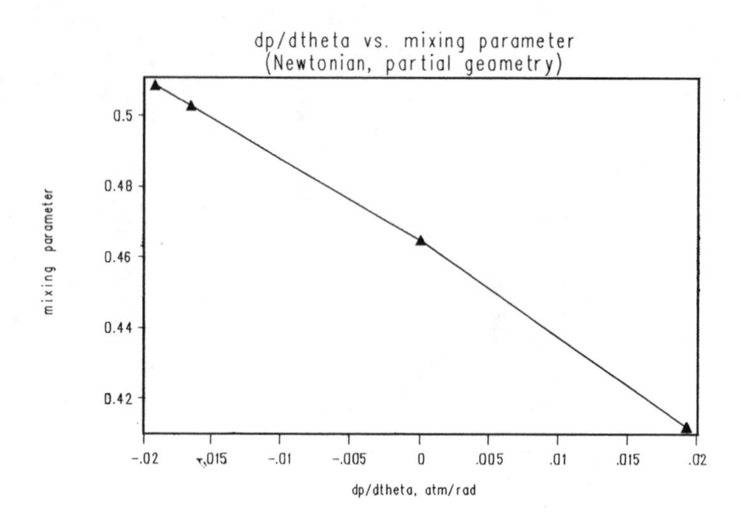

Figure 4.27 Effect of the pressure gradient on the mixing parameter (using 360 elements).

Finally, the result from both simulations were compared. Figure 4.28 shows the comparison of the results of these simulations in terms of the mixing parameter for different power-law

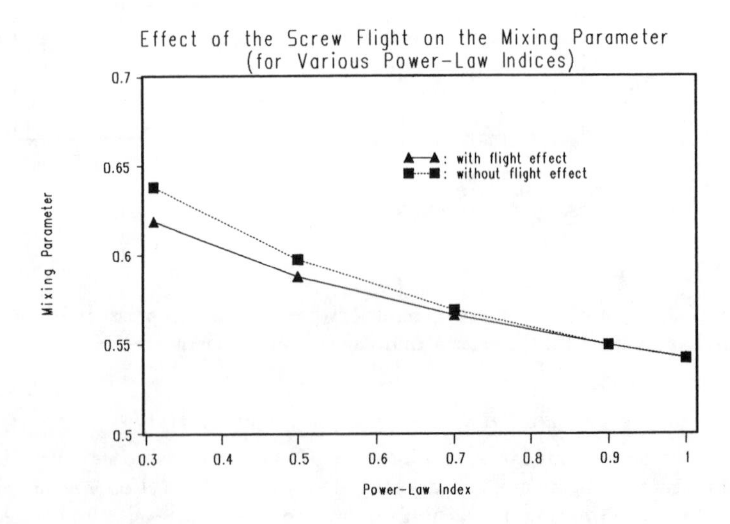

Figure 4.28 Variation of the mixing parameter, x_f, with the power-law index, n, with and without the flight effect.

indices of the material. For each pair of power-law indices, the same u, v, and w velocities were applied as the inlet boundary conditions for both cases. As can be seen here, the effect of the screw flight on the mixing parameter was negligible. Similarly, the effect was also found to be negligible in terms of the velocity and pressure fields. In this regard, the discussions in the preceding sections are still valid even if they ignored the screw flight effect.

4.7 MODELING OF THE TRANSPORT IN A TWIN-SCREW EXTRUDER

A cross section of a twin-screw extruder is shown schematically in Figure 4.1. The fluid flow at any cross section in a twin-screw extruder can be obtained by an appropriate coupling of the single-screw model with that of the intermeshing zone. As shown in Figure 4.1, the cross section is divided in two zones: the Translation Zone and the Mixing Zone. The flow in the translation zone is obtained by employing the single-screw model and the flow in the mixing zone is obtained by employing the model for the intermeshing zone. Both of these models have been discussed in detail in the preceding sections.

The complexity of the twin-screw extrusion process arises from the cyclic flow pattern of material between the two screws and the figure-eight-shaped barrel. The flow is transient in nature when the material undergoes structural changes during the process. In twin-screw extrusion, there is no frame of reference which enables the flow pattern to be approximated as steady state, unlike that for the single-screw extrusion in the moving barrel formulation. The difficulty posed by this transient cyclic behavior might be alleviated by introducing an assumption that the changes in material characteristics and flow behavior during one turn of the screws, are negligible. That is, during one turn of the screws, the flow is considered as quasi-steady. Based upon this quasi-steady approximation, the analysis of twin-screw extruders can be achieved by coupling the two zones in a sequence. The two zones will be coupled together appropriately at each screw turn, as described below.

The outflow conditions of the translation zone, obtained from the single-screw model with the moving barrel formulation, will be transformed to the velocity profiles with the moving screw formulation and then will be used as the boundary conditions for the twin-screw finite element model (moving screw formulation) for the mixing zone. Subsequently, the velocity and pressure fields of the mixing zone will be obtained, which then will become the inlet conditions for the translation zone in the next screw turn. This procedure will be repeated in the down-channel direction for the full analysis of the twin-screw extrusion, as depicted in Figure 4.2.

Since this chapter was written, significant amount of further work has been carried out by this group of investigators. Detailed study of heat transfer and characteristics of a single-screw extruder have been carried out [63,64]. The mixing region has been studied in greater detail [65,66] to characterize the effect of various parameters on the mixing. The results have also been compared with experiments [67] and a good agreement between the model and the experimental results was obtained. The coupling of the two models described earlier has been successfully carried out [68] to simulate a full twin-screw extruder.

4.8 CONCLUSIONS

A preliminary model for the flow and heat transfer in a twin-screw extruder is developed. The results are based upon the calculations for the translation and the mixing zones. Several important and interesting results have been obtained. In particular, the mathematical model for single-screw extruder was employed to study the flow and the thermal fields in the extruder for a very wide range of operating conditions such as rpm, throughput, and material properties. The numerical results were obtained in dimensionless terms in order to allow the characterization of the diversity of design, dimensions, materials, and operating parameters generally encountered in practice. The shear, pressure, and temperature variation undergone by a polymer material as it progresses through an extruder were investigated in detail. The residence time distribution (RTD) was computed and found to agree closely with earlier experimental and numerical studies.

The simulation of the intermeshing or mixing zone has also yielded important and valuable results. The mixing parameter that characterizes the resulting mixing in this region was determined as function of material properties and inlet conditions. It was found that for a given throughput or volumetric flow rate the mixing parameter increases as the power-law index n is decreased. It increases with an increase of throughput, for a given power-law index. It was also found to increase as the depth of the screw channel is decreased, for a given material.

The results obtained are valuable for the design, operation, and optimization of polymer extruders. This work is relevant to a consideration of the temperature, shear, and pressure a given plastic material undergoes to as it moves through the extruder, along with the time spent. The residence time distribution is an important result, along with the mixing parameter calculations. All these results pertain to actual materials, typical operating conditions and common extruders. Thus, the results are valuable to the scientific field as well as to industry. The basic aspects studied include the nature of the flow and thermal fields, mixing mechanisms, effect of operating conditions on the process and the proper imposition of the boundary conditions. From a practical standpoint, the effect of operating conditions on the thermal process undergone by the material is important. Also, of crucial importance is the mixing that arises in the intermeshing zone and its dependence on geometry, material and operating conditions. Several quantitative results have been obtained that can be employed to answer fundamental questions that arise in the process under consideration and to indicate important variables and their effects in practical extruders.

ACKNOWLEDGEMENTS

This work has been supported by a grant from the Center for Advanced Food Technology (CAFT), Rutgers University. The authors would like to thank Prof. V. Sernas and Mr. M. Esseghir for discussions throughout this work.

NOMENCLATURE

b:	Temperature coefficient of viscosity, Eq (8)
C:	Specific heat of the fluid
D_{ij}:	Rate of deformation tensor
G:	Griffith number, $G = \bar{\mu} V_{bz}^2/k(T_b - T_i)$
H:	Height of the screw channel
k:	Thermal conductivity of the fluid
L:	Axial screw length
n:	Power law index, Eq (8)
n_x:	Normal vector along x direction in the FEM model

n_y: Normal vector along y direction in the FEM model
p: Pressure
$\bar{p}$: Reference pressure, $\bar{p} = \bar{\mu} V_{bz}/H$
Pe: Peclet number, $Pe = V_{bz}H/\alpha$
Q: Total volumetric flow rate
r: Radial distance from the center of the screw in the FEM model
S: Slab thickness, Eq (30)
S_v: Velocity boundary surface in the FEM model, Eq (20)
S_t: Traction boundary surface in the FEM model, Eq (21)
$\bar{t}_x, \bar{t}_y$: Prescribed traction in x and y directions, respectively in the FEM model
T: Temperature
T_i: Inlet temperature
T_o: Reference temperature, Eq (8)
u: Velocity component in x direction
u_θ: Tangential velocity in the FEM model
$\bar{u}$: Prescribed u velocity in the FEM model
v: Velocity component in y direction
$\bar{v}$: Prescribed v velocity in the FEM model
V_b: Tangential velocity of the barrel
V_{bx}: Component of V_b along x, $V_{bx} = V_b \cos(\phi)$
V_{bz}: Component of V_b along z, $V_{bz} = V_b \sin(\phi)$
w: Velocity component in z direction
$\bar{w}$: Prescribed w velocity in the FEM model
W: Width of the screw channel
x: Coordinate axis normal to screw flights in the single-screw model
Horizontal coordinate axis normal to the screw axis in the FEM model
x_f: Mixing parameter
y: Coordinate axis normal to the screw root in the single-screw model
Vertical coordinate axis normal to the screw axis in the FEM model
z: Coordinate axis along the screw channel in the single-screw model
$\Delta z^*, \Delta y^*$: Grid spacing along z and y directions, respectively, in the finite difference model
α: Thermal diffusivity, $\Delta = k/\rho C$
β: Dimensionless b, $\beta = b(T_b - T_i)$
ϕ: Screw helix angle
$\dot{\gamma}$: Strain rate
$\dot{\gamma}_o$: Reference strain rate, Eq (8)
μ: Dynamic viscosity
$\bar{\mu}$: Reference viscosity, $\bar{\mu} = \mu_o \dot{\gamma}^{1-n} (V_{bz}/H)^{n-1} \exp\{-b(T_i - T_o)\}$
θ^*: Dimensionless temperature, $\theta^* = (T - T_i)/(T_b - T_i)$
θ: Angular coordinate in the FEM model
ρ: Density of the fluid
τ: Shear stress

subscripts

b: Barrel
dev: Developed
i: Inner, in the FEM model

o: Reference quantity in the single-screw model
Outer, in the FEM model
s: Screw
t: Total

superscripts

*: Dimensionless quantity

REFERENCES

1. Griffith, R.M., *Ind. Eng. Chem. Fund.*, *1*, 181 (1962).
2. Zamodits, H.J., Pearson, J.R.A., *Trans. Soc. Rheol.*, *13*, 357 (1969).
3. Rauwendaal, C., *SPE Tech. Papers*, *31*, 30 (1985).
4. Lappe, H., Potente, H., *SPE Tech. Papers*, *29*, 174 (1983).
5. Klein, I., *SPE Tech. Papers*, *22*, 444 (1976).
6. Klein, I., Klein, R., *SPE Tech. Papers*, *24*, 519 (1978).
7. Derezinski, S.J., *SPE Tech. Papers*, *31*, 69 (1985).
8. Peng, Y., Cheng, J., Wu, Y., *SPE Tech. Papers*, *31*, 73 (1985).
9. Donovan, R.C., *Polym. Eng. Sci.*, *14*, 101 (1974).
10. Lindt, J.T., *SPE Tech. Papers*, *22*, 429 (1976).
11. Chung, K.H., Chung, C.I., *Polym. Eng. Sci.*, *23*, 191 (1983).
12. Lidor, G., Tadmor, Z., *Polym. Eng. Sci.*, *16*, 450 (1976).
13. Bigg, D.M., Middleman, S., *Ind. Eng. Chem. Fund.*, *13*, 66 (1974).
14. Tadmor, Z., Gogos, C., "Principles of Polymer Processing," John Wiley & Sons, New York (1979).
15. Fenner, R.T., "The Design of Large Hot Melt Extruders," conf.: Eng. Design of Plastics Processing Machinery, University of Bradford, England, April, 1974.
16. Fenner, R.T., *Polymer*, *18*, 617 (1977).
17. Fenner, R.T., "Principles of Polymer Processing," Chem. Publ., New York (1979).
18. Agur, E.E., Vlachopolous, J., *Polym. Eng. Sci.*, *22*, 1084 (1982).
19. Fukase, H., Kunio, T., Shinya, S., Nomura, A., *Polym. Eng. Sci.*, *22*, 578 (1982).
20. Elbirli, B., Lindt, J.T., *SPE Tech. Papers*, *28*, 469 (1982).
21. Elbirli, B., Lindt, J.T., *SPE Tech. Papers*, *28*, 472 (1982).
22. Mokhtarian, F., Erwin, L., *SPE Tech. Papers*, *28*, 476 (1982).
23. Erwin, L., Mokhtarian, F., *Polym. Eng. Sci.*, *23*, 49 (1983).
24. Baird, D.J., Joseph, E.G., Luxenburg, L.A., "Advances in Rheolohy," *4*, 145, Proceed. 9th International Congress Rheology, Acapulco, Mexico (1984).
25. Chen, A.H., Jao, Y.C., Larkin, J.W., Goldstein, W.E., *J. Food Process Eng.*, *2*, 337 (1978).
26. Harper, M., Rhodes, T.P., Wanninger, Jr., L.A., *Chem. Eng. Prog. Sym. Series*, *67*, 108, 40 (1967).
27. Bruin, S., Van Zuilichem, D.J., Stolp, W., *J. Food Process Eng.*, *2*, 1 (1978).
28. Mohamed, I.O., Morgan, R.G., "Average Heat Transfer Coefficients in Single-Screw Extruders of Non-Newtonian Food Materials," Fundamentals of Food Extrusion and Forming Symposium, AIChE, Summer, Boston, MA (1986).
29. White, J.L., Szydlowski, W., Min, K., Kim, M.H., *Adv. Polym. Technol.*, *7*, 295 (1987).

30. Burkhardt, K., Herrmann, H., Jacopin, S., *SPE Tech. Papers*, *24*, 498 (1978).
31. Nichols, R.J., *SPE Tech. Papers*, *29*, 130 (1983).
32. Nichols, R.J., *SPE Tech. Papers*, *30*, 6 (1984).
33. Nichols, R.J., Yao, J., *SPE Tech. Papers*, *28*, 416 (1982).
34. Howland, C., Erwin, L., *SPE Tech. Papers*, *29*, 113 (1983).
35. Janssen, L.P.B.M., "Twin-Screw Extrusion," Elsevier Scientific Publishing Company, Amsterdam (1978).
36. Secor, R.M., *Polym. Eng. Sci.*, *26*, 647 (1986).
37. Wyman. C.E., *Polym. Eng. Sci.*, *15*, 601 (1975).
38. Yacu, W.A., *J. Food Eng.*, *8*, 1 (1985).
39. Hold, P., *Adv. Polym. Technol.*, *4*, 281 (1984).
40. Illing, G., *Modern Plastics*, 70, August 1969.
41. Jakopin, S., *Adv. Chem.*, *134*, 114 (1974).
42. Pinto, G., Tadmor, Z., *Polym. Eng. Sci.*, *10*, 279 (1970).
43. Walk, C.J., *SPE Tech. Papers*, *28*, 423 (1982).
44. Todd, D.B., *Polym. Eng. Sci.*, *15*, 437 (1975).
45. Kao, S.V., Allison, G.R., *Polym. Eng. Sci.*, *24*, 645 (1984).
46. Wolf, D., Holin, N., White, D.H., *Polym. Eng. Sci.*, *26*, 640 (1986).
47. Herrmann, H., Eise, K., *SPE Tech. Papers*, *27*, 614 (1981).
48. Booy, M.L., *Polym. Eng. Sci.*, *18*, 973 (1978).
49. Booy, M.L., *Polym. Eng. Sci.*, *20*, 1220 (1980).
50. Herrmann, H., Jacopin, S., *SPE Tech. Papers*, *23*, 481 (1977).
51. Eise, K., Herrmann, H., Werner, H., Burkhardt, U., *Adv. Plastics Technol.*, *1*, 18 (1981).
52. Denson, C.D., Hwang, Jr., P.K., *SPE Tech. Papers*, *26*, 107 (1980).
53. Denson, C.D., Hwang, Jr., P.K., *Polym. Eng. Sci.*, *20*, 965 (1980).
54. Hwang, B.K., "Fluid Flow Studies in Twin-Screw Extruders," Ph.D. Thesis, University of Delaware (1982).
55. Maheshri, J.C., Wyman, C.E., *Polym. Eng. Sci.*, *20*, 601 (1980).
56. Schlichting, H., "Boundary Layer Theory," 7th Ed., McGraw-Hill, New York (1979).
57. Kays, W.M., "Convective Heat and Mass Transfer," McGraw-Hill, New York (1966).
58. Karwe, M.V., Jaluria, Y., Trans. ASME, *J. Heat Transfer*, *108*, 728 (1986).
59. Eckert, E.R.G., Drake, R.M., Jr., "Analysis of Heat and Mass Transfer," McGraw-Hill, New York (1972).
60. Crochet, M.J., Davies, A.R., Walters, K., "Numerical Simulation of Non-Newtonian Flow," Elsevier Publ. Co., New York (1984).
61. Jaluria, Y., Torrance, K.E., "Computational Heat Transfer," Hemisphere Publishing Corp., New York (1986).
62. Kwon, T.H., Shen, S.F., Wang, K.K., *Polym. Eng. Sci.*, *26*, 214 (1986).
63. Karwe, M.V., Jaluria, Y., *Numerical Heat Transfer*, Part A, *17*, 167 (1990).
64. Gopalakrishna, S., Karwe, M.V., Jaluria, Y., *NUMIFORM 89*, 205 (1989).
65. Sastrohartono, T., Kwon,.H., "Numerical Methods in Industrial Forming Processes," Proceed. 3rd Int. Conf., Fort Collins, Colorado, 277 (1989).
66. Sastrohartono, T., Kwon, T.H., *Int. J. Num. Meth. Eng.*, (1990), to appear.
67. Sastrohartono, T., Esseghir, M., Kwon, T.H., Sernas, V., *Polym. Eng. Sci.*, (1990), to appear.
68. Sastrohartono, T., Karwe, M.V., Jaluria, Y., "Extrusion and Rheology of Foods," Proceed. 2nd Int. Symp., Rutgers University, New Brunswick (1990).

CHAPTER 5

FINITE ELEMENT SIMULATION OF THERMOFORMING AND BLOW MOLDING

by H.G. deLorenzi and H.F. Nied

General Electric Corporate Research and Development
Schenectady, New York 12301
U.S.A.

ABSTRACT
5.1 INTRODUCTION
5.2 FINITE ELEMENT FORMULATION
5.2.1 Theoretical Background
5.2.2 The Finite Element Concept
5.2.3 The Equilibrium Equations
5.2.4 The Solution Procedure
5.3 MATERIAL BEHAVIOR
5.3.1 Background and Experimental Observations
5.3.2 Nonlinear Elastic Constitutive Models
5.3.3 Experimental Techniques for Obtaining Material Data
5.4 THERMOFORMING EXAMPLES
5.4.1 Free Inflation of a Flat Membrane
5.4.2 Comparison with Experimental Results for a Deep Vacuum Formed Cylinder
5.4.3 Forming a Deep Cylinder With a Male Mold
5.4.4 Thermoforming a 3-D Box
5.5 BLOW MOLDING EXAMPLES
5.5.1 Blow Molded Jar
5.5.2 Blow Molding of a Rectangular Box
5.6 DESIGN METHODOLOGY
5.6.1 Design Philosophy
5.6.2 Example of Iterative Design Procedure
5.7 SOME UNRESOLVED ISSUES
5.8 CONCLUSIONS
ACKNOWLEDGEMENT
REFERENCES

ABSTRACT

The chapter provides step-by-step description of the finite element method as used to simulate thermoforming and blow molding processes. After reviewing the literature, the problems associated with use of the finite element method to processes involving large strains, nonlinear material behavior, contact between hot melt and cold walls, and in some cases physical instabilities during forming, are discussed. The iterative solution incorporates an assumption that the polymer in contact with the cold mold surface stops deforming. The polymer is considered to behave as nonlinear elastic body, described by either Mooney-Rivlin or Ogden models, with the parameters determined experimentally. The results from the finite element calculations of a flat membrane inflation were compared with classical analytical solutions. Furthermore, the finite element simulations of variety of thermoforming and blow molding problems (both axisymmetric and general 3-D) were compared with the experimental measurements of wall thickness distribution. Good agreement was obtained. One of the most useful feature of the finite element method is its ability for tracking the material from the final configuration back to the original parison.

5.1 INTRODUCTION

In this chapter the use of the finite element method to simulate thermoforming and blow molding processes will be examined. Since finite element simulation of these forming processes is relatively new, many of the modeling assumptions are in a current state of flux. Indeed, the underlying physical mechanisms which govern these forming processes are poorly understood. It is anticipated that this situation will improve dramatically based on the results of current research initiatives at universities and in industry. Despite the many large gaps in fundamental knowledge concerning large deformation polymer behavior as it relates to these forming processes, finite element simulation can provide crucial insight into how one should proceed to improve current processing technology.

Figures 5.1 - 5.3 depict examples from the two families of forming processes that will be examined in this chapter. Thermoforming is considered a "secondary" forming process in which a previously extruded sheet of thermoplastic is reheated and "inflated" into a mold cavity. This simplest version of thermoforming, i.e. straight vacuum forming, is illustrated in Figure 5.1. In this process, the clamped sheet of plastic is heated (Fig. 5.1a) and then formed into the mold by using either positive or negative (vacuum) pressure (Fig. 5.1b). The final part (Fig. 5.1c) usually exhibits considerable thickness variations, with the corners typically ending up as the thinnest regions. There are many variations on this same theme [1,2], for example the plug-assist thermoforming shown in Figure 5.2. In plug-assist thermoforming, the hot plastic sheet (Fig. 5.2a) is mechanically stretched with the assistance of a plug (Fig. 5.2b) prior to the application of vacuum or pressure (Fig. 5.2c). This procedure typically results in a part with thicker corners and thinner side walls than obtained in straight thermoforming (Fig. 5.2d). In contrast to thermoforming, the primary steps in the extrusion-blow molding cycle are illustrated in Figure 5.3. In this process, a tube or parison of hot plastic is extruded (Fig. 5.3a), pinched-off at top and bottom (Fig. 5.3b), and then inflated into the mold cavity (Fig. 5.3c). As in thermoforming, there are many variations of the simple basic blow molding cycle depicted in Figure 5.3. These include injection-blow molding (the initial parison is injection molded) and stretch-blow molding (the parison is stretched with a plug prior to inflation).

Thermoforming and blow molding are two closely related processes that are characterized by the "inflation" of a hot plastic membrane into a mold cavity. Both processes

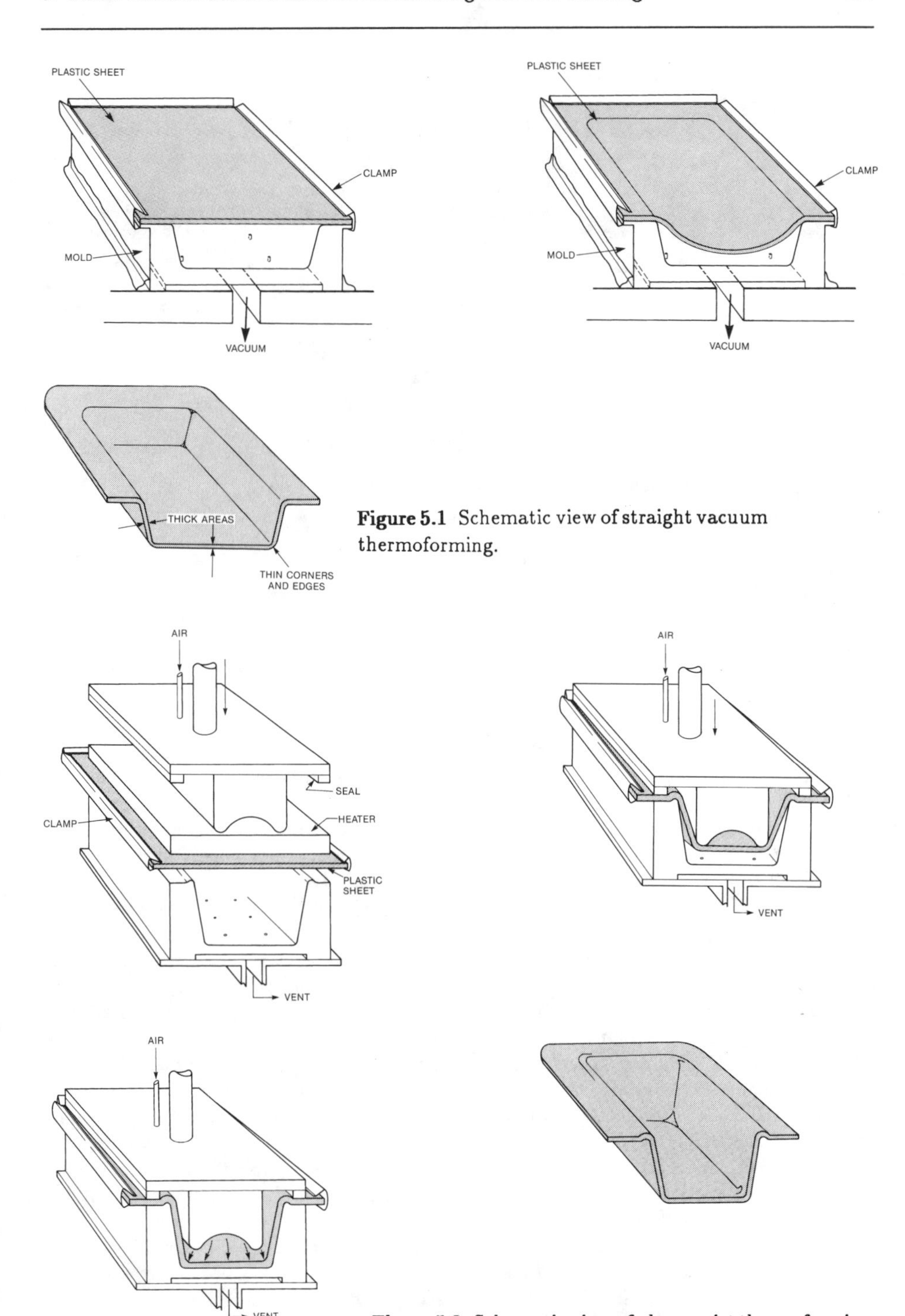

Figure 5.1 Schematic view of straight vacuum thermoforming.

Figure 5.2 Schematic view of plug-assist thermoforming.

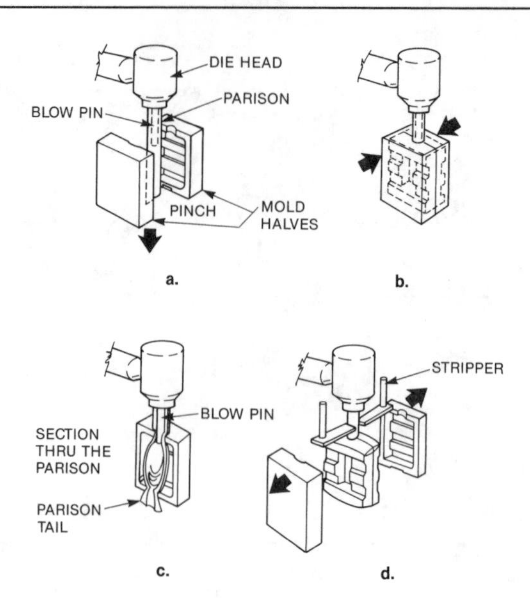

Figure 5.3 Basic blow molding cycle (a) Mold is open; parison is dropped. (b) Mold closes; parison is pinched in 2 places. (c) Air is blown into parison (inflation). (d) Mold opens; part is stripped off.

offer sufficient process control to accurately form thin shell type structures which have acceptable wall thicknesses in critical load bearing locations. For example, in blow molding, the parison can be "programmed" with an intentional variation in thickness, to compensate for thinning of material that would be stretched into edges or corners. In thermoforming, final part thickness can be controlled by utilizing differential heating on the plastic sheet and with plug assist. Unfortunately, the critical parameters associated with these techniques for controlling final part thicknesses are generally determined by trial and error. Indeed, mold development and material selection for these two processes has historically been a "trial and error" process, and as a result, the development of new mold designs and thermal process parameters has been inefficient and expensive. In addition, this method of process development does not allow a quick comparison of competing designs and different materials.

The objective of computer simulation for these hot forming processes is to provide a rational means for mold design and to permit design of the "optimal" final parts using the minimum amount of material. This is done by comparing the simulated behavior of different polymeric materials in candidate mold configurations under different processing conditions. For example, by changing the initial parison geometry in a blow molding simulation, the effects on the thickness distribution in the final part can be predetermined. If the thicknesses are not satisfactory, the analysis can be repeated with different initial thickness distributions in the parison and/or altered process parameters. This procedure is repeated iteratively, until acceptable thicknesses are obtained. In addition, if the part is to be used in a highly stressed structural application, the wall thicknesses predicted by the simulation can be used as input to a structural analysis program. The part can be analyzed for the expected loads and the working stresses can be predicted. If the stresses are not within the desired range, the tooling or process conditions can be changed in the forming simulation and the iterative procedure can be repeated until the stresses are within an acceptable range.

Initial efforts at predicting wall thickness distributions in thermoformed parts concentrated on developing models which did not depend on material behavior and were applicable only for very simple geometries [3-6]. These models assume that the "bubble" shape during free inflation is known (usually spherical) and that the sheet solidifies upon contact with the mold walls. Thus, the final thickness distribution is determined from a simple mass balance and the imposed boundary constraints. In the case of a truncated cone, comparison of these calculations with experimental data indicates that the predicted thicknesses can be 10-45% lower than the measured values [2,4,5], with the predictions for the shallow truncated cone configuration yielding the best agreement with experimental measurements. As noted, the predictions obtained from these simple "conservation of mass" models are independent of the material used. In [7], an attempt was made to incorporate a Neo-Hookean rubber constitutive relationship into the model. However, the constitutive relationship was used only to estimate the forming pressure, while thickness was determined from an expression based on mass conservation and the approximation that the deformation is everywhere equibiaxial. Thus, as in [3-6] the final thickness distribution is independent of material properties. In blow molding analysis, substantial advances have been made modeling simple geometric configurations using viscous fluid constitutive relationships [8-10]. However, the parison inflation analysis examined in [8-10] assumes that the parison has a uniform thickness at the end of free inflation, and yet, as noted in [9]; "the experimental results indicate that the thickness is nonuniform." The simultaneous inflation and extension of a viscoelastic tube has also been examined [11-12], but no attempt has been made to extend this work to include contact with mold surfaces. Moreover, the formidable experimental measurements needed to construct large deformation viscoelastic constitutive equations have not yet been performed for most thermoplastics of interest.

Recent advances in computer simulation of thermoforming and blow molding have relied on the finite element method [13-16]. The advantages of using a finite element formulation for analysis of thermoforming and blow molding are twofold: first, the formulation is not restricted to any particular geometry, and second, the highly nonlinear behavior typically associated with large strains and nonlinear polymer material behavior can be directly accommodated.

If an analytic model is to accurately reflect actual hot polymer behavior in complex mold configurations it should incorporate certain basic processing observations. First, the model should simulate the nonuniform thinning of the polymer which occurs during the free inflation stage of the forming process prior to contact with the mold surfaces. This is especially critical if there is considerable inflation, i.e., large deformation, prior to contact with the mold surface. Secondly, the model should incorporate the observation that upon contact with the mold surface, sudden cooling and adhesion between the polymer and the mold surfaces rarely permits any additional stretching of the polymeric material in contact with the mold. The actual contact behavior between hot polymer and mold surface is poorly understood and it has not yet been determined how the contact conditions change as a function of temperature and mold surface texture. Observations of actual thermoforming and blow molding processes have verified that the "sticking" assumption is reasonably good. There is some evidence which indicates that under certain special processing conditions the polymer can be "dragged" over shallow indentations and smooth curves during processing, but this is not seen in most circumstances. In any case, the experimental measurements needed to correctly model frictional sliding and plastic/mold adhesion at elevated temperatures have not yet been performed. Detailed examination of this is a secondary issue since most thermoformed and blow molded parts do not exhibit any significant sliding between the plastic and the mold surfaces.

The finite element approach described in this chapter contains assumptions that simplify the formulation of the problem while still incorporating the dominant physical phenomena that have been observed in the actual processes. Since most of the items formed by either of these processes consist of thin walled shell structures, the hot polymer is modeled as a membrane. Thus, the bending resistance of the hot polymer is neglected and the material thickness is assumed to be small when compared to other dimensions of the structure. The membrane assumption would appear to be quite reasonable for the bulk of the structure with the possible exception of material next to clamping rings and "holding" fixtures. In these locations it is possible that substantial cooling may exist which could impart significant bending strength to the plastic.

The mold surface in this finite element model is a rigid predefined boundary through which the plastic membrane is not permitted to penetrate. During the analysis, collision calculations are continuously performed to determine whether contact has occurred between the plastic membrane and the rigid mold surface. When contact does occur, the plastic membrane is permanently fixed to the mold surface at the point of contact. Since the material which contacts the mold surface is not permitted to move after contact occurs, the final thickness of the plastic in this location after contact remains constant. In essence, the thinning of the membrane is determined solely by the stretching which occurs during inflation and the sequence in which material elements contact the mold surface. An additional benefit of the current contact algorithm is the fact that as contact between plastic and mold occurs, degrees of freedom, and thus unknowns, can be eliminated. This means that the number of equations that have to be solved during the analysis decreases as more and more of the polymer contacts the mold surface. This unique characteristic of the formulation is particularly beneficial when it comes to solving three-dimensional problems which would otherwise have excessively long computation times.

The plastic membrane itself is modeled as a "rubbery," i.e., nonlinear, elastic, incompressible material and therefore does not exhibit time-dependent behavior. This constitutive relationship is often called hyperelastic and there seems to be strong experimental evidence [17-19] that such a constitutive relationship is applicable for modeling these plastic forming processes. Thermoplastics are generally considered to be viscoelastic materials. Thus rigorous analysis valid for arbitrary strain rates and deformation history should incorporate a nonlinear viscoelastic constitutive relationship suitable for large biaxial deformations. However, considerable experimental work remains before such viscoelastic models can be constructed for polymers at the processing temperatures of interest. Furthermore, since most industrial processing is conducted at relatively high strain rates and in the temperature range where the polymer often exhibits "rubberlike" behavior, the hyperelastic material model seems to be the most reasonable constitutive relationship available based on current knowledge of high temperature thermoplastic behavior.

The following section contains a brief review of the theoretical foundation connected with finite deformation analysis. This is followed by the details of the finite element formulation and material relationships used to solve the thermoforming and blow molding examples presented later in this chapter.

5.2 FINITE ELEMENT FORMULATION

Modeling thermoforming and blow molding inflation processes with the finite element method requires that many of the most difficult aspects of the finite element method be addressed in the analysis. These difficulties arise because of large strains, large deformations, nonlinear material behavior, contact between polymer and mold wall, and in certain instances physical instability during inflation of the polymer. All these complications lead to a set of nonlinear equations which have to be solved in an iterative manner. The final equations pertinent to the formulation used for thermoforming and blow molding analyses will be presented, but details concerning the derivations of these equations will not be given in this chapter. General descriptions of the finite element method may be found in [20-23], while a detailed description of the axisymmetric finite element implementation for blow molding and thermoforming is given in [16].

5.2.1 Theoretical Background

Before outlining the finite element implementation, it is necessary to give a brief introduction to some theoretical concepts used in describing large deformations of thin membranes. A more rigorous treatment of this subject may be found in [24-26]. For the following discussions it is necessary to introduce the concept of a material coordinate system which is characterized by an R-axis and an S-axis. The axes of the material coordinate system may be visualized by two lines drawn on the membrane such that the two lines are crossing each other (Fig. 5.4). Since the membrane may be curved, the two lines will in general not be straight lines nor will they in general be perpendicular to each other. The axes have unit divisions which may be different from one axis to the other. A point P on the membrane can now be defined by its material coordinates, which are determined by the intersections between the axes and lines drawn "parallel" to the axes through P (Fig. 5.4). Since the coordinate system is embedded in the membrane, all coordinate lines will deform with the membrane with the result that the point P will always have the same coordinates in this system independent of the deformation of the membrane.

In addition to the material coordinate system, a Cartesian system, x-y-z, will also be introduced. A point P with the material coordinates (R, S) may equally well be characterized by its Cartesian coordinates (x, y, z), which are functions of the location of the point P, i.e.,

$$x = x(R, S), \quad y = y(R, S), \quad z = z(R, S) \tag{1}$$

A point P' with the material coordinates (R + dR, S + dS) will then have the Cartesian coordinates (x + dx, y + dy, z + dz) where

$$dx = \frac{\partial x}{\partial R}dR + \frac{\partial x}{\partial S}dS, \quad dy = \frac{\partial y}{\partial R}dR + \frac{\partial y}{\partial S}dS, \quad dz = \frac{\partial z}{\partial R}dR + \frac{\partial z}{\partial S}dS \tag{2}$$

When studying the local deformations around the point P for a certain deformed state, the Cartesian system will be placed such that the origin of this coordinate system is at the point P and the system will be oriented so that the z-axis coincides with the normal to the membrane at P. With this orientation the x-y plane coincides with the tangent plane at P with the result that the calculation of the infinitesimal distance PP' will not involve the z-coordinate. The

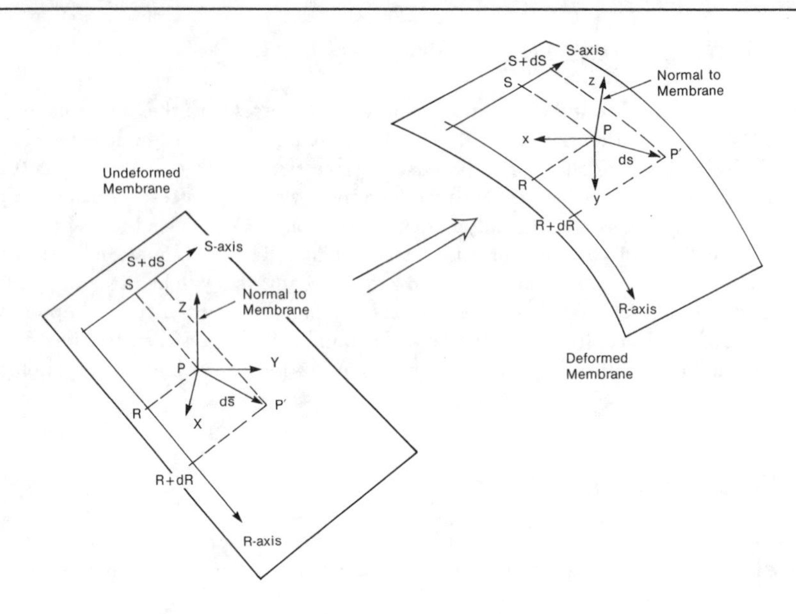

Figure 5.4 Coordinate systems associated with membrane.

last equation in Eq (2) is, therefore, not needed and any reference to the z-coordinate will be dropped. By introducing the Jacobian matrix:

$$\{j\} = \begin{Bmatrix} \dfrac{\partial x}{\partial R} & \dfrac{\partial x}{\partial S} \\ \dfrac{\partial y}{\partial R} & \dfrac{\partial y}{\partial S} \end{Bmatrix} \tag{3}$$

Eq (2) can be rewritten as

$$\begin{Bmatrix} dx \\ dy \end{Bmatrix} = \{j\} \begin{Bmatrix} dR \\ dS \end{Bmatrix} \tag{4}$$

The length ds of the line element PP' (Fig. 5.4) determined by the vector $\{dx \quad dy\}^T$ is then given by

$$ds^2 = \{dx \quad dy\} \begin{Bmatrix} dx \\ dy \end{Bmatrix} = \{dR \quad dS\} \{j\}^T \{j\} \begin{Bmatrix} dR \\ dS \end{Bmatrix} \tag{5}$$

The product of the transpose of the Jacobian matrix and the Jacobian matrix itself is called the metric tensor and is denoted by $\{g\}$, i.e., $\{g\} = \{j\}^T \{j\}$.

So far the discussion has concentrated on calculating the length of PP' at a certain fixed state of deformation. To get a measure of the absolute deformations, it is, however, necessary to compare the length of PP' with its original length. For this purpose a second Cartesian coordinate system X-Y-Z is introduced. This coordinate system is placed on the membrane when it is in its original undeformed position. At this point it again has its origin at the material point P and has the Z-axis coinciding with the normal of the membrane in this initial position. In this configuration the surface of the membrane can be described by the parametric form $X = X(R, S)$, $Y = Y(R, S)$ and the vector PP' and the length $\overline{ds}$ of PP' can similarly be written as

$$\begin{Bmatrix} dX \\ dY \end{Bmatrix} = \{J\} \begin{Bmatrix} dR \\ dS \end{Bmatrix} \tag{6}$$

$$\overline{ds}^2 = \{dR \quad dS\} \{J\}^T \{J\} \begin{Bmatrix} dR \\ dS \end{Bmatrix} \tag{7}$$

where the Jacobian matrix for this initial position is given by

$$\{J\} = \begin{bmatrix} \frac{\partial X}{\partial R} & \frac{\partial X}{\partial S} \\ \frac{\partial Y}{\partial R} & \frac{\partial Y}{\partial S} \end{bmatrix} \tag{8}$$

The metric tensor is in this case given by $\{G\} = \{J\}^T \{J\}$. The difference between the square of the length of PP' in the deformed position and the square of the length in the initial position (Fig. 5.4) is then

$$ds^2 - \overline{ds}^2 = \{dR \quad dS\} \left[\{j\}^T \{j\} - \{J\}^T \{J\} \right] \begin{Bmatrix} dR \\ dS \end{Bmatrix} \tag{9}$$

or by inverting Eq (6) and substituting into Eq (9)

$$ds^2 - \overline{ds}^2 = \{dX \quad dY\} (\{C\} - \{I\}) \begin{Bmatrix} dX \\ dY \end{Bmatrix} \tag{10}$$

where $\{I\}$ is the identity matrix and $\{C\}$ is Green's deformation tensor given by

$$\{C\} = \left[\{j\} \{J\}^{-1} \right]^T \left[\{j\} \{J\}^{-1} \right] \tag{11}$$

The quantity $\{j\} \{J\}^{-1}$ is called the deformation gradient.

In the special case, where P' has the coordinates $(X + dX, Y)$ and P has the coordinates (X, Y) in the X-Y system, the length of the vector PP' in the undeformed state is $\overline{ds}^2 = dX^2$. From Eq (10) $ds^2 - \overline{ds}^2 = (C_{11} - 1)dX^2$, hence the final length of PP' is determined by $ds^2 = C_{11}\overline{ds}^2$. The stretch λ_X in the X-direction is defined as the final length of this line element divided by its original length in the X-direction, i.e.,

$$\lambda_X = \frac{ds}{\overline{ds}} = \sqrt{C_{11}} \tag{12}$$

The stretch in the Y-direction is similarly found to be $\lambda_Y = \sqrt{C_{22}}$. In general, all elements in the matrix $\{C\}$ are nonzero, but by rotating the X-Y-Z system around the Z-axis, a position may be reached where $C_{12} = C_{21} = 0$. The stretches in this particular orientation of the X- and Y-axes are called the principal stretches and are denoted by λ_1 and λ_2. The stretch in the membranes thickness direction is $\lambda_3 = h/H$ where h and H are the current and original membrane thicknesses, respectively, at the point under consideration. If the membrane material is incompressible, it can be shown that

$$\lambda_3 = \frac{h}{H} = \frac{1}{\sqrt{\det\{C\}}} \tag{13}$$

where $\det\{C\}$ is the determinant of $\{C\}$.

The traction in a cross section of the membrane is defined as the force per unit area, on the cross section. The tractions $\{t\}$ can be derived from the stress tensor

$$\{\sigma\} = \begin{Bmatrix} \sigma_{11} \, \sigma_{12} \\ \sigma_{21} \, \sigma_{22} \end{Bmatrix} \tag{14}$$

through

$$\{t\} = \{\sigma\} \begin{Bmatrix} n_x \\ n_y \end{Bmatrix} \tag{15}$$

where σ is the Cauchy stress tensor which is equivalent to the stress tensor used in small displacement analyses and (n_x, n_y) is the normal to the cross section (not the normal to the membrane). The normal is here defined in the local Cartesian system described previously, and since the membrane lies in the x-y plane at the point under consideration, the normal to the cross section will also lie in the x-y plane. For the discussions of the finite element implementation and for the description of the material behavior it is also necessary to introduce the 2nd Piola-Kirchhoff stress tensor, $\{T\}$ [24-25], which under the assumption of an incompressible material is defined by

$$\{T\} = \left[\{J\} \{j\}^{-1} \right] \{\sigma\} \left[\{J\} \{j\}^{-1} \right]^T \tag{16}$$

The Cauchy stress can be determined from the 2nd Piola-Kirchhoff stress from

$$\{\sigma\} = \left[\{j\} \{J\}^{-1} \right] \{T\} \left[\{j\} \{J\}^{-1} \right]^T \tag{17}$$

5.2.2 The Finite Element Concept

In the finite element method, a body to be analyzed is divided into a number of small subdivisions, or finite elements (Fig. 5.5). The elements are defined by a number of nodal points, which for the elements considered in this chapter simply are the corner points of the elements. The displacements of the node points are the unknowns in the finite element analysis. Through energy considerations, a set of equilibrium equations are derived which relate the deflections of these nodes to the loads applied to the body.

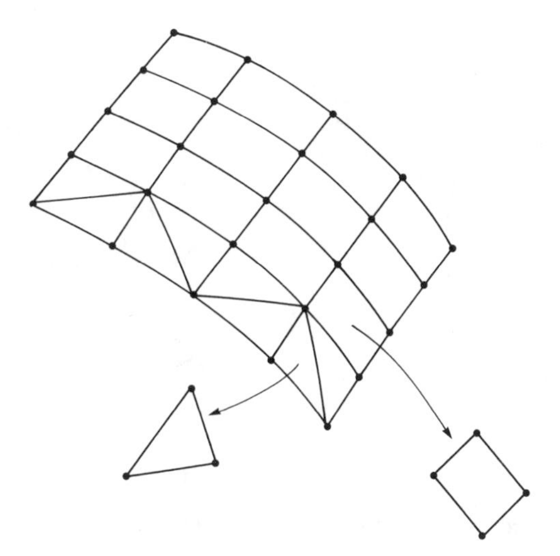

Figure 5.5 A membrane divided into finite elements.

Inside an element, the displacements are interpolated between the nodes of the element using a polynomial interpolation function. For all elements considered in this chapter it is assumed that the displacements within the elements vary linearly along the edges. In conventional finite element analyses, it is now customary to use elements which have a higher order interpolation function than this, but in the thermoforming and blow molding processes, the polymer can be severely deformed during contact with the mold surface resulting in abrupt slope changes where the polymer contacts the mold. Elements with a higher order interpolation function can be overly deformed when subjected to these conditions and often give rise to numerical instabilities. On the other hand, a linear interpolation of the deformations inside the element has been shown to be more stable for analyses involving these kinds of slope discontinuities. The following description will, therefore, only deal with linear elements.

As mentioned in the introduction, since most of the items made by the thermoforming and blow molding processes consist of thin walled structures, the polymer will be modeled as a membrane. Thus, the material thickness will be assumed to be small compared to the other dimensions of the finished part and only the stretching of the polymer is assumed to be important during the inflation, while the bending resistance of the polymer will be neglected. With this approximation the elements in Figure 5.5 can be used to model a general membrane. The 1D elements shown in Figure 5.6 may be used to model a cross section of a membrane for axisymmetric deformations and for deformations of long sheets where the deformation is uniform in one direction.

5.2.3 The Equilibrium Equations

For simplicity, the outline of the finite element formulation will be given for the 4-noded element shown in Figure 5.5 and at the end of this subsection the axisymmetric formulation for the 2-noded element of Figure 5.6 will be discussed briefly.

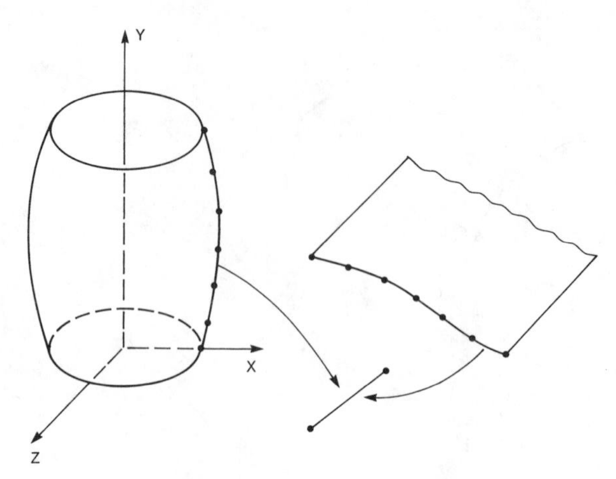

Figure 5.6 1-D elements used to model axisymmetric and long membranes.

Since all motions are slow enough for dynamic effects to be negligible, the equilibrium condition for the membrane can be derived by comparing the rate of work performed by the pressure on the membrane with the rate of increase in internal energy in the membrane [23], or:

$$\int_a p\,\mathbf{n}\cdot\mathbf{v}\,da = \frac{\partial}{\partial t}\int_v \rho\,\varepsilon\,dv \tag{18}$$

where p is the pressure on the membrane, **n** is the normal to the membrane, **v** is its velocity, ρ is the material density, ε is the internal energy density, a is the current surface area of the membrane, and v its volume. The integration in Eq (18) is performed element by element in a local element coordinate system which corresponds to the material coordinate system discussed in Section 5.2.1 and is shown schematically in Figure 5.7 for the 4-noded element.

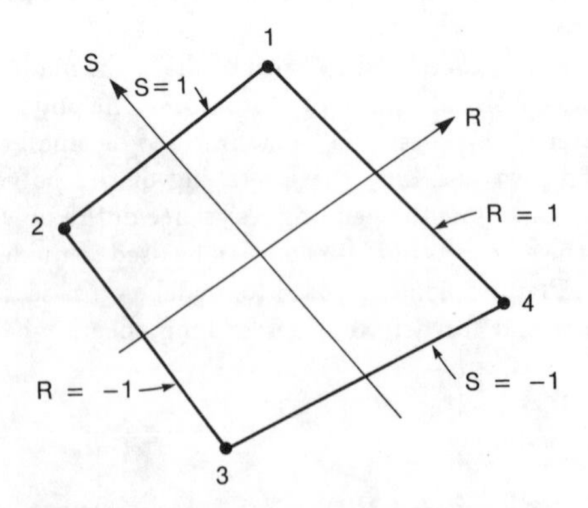

Figure 5.7 Local element coordinate system and local node numbering scheme for 4-noded element.

This coordinate system is fixed to the element and moves with it during the deformations. This coordinate system is, therefore, in general not a Cartesian system. A point with the coordinates (R, S) (Fig. 5.7), with $R = \pm 1$ or $S = \pm 1$, corresponds to a point on the sides of the element. In a global Cartesian coordinate system the coordinates of the node "i" will be denoted (x_i, y_i, z_i). The global coordinates of node "i" change constantly during the deformation and for reference the coordinates in the undeformed configuration will be denoted by capital letters (X_i, Y_i, Z_i). It should be noted that the global Cartesian coordinate system described here is different from the special purpose Cartesian systems outlined in Section 5.2.1.

As mentioned above, linear interpolation functions are used for the deformations inside the element. In matrix notation this can be expressed as:

$$\{x\} = \{N\}\{x^e\} \tag{19}$$

where

$$\{x\} = \{x\,y\,z\}^T \tag{20}$$

$$\{x^e\} = \{x_1 y_1 z_1 x_2 y_2 z_2 x_3 y_3 z_3 x_4 y_4 z_4\}^T \tag{21}$$

$$\{N\} = \begin{Bmatrix} N_1 & 0 & 0 & N_2 & 0 & 0 & N_3 & 0 & 0 & N_4 & 0 & 0 \\ 0 & N_1 & 0 & 0 & N_2 & 0 & 0 & N_3 & 0 & 0 & N_4 & 0 \\ 0 & 0 & N_1 & 0 & 0 & N_2 & 0 & 0 & N_3 & 0 & 0 & N_4 \end{Bmatrix} \tag{22}$$

In Eqs (20) - (21), (x, y, z) are the coordinates of a point inside an element, and $(x_i\, y_i\, z_i)$ are the coordinates of the "i-th" node of the element (Fig. 5.7). The interpolation functions N_i are given by:

$$N_1 = \frac{1}{4}(1+R)(1+S) \qquad N_2 = \frac{1}{4}(1-R)(1+S)$$
$$N_3 = \frac{1}{4}(1-R)(1-S) \qquad N_4 = \frac{1}{4}(1+R)(1-S) \tag{23}$$

Hence, the function N_1 takes on the value 1 at the node where $R = S = 1$ and the value 0 at all other nodes. Each of the other functions similarly take on the value 1 at one node and zero at all the other nodes. For a fixed value of R (or S), Eq (19), therefore, defines a linear interpolation between opposing element edges.

As mentioned above, the equilibrium condition for the membrane may be derived by comparing the rate of increase in internal energy with the rate of work done by the pressure on the membrane. This results in the equilibrium condition given by Eq (18). After some manipulation [16], this equation may also be formulated as:

$$\int_A p\sqrt{\det\{g\}}\,\mathbf{n}\cdot\dot{\mathbf{x}}\,dRdS = \int_A \frac{1}{2}H\sqrt{\det\{G\}}\,T_{IJ}\frac{\partial C_{IJ}}{\partial t}\,dRdS \tag{24}$$

where the index notation has been used for repeated indices, T_{IJ} are the components of the 2nd Piola-Kirchhoff stress tensor defined in Section 5.2.1, C_{IJ} are the components of Green's deformation tensor, H is the initial thickness of the polymer, {G} and {g} are the metric tensors defined in Section 5.2.1, and p is the pressure acting on the surface of the membrane. It is important to note, that all integrations in Eq (24) are performed over the undeformed surface area A. In matrix notation the resulting equilibrium equation may be written as:

$$\{K\}\{x\} = \{F\} \tag{25}$$

where the unknown $\{x\}$ is the vector of all nodal coordinates in the model, i.e.,

$$\{x\} = \{x_1 y_1 z_1 x_2 y_2 z_2 \cdots x_n y_n z_n\}^T \tag{26}$$

The global stiffness matrix $\{K\}$ and the global load vector $\{F\}$ are assembled from the individual element stiffness matrices and load vectors as described in [21]. The stiffness matrix $\{k\}$ and load vector $\{f\}$ for the m-th element are given by

$$\{k\} = \int_{A_m} H\sqrt{\det\{G\}} \left\{\frac{\partial N}{\partial X}\right\}^T \{T^*\} \left\{\frac{\partial N}{\partial X}\right\} dR\, dS \tag{27}$$

$$\{f\} = \int_{A_m} p\sqrt{\det\{g\}}\, \{nN\}\, dR\, dS \tag{28}$$

where A_m is the area of the m-th element. The matrix expressions in Eqs (27) - (28) are given by:

$$\left\{\frac{\partial N}{\partial X}\right\} = \left\{\begin{matrix} \frac{\partial N_1}{\partial X} & 0 & 0 & \frac{\partial N_2}{\partial X} & 0 & 0 & \cdots & 0 \\ 0 & \frac{\partial N_1}{\partial X} & 0 & 0 & \frac{\partial N_2}{\partial X} & 0 & \cdots & 0 \\ 0 & 0 & \frac{\partial N_1}{\partial X} & 0 & 0 & \frac{\partial N_2}{\partial X} & \cdots & \frac{\partial N_n}{\partial X} \\ \frac{\partial N_1}{\partial Y} & 0 & 0 & \frac{\partial N_2}{\partial Y} & 0 & 0 & \cdots & 0 \\ 0 & \frac{\partial N_1}{\partial Y} & 0 & 0 & \frac{\partial N_2}{\partial Y} & 0 & \cdots & 0 \\ 0 & 0 & \frac{\partial N_1}{\partial Y} & 0 & 0 & \frac{\partial N_2}{\partial Y} & \cdots & \frac{\partial N_n}{\partial Y} \end{matrix}\right\} \tag{29}$$

$$\{T^*\} = \left\{\begin{matrix} T_{11} & 0 & 0 & T_{12} & 0 & 0 \\ 0 & T_{11} & 0 & 0 & T_{12} & 0 \\ 0 & 0 & T_{11} & 0 & 0 & T_{12} \\ T_{21} & 0 & 0 & T_{22} & 0 & 0 \\ 0 & T_{21} & 0 & 0 & T_{22} & 0 \\ 0 & 0 & T_{21} & 0 & 0 & T_{22} \end{matrix}\right\} \tag{30}$$

$$\{nN\} = \{n_x N_1\, n_y N_1\, n_z N_1\, n_x N_2\, n_y N_2\, n_z N_2 \cdots n_x N_4\, n_y N_4\, n_z N_4\}^T \tag{31}$$

where (n_x, n_y, n_z) are the components of the element normal vector. In Eq (30) T_{IJ} are the components of the 2nd Piola-Kirchhoff stress tensor defined in Section 5.2.1. The components of stress are functions of the deformation gradients and are determined from a suitable constitutive relationship (Section 5.3.2). In Eq (29), the derivatives are with respect to a

Cartesian coordinate system which has been rotated such that X and Y are in the plane of the polymer membrane and Z is in the direction of the normal while the membrane is in the undeformed state.

Since the stresses in Eq (30) depend on the deformation state in a nonlinear manner, the stiffness matrix, Eq (27), is a nonlinear function of the deformations of the membrane and consequently the equilibrium condition, Eq (25), represents a set of nonlinear equations in $\{x\}$. By assuming that an equilibrium position is known for some pressure p_0, the equilibrium Eq (25) can be linearized by expanding the left hand side in a Taylor series about this position. This leads to the following equation for the determination of the increment $\{\Delta x\}$ in the nodal coordinates:

$$\{K\}_0\{x\}_0 + \left[\{K\} + \left\{ \frac{\partial K}{\partial x} \right\} \{x\} \right]_0 \{\Delta x\} = \{F\} \tag{32}$$

where:

$$\{\Delta x\} = \{\Delta x_1\, \Delta y_1\, \Delta z_1\, \Delta x_2\, \Delta y_2\, \Delta z_2 \cdots \Delta x_n\, \Delta y_n\, \Delta z_n\}^T \tag{33}$$

The product $\{K\}_0\{x\}_0$ is commonly referred to as the residual load vector, while the pre-multiplier for the vector $\{\Delta x\}$ is called the tangent stiffness matrix. Equation (32) may, therefore, be written as:

$$\{K_t\}\{\Delta x\} = \{F\} - \{R\} \tag{34}$$

where the element tangent stiffness matrix $\{k_t\}$ and the residual load vector $\{r\}$ are given by:

$$\{k_t\} = \int_{A_m} H \sqrt{\det\{G\}} \left\{ \frac{\partial N}{\partial X} \right\}^T \left[\left\{ \frac{\partial x}{\partial X} \right\}^T \left\{ \frac{\partial T}{\partial C} \right\} \left\{ \frac{\partial x}{\partial X} \right\} + \{T^*\} \right]_0 \left\{ \frac{\partial N}{\partial X} \right\} dR\, dS \tag{35}$$

$$\{r\} = \int_{A_m} H \sqrt{\det\{G\}} \left[\left\{ \frac{\partial N}{\partial X} \right\}^T \left\{ \frac{\partial x}{\partial X} \right\}^T \{T\} \right]_0 dR\, dS \tag{36}$$

The load vector $\{F\}$ in Eq (34) is still assembled from the individual load vectors given by Eq (28). In the above expressions:

$$\left\{ \frac{\partial x}{\partial X} \right\} = \begin{bmatrix} \frac{\partial x}{\partial X} & \frac{\partial y}{\partial X} & \frac{\partial z}{\partial X} & 0 & 0 & 0 \\ 0 & 0 & 0 & \frac{\partial x}{\partial Y} & \frac{\partial y}{\partial Y} & \frac{\partial z}{\partial Y} \\ \frac{\partial x}{\partial Y} & \frac{\partial y}{\partial Y} & \frac{\partial z}{\partial Y} & \frac{\partial x}{\partial X} & \frac{\partial y}{\partial X} & \frac{\partial z}{\partial X} \end{bmatrix} \tag{37}$$

$$\left\{\frac{\partial T}{\partial C}\right\} = 2\left\{\begin{matrix} \frac{\partial T_{11}}{\partial C_{11}} & \frac{\partial T_{11}}{\partial C_{22}} & \frac{\partial T_{11}}{\partial C_{12}} \\ \frac{\partial T_{22}}{\partial C_{11}} & \frac{\partial T_{22}}{\partial C_{22}} & \frac{\partial T_{22}}{\partial C_{12}} \\ \frac{\partial T_{12}}{\partial C_{11}} & \frac{\partial T_{12}}{\partial C_{22}} & \frac{\partial T_{12}}{\partial C_{12}} \end{matrix}\right\} \tag{38}$$

$$\{T\} = \{T_{11}\, T_{22}\, T_{12}\}^T \tag{39}$$

and $\{C\}$ is the Green deformation tensor from Section 5.2.1.

For axisymmetric conditions, the cross section of the membrane is assumed to be given in the x-y plane with the y-axis as the axis of symmetry (Fig. 5.6). This cross section is modeled with the 2-noded line elements. The equilibrium equation, Eq (34), and the expressions for the tangent stiffness matrices, Eq (35), the load vectors, Eq (28), and the residual load vectors, Eq (36), are still valid with the following simplifications for the matrices involved:

$$\left\{\frac{\partial N}{\partial X}\right\} = \left\{\begin{matrix} \frac{\partial N_1}{\partial X} & 0 & \frac{\partial N_2}{\partial X} & 0 \\ 0 & \frac{\partial N_1}{\partial X} & 0 & \frac{\partial N_2}{\partial X} \\ \frac{N_1}{r_0} & 0 & \frac{N_2}{r_0} & 0 \end{matrix}\right\} \tag{40}$$

$$\left\{\frac{\partial x}{\partial X}\right\} = \left\{\begin{matrix} \frac{\partial x}{\partial X} & \frac{\partial y}{\partial X} & 0 \\ 0 & 0 & \frac{x}{r_0} \end{matrix}\right\} \tag{41}$$

$$\left\{\frac{\partial T}{\partial C}\right\} = \left\{\begin{matrix} \frac{\partial T_{11}}{\partial C_{11}} & \frac{\partial T_{11}}{\partial C_{22}} \\ \frac{\partial T_{22}}{\partial C_{11}} & \frac{\partial T_{22}}{\partial C_{22}} \end{matrix}\right\} \tag{42}$$

$$\{T^*\} = \left\{\begin{matrix} T_{11} & 0 & 0 \\ 0 & T_{11} & 0 \\ 0 & 0 & T_{22} \end{matrix}\right\} \tag{43}$$

$$\{nN\} = \{n_x N_1\, n_y N_1\, n_x N_2\, n_y N_2\}^T \tag{44}$$

$$\{T\} = \{T_{11}\, T_{22}\}^T \tag{45}$$

where r_0 is the initial radius at the point under consideration. In the integration of Eq (28) and Eqs (35) - (36), $\sqrt{\det\{G\}}$ and $\sqrt{\det\{g\}}$ should now be replaced by $r_0\sqrt{G_{11}}$ and $r_0\sqrt{g_{11}}$ and the integration only needs to be performed in one dimension, i.e., along the axial arc length of the membrane.

5.2.4 The Solution Procedure

In the previous sections, the equations have been set up so that the deformation increments can be calculated for a small increase in pressure when an equilibrium shape is known. Since the initial state without any loads is an equilibrium state, the analysis is started from the zero load condition and a small load is applied. For this small load, the deformation increments can be found. This furnishes a new equilibrium state and another small load increment can be applied. Hence, by successive applications of small load increments the final deformation of the polymer can be found.

However, even for a small load increment the deformations often change enough for the tangent stiffness matrix to change significantly, with the result that the solution for each pressure increment has to be obtained in an iterative manner. This is done by first incrementing the pressure, calculating the tangent stiffness matrix and load vector, and finding the increments in the deformation of the membrane. With this updated geometry as the starting point and keeping the pressure constant, the tangent stiffness matrix, the load vector, and the residual load vector are recalculated and a new change in the deformations is found. After several iterations with the pressure kept constant, the deformation increments eventually get so small that the solution can be considered to have converged for this particular pressure. As mentioned above, even though the pressure remains constant, the load vector still has to be recalculated during each iteration since the load includes an area integration and the shape and orientation of the membrane changes continuously.

During the sequence of load steps and successive iterations the polymer might also contact the mold surface, at which time that portion of the polymer will stop deforming. The mold surface is defined as a rigid surface through which the nodes of the finite element model are not permitted to penetrate. During the analysis, collision calculations are continuously performed to determine whether contact has occurred between the finite element nodes and the mold. When contact occurs, the nodes that have contacted are permanently fixed to the mold surface at the contact points and equilibrium is reestablished. Since an element which has all of its nodes fixed to the mold surface can not deform any further, the degrees of freedom for the nodes that have contacted the mold can be eliminated. This results in an increasingly smaller set of equations to be solved and greatly speeds up the computation time as more and more of the membrane contacts the mold.

5.3 MATERIAL BEHAVIOR

5.3.1 Background and Experimental Observations

It is generally recognized that thermoplastics exhibit viscoelastic behavior at elevated temperatures. In viscoelastic materials, the application of a stress results in a strain state which depends upon the manner in which the stress is applied; i.e., whether the load is applied rapidly or slowly. Thus, the strains in a viscoelastic body depend on the history of loading as well as on the actual magnitude of the loads.

Fortunately, at temperatures above the glass transition temperature of a polymer, strong experimental evidence exists, which suggests that under certain conditions, polymer behavior can be adequately modeled using constitutive models originally developed for rubberlike materials, i.e., nonlinear elastic behavior [4,17,27]. When extended at relatively high strain rates, most hot polymers exhibit some degree of reversible "rubberlike" elastic behavior. For example, in the experiments performed by Schmidt and Carley [17,18], circular sheets of heat softened polymer were rapidly inflated to large bubble domes. Upon rupture, these polymer bubbles contracted back to their original shape in less than 1/700 s. This behavior is not totally surprising, since at temperatures above the glass transition temperature, chain rotation, and uncoiling of the long polymer molecules can take place with relatively little viscous effect, so that rubberlike behavior is observed.

The assumption of nonlinear elastic behavior greatly simplifies the formulation of the finite element equations and will be used throughout this chapter. However, formulations incorporating viscoelastic effects can and should be utilized in processing simulations, when the experimental measurements become available to develop constitutive relationships which adequately characterize the large deformation, viscoelastic behavior of hot polymers. Section 5.7 contains a further discussion of this issue.

5.3.2 Nonlinear Elastic Constitutive Models

Extensive classical treatises on the mechanics of nonlinear finite elasticity can be found in [24-26,28]. In [29], Treloar reviews the various constitutive models and includes experimental data for rubber. In this section two constitutive relationships will be examined in detail. They are the so-called (1) Mooney-Rivlin formulation and the (2) Ogden formulation. Both of these constitutive relationships have been developed for an ideally elastic solid that possesses a stress potential. This special elastic solid is often called a hyperelastic solid. The Mooney-Rivlin formulation has been used extensively to obtain solutions for a large class of elasticity problems and Mooney-Rivlin material constants have been obtained experimentally for a number of elastomers [29] and thermoplastics [18]. Thus, this constitutive model is particularly useful for comparing results from finite element calculations with classical solutions obtained using other techniques. The more recently developed Ogden model has certain advantages when it comes to curve fitting actual experimental data and physical interpretation. However, neither model is intrinsically better than the other and both are outlined below.

For a hyperelastic material, the 2nd Piola-Kirchhoff stress tensor $\{T\}$, Eq (16), can be expressed as the derivative of the strain energy function, W. The form of W, as a function of the deformation gradients, is subject to certain restrictions. One restriction is that superposition on the assumed deformation of a rigid rotation leaves the strain energy function W unchanged. The strain energy depends only on the changes in distance between neighboring particles and so depends solely on the components of the deformation tensor, i.e., $W = W(C_{IJ})$, where the C_{IJ}'s are the components of Green's deformation tensor, Eq (11). The components of the 2nd Piola-Kirchhoff stress tensor $\{T\}$ can, therefore, be expressed as:

$$T_{IJ} = \frac{\partial W}{\partial C_{IJ}} + \frac{\partial W}{\partial C_{JI}} = 2\frac{\partial W}{\partial C_{IJ}}, \quad I, J = 1, 3 \tag{46}$$

where the last equality follows if W is written as a symmetric function in C_{IJ}.

For an isotropic material, i.e. a material which does not have any preferred directions, the strain energy function W cannot depend on the initial material orientation. Since this is the case, the strain energy function W for an isotropic material can be expressed in terms of the invariants of the deformation gradient $\{C\}$. These invariant quantities, I_1, I_2 and I_3 are given by:

$$I_1 = C_{II} \tag{47}$$

$$I_2 = \frac{1}{2}\left[I_1^2 - C_{IJ}C_{JI}\right] \tag{48}$$

$$I_3 = \det\{C\} \tag{49}$$

In Eqs (47) - (49) the index notation is used, with the usual convention that expressions with repeated indices are summed from 1 to 3. For example, $C_{II} = C_{11} + C_{22} + C_{33}$.

Materials which undergo little change in volume at stress levels that cause severe deformations can be treated as incompressible. From Eq (12) it can be seen that in terms of the principal stretches, this incompressibility condition is given by

$$\lambda_1\lambda_2\lambda_3 = 1 \tag{50}$$

hence $I_3 = 1$. The strain energy function for an incompressible material is thus given by $W = W(I_1, I_2)$. It has been shown experimentally, that the mechanical behavior of many polymers above the glass transition temperature is essentially incompressible [27,30], and thus incompressible behavior is a reasonable assumption.

In the Mooney-Rivlin formulation it is assumed that W can be expressed as a polynomial function of the invariants I_1 and I_2, or rather as a polynomial function of $I_1 - 3$, $I_2 - 3$, so that the stresses will be zero when there is no strain. The generalized Mooney-Rivlin form of the strain energy function is given by [25]:

$$W = \sum_{i=0}^{M}\sum_{j=0}^{N} A_{ij}(I_1 - 3)^i (I_2 - 3)^j \tag{51}$$

where A_{ij} are empirically determined constants. Since $W = 0$ when the body is undeformed, $A_{00} = 0$. If in expression (51), only the first two terms are retained, i.e., A_{10} and A_{01}, then the standard Mooney-Rivlin form for the strain energy function is obtained and is given by:

$$W = A_{10}(I_1 - 3) + A_{01}(I_2 - 3) \tag{52}$$

Using Eq (46), assuming that the magnitude of the stress normal to the membrane is negligible, and assuming incompressibility, the components of the 2nd Piola-Kirchhoff stress tensor can be expressed as:

$$T_{11} = 2\left[1 - C_{22}C_{33}^2\right]\left[\frac{\partial W}{\partial I_1} + C_{22}\frac{\partial W}{\partial I_2}\right] \tag{53}$$

$$T_{22} = 2\left[1 - C_{11}C_{33}^2\right]\left[\frac{\partial W}{\partial I_1} + C_{11}\frac{\partial W}{\partial I_2}\right] \tag{54}$$

$$T_{12} = 2\left[C_{21}C_{33}^2\right]\frac{\partial W}{\partial I_1} + 2\left[-C_{12} + C_{33}^2 C_{21}\left[C_{11} + C_{22}\right]\right]\frac{\partial W}{\partial I_2} \tag{55}$$

For a membrane loaded in biaxial tension, Eqs (53) - (54) can be reduced to the following simple expression in terms of the principal stretches λ_1, λ_2:

$$T_{11} = 2\left[1 - \frac{1}{\lambda_1^4 \lambda_2^2}\right]\left\{\frac{\partial W}{\partial I_1} + \lambda_2^2 \frac{\partial W}{\partial I_2}\right\} \tag{56}$$

$$T_{22} = 2\left[1 - \frac{1}{\lambda_1^2 \lambda_2^4}\right]\left\{\frac{\partial W}{\partial I_1} + \lambda_1^2 \frac{\partial W}{\partial I_2}\right\} \tag{57}$$

Alternatively, Eqs (56) - (57) can be transformed using Eq (17) to yield the Cauchy stresses:

$$\sigma_{11} = \lambda_1^2 T_{11} = 2\left[\lambda_1^2 - \frac{1}{\lambda_1^2 \lambda_2^2}\right]\left\{\frac{\partial W}{\partial I_1} + \lambda_2^2 \frac{\partial W}{\partial I_2}\right\} \tag{58}$$

$$\sigma_{22} = \lambda_2^2 T_{22} = 2\left[\lambda_2^2 - \frac{1}{\lambda_1^2 \lambda_2^2}\right]\left\{\frac{\partial W}{\partial I_1} + \lambda_1^2 \frac{\partial W}{\partial I_2}\right\} \tag{59}$$

For equibiaxial stretching ($\lambda_1 = \lambda_2 = \lambda$) the stress vs. stretch relationship takes the form:

$$\sigma_{11} = \sigma_{22} = 2\left[\lambda^2 - \frac{1}{\lambda^4}\right]\left[\frac{\partial W}{\partial I_2} + \lambda^2 \frac{\partial W}{\partial I_1}\right] \tag{60}$$

Finally, it can be shown that for simple elongation, i.e. $\sigma_{22} = \sigma_{33} = 0$ and $\lambda_1 = \lambda$, $\lambda_2 = 1/\sqrt{\lambda}$, the Cauchy stress is given by:

$$\sigma_{11} = 2\left[\lambda^2 - \frac{1}{\lambda}\right]\left[\frac{\partial W}{\partial I_1} + \frac{1}{\lambda}\frac{\partial W}{\partial I_2}\right] \tag{61}$$

It should be noted that depending on the form of the derivatives of W with respect to I_1 and I_2 (for the Mooney-Rivlin form Eq (52) these will be the constants A_{10} and A_{01} respectively), the expression for the stress is a highly nonlinear function of the stretch λ, even for simple elongation.

An alternative constitutive relationship that is examined in this section, is the Ogden model. The Ogden model is a hyperelastic material model based on a strain energy function W which is expressed as a function of the principal stretches as opposed to the strain invariants. In the Ogden model, the strain energy is written as an expansion in the principal stretches, λ_1, λ_2, λ_3, and has the form [31]:

$$W = \sum_{n=1}^{r} \frac{\mu_n}{\alpha_n}\left[\lambda_1^{\alpha_n} + \lambda_2^{\alpha_n} + \lambda_3^{\alpha_n} - 3\right] \tag{62}$$

where μ_n and α_n are experimentally determined constants. The constants, μ_n and α_n, can be noninteger and negative, with the only restriction being that the total summation in Eq (62) results in a positive strain energy function. A description of the advantages of the Ogden model, comparison with other hyperelastic relationships and correlation with experimental measurements is given in [27,29,31]. Since the Ogden model is represented directly in terms of the stretches, λ_i, instead of the less tangible invariants I_1 and I_2, the physical interpretation of stress-stretch relationships is often much clearer using the Ogden formulation.

The 2nd Piola-Kirchhoff stress tensor components for the Ogden model are determined from Eqs (46) and (62). The resulting relationship between stress and stretch for an incompressible membrane subjected to in-plane stretching is given by:

$$T_{11}=\sum_{n=1}^{r}\frac{\mu_n}{2}\left[\lambda_1^{\alpha_n-2}(1+D)+\lambda_2^{\alpha_n-2}(1-D)-2C_{22}\lambda_3^{\alpha_n+2}\right] \tag{63}$$

$$T_{22}=\sum_{n=1}^{r}\frac{\mu_n}{2}\left[\lambda_1^{\alpha_n-2}(1-D)+\lambda_2^{\alpha_n-2}(1+D)-2C_{11}\lambda_3^{\alpha_n+2}\right] \tag{64}$$

$$T_{12}=\sum_{n=1}^{r}\frac{\mu_n}{2}C_{21}\left[\frac{\left[\lambda_1^{\alpha_n-2}-\lambda_2^{\alpha_n-2}\right]}{B}+2\lambda_3^{\alpha_n+2}\right] \tag{65}$$

where:

$$D=(C_{11}-C_{22})/2B \tag{66}$$

$$\lambda_3=(C_{33})^{1/2}=(C_{11}C_{22}-C_{12}C_{21})^{-1/2} \tag{67}$$

$$\lambda_{1,2}=(A\pm B)^{1/2} \tag{68}$$

$$A=\frac{C_{11}+C_{22}}{2} \tag{69}$$

$$B=\left[\frac{1}{4}(C_{11}-C_{22})^2+C_{12}C_{21}\right]^{1/2} \tag{70}$$

As in the Mooney-Rivlin formulation, the expressions for the stress components in terms of the Ogden constants simplify considerably for the special case when shear deformations are absent. For this special case $D = 1$. For equibiaxial stretching with $\lambda_1 = \lambda_2 = \lambda$ and therefore the thickness stretch $\lambda_3 = \lambda^{-2}$, the pertinent Cauchy stress components are given by:

$$\sigma_{11}=\sigma_{22}=\lambda^2T_{11}=\lambda^2T_{22}=\sum_{n=1}^{r}\mu_n\left[\lambda^{\alpha_n}-\lambda^{-2\alpha_n}\right] \tag{71}$$

In a likewise manner the Cauchy stress components for simple extension $\sigma_{22} = \sigma_{33} = 0$, can be expressed as:

$$\sigma_{11}=\lambda^2T_{11}=\sum_{n=1}^{r}\mu_n\left[\lambda^{\alpha_n}-\lambda^{-\alpha_n/2}\right] \tag{72}$$

The slope of the stress vs. stretch behavior in simple extension is obtained by taking the derivative of Eq (72) with respect to λ resulting in:

$$\frac{d\sigma_{11}}{d\lambda} = \sum_{n=1}^{r} \frac{\mu_n \alpha_n}{2} \left[2\lambda^{\alpha_n - 1} + \lambda^{-\left(\frac{\alpha_n}{2}+1\right)} \right] \tag{73}$$

Taking the limit of the derivative as $\lambda \to 1$, yields:

$$\lim_{\lambda \to 1} \frac{d\sigma_{11}}{d\lambda} = \frac{3}{2} \sum_{n=1}^{r} \mu_n \alpha_n \tag{74}$$

Thus, for small values of stretch, the quantity in Eq (74) corresponds to the elastic modulus for infinitesimal deformations. For an incompressible material subjected to infinitesimal deformations, the shear modulus can be given by $\Sigma \mu_n \alpha_n/2$, i.e., 1/3 the extensional modulus.

5.3.3 Experimental Techniques for Obtaining Material Data

Both the Mooney-Rivlin and Ogden constitutive models contain constants which must be obtained empirically. It is highly desirable that the experimental measurements used to obtain these constants simulate the actual deformation modes observed in thermoforming and blow molding. This requirement means that extensional measurements must be made on the hot polymer. Unfortunately, extensional measurements are not easy to perform on plastics at high temperatures, especially for biaxial deformations. In rubber elasticity, three special types of experiments have historically been used to determine material constants. These tests are: (1) simple uniaxial tension (Eqs (61), (72)), where $\sigma_{22} = \sigma_{33} = 0$ and $\lambda_{22} = \lambda_{33} = \lambda^{-1/2}$, (2) uniaxial stretching, sometimes called "pure shear," where the material is stretched in the axial or 1 direction, while it is prevented from contracting in the lateral or 2 direction, i.e., $\lambda_{22} = 1$ and thus $\lambda_{33} = \lambda^{-1}$, and finally (3) equibiaxial stretching (Eqs (60),(71)).

In the special case of uniaxial stretching or so-called "pure shear" the relationship between the axial Cauchy stress and the stretch λ for an Ogden material is given by

$$\sigma_{11} = \sum_{n=1}^{r} \mu_n \left[\lambda^{\alpha_n} - \lambda^{-\alpha_n} \right] \tag{75}$$

Figures 5.8a,b depict a technique from [32] for performing the "pure shear" test on rubber at room temperature. In this test, a narrow strip is pulled with the special grips shown. Because of the constraint provided by the surrounding material, the material in the center of the specimen is prevented from contracting in the lateral or 2 direction, i.e., in the central region $\lambda_{22} = 1$. Equibiaxial measurements can be performed by inflating a circular membrane and measuring the stretch at the pole. For rubberlike materials, load vs. stretch measurements using these three tests, can yield a sufficient amount of information to construct multiple term constitutive relationships, using either the Mooney-Rivlin or Ogden models, which accurately characterize material behavior when subjected to relatively general biaxial deformation. However, a large number of terms is not practical and most of the experimental data for rubber has been used to obtain the two constants needed for the standard Mooney-Rivlin expression, Eq (52), or for a three term Ogden series expansion, i.e., r = 3 in Eqs (62) - (65). For example, from experiments conducted by Treloar on rubber, Ogden [31] obtained a three term strain

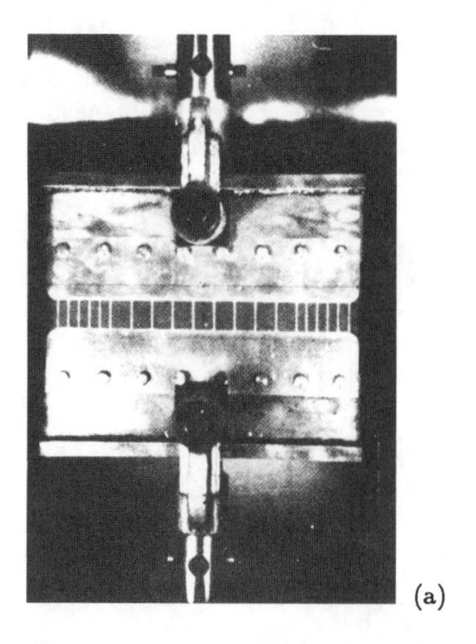

(a)

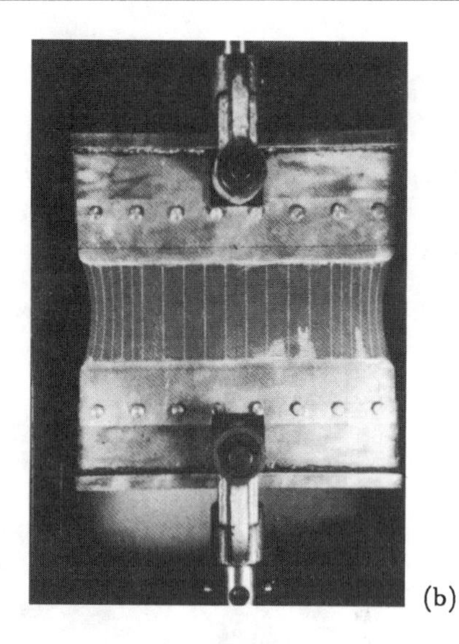

(b)

Figure 5.8 (a) Strip-biaxial tension or "pure shear" at $\lambda = 1$. (b) Strip-biaxial tension or "pure shear" at $\lambda = 3.5$ (from [32]).

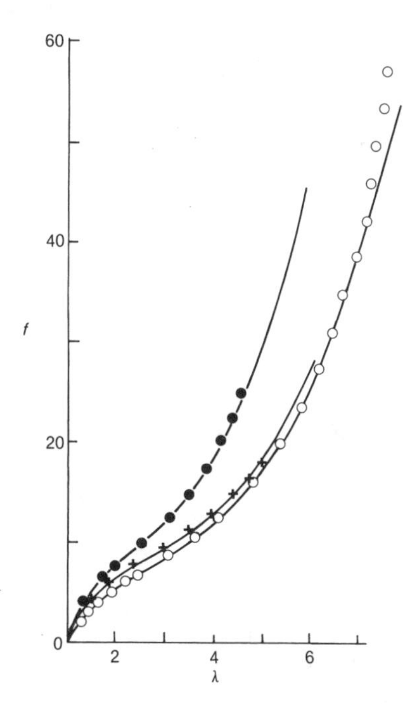

Figure 5.9 Three-term theory compared with Treloar's data in simple tension (O), pure shear (+) and equibiaxial tension (O) ($\mu_1 = 6.3$ kg/cm^2, $\alpha_1 = 1.3$, $\mu_2 = 0.012$ kg/cm^2, $\alpha_2 = 5.0$, $\mu_3 = -0.1$ kg/cm^2, $\alpha_3 = -2.0$, from [31]).

energy function which very closely fits Treloar's data in simple tension, "pure shear" and equibiaxial tension, see Figure 5.9. The values he obtained contain both positive and negative values for the constants μ_n and α_n. The three term Ogden expansion accurately fits the load vs. stretch data for all three of these different extensional experiments up to very large values of stretch λ.

Substantial difficulties are encountered when extensional measurements are performed on hot polymers. Various types of specialized test equipment and procedures have been developed to perform extensional measurements on polymers at high temperatures [17,19,32-40]. The approach taken by Meissner et al. [33-34], using special corrugated rollers to pull the hot plastic in multiple directions (see Fig. 5.10), seems to be capable of yielding very

Figure 5.10 View of rheometer with eight rotary clamps for homogeneous equibiaxial elongation of polymer sheet (courtesy Professor J. Meissner from [33]).

accurate biaxial stretch vs. force data for polymer sheets. However, biaxial measurements using Meissner's approach have not yet been performed on the hot polymers that are typically used in thermoforming or blow molding applications and at the stretch and strain rates needed to adequately simulate thermoforming and blow molding. The use of bubble and tube inflation to obtain biaxial material measurements also seems to be quite promising [35-41], but temperature control is very difficult to maintain during the inflation. In addition, it is not a simple task to measure stretch on the hot membrane surface during inflation. Figure 5.11 shows how DeVries et al. [19] used a special extensometer to measure equibiaxial stretch at the pole of an inflated polystyrene dome, in a range of temperatures above the glass transition region. From such experiments DeVries et al. were able to produce biaxial stress vs. stretch curves at different strain rates and temperatures for polystyrene (see Fig. 5.12). The rubberlike behavior of a number of polymers was confirmed in [19] by the "fact that both uni- and biaxial strains were completely recoverable after stress release." However, unlike rubber, the relationship between stress and strain for amorphous polymers, in the rubbery state, depends

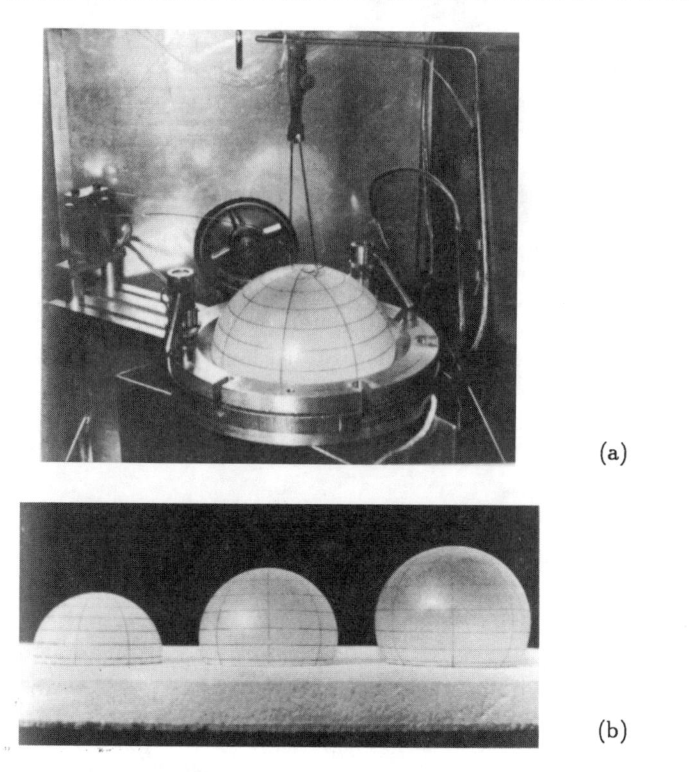

Figure 5.11 (a) Inflated bubble in biaxial extension. (b) Effect of increasing degree of biaxial extension on bubble shape (from [19]).

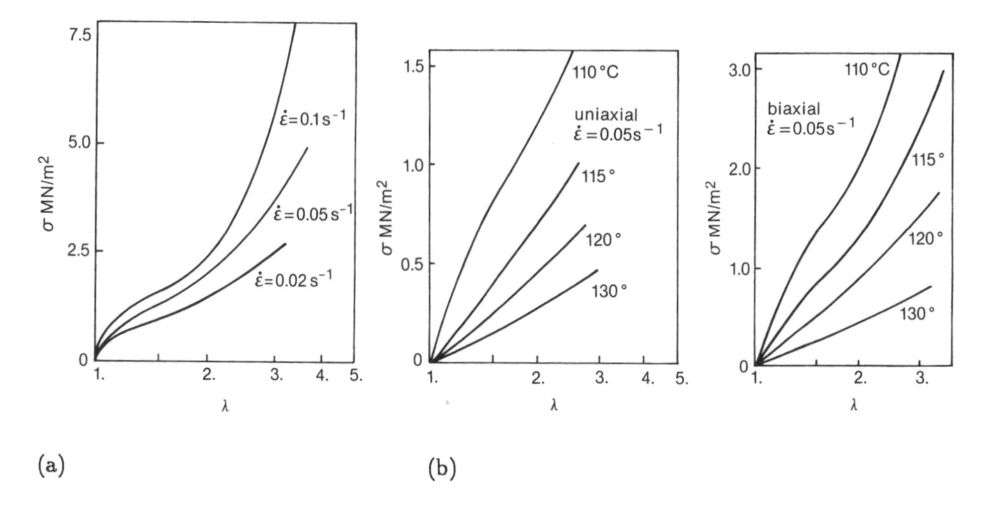

Figure 5.12 (a) Effect of strain rate on biaxial extension of polystyrene at 110° C. (b) Effect of temperature on uni- and biaxial extension of polystyrene (from [19]).

on strain rate and most importantly, on temperature (Fig. 5.12). It is interesting to note when comparing the uniaxial behavior to the equibiaxial measurements in Figure 5.12, that for polystyrene, the shape of the stress-stretch curves is more or less identical at the same temperature and strain rate. However, the tensile stress in the biaxial case is approximately twice as large as the tensile stress in the uniaxial case.

At present, the type of biaxial experiments needed to adequately characterize the behavior of hot polymers for thermoforming and blow molding simulation, are not routinely performed over a wide range of temperatures and strain rates. This contrasts with uniaxial extension tests which can be more or less routinely performed on thermoplastics at elevated temperatures. By photographing the specimen while it is being stretched axially, it is possible to generate true stress vs. stretch curves for polymers at elevated temperatures [42–43]. Of course, one should be leary of exclusively using uniaxial data to obtain material constants for constitutive relationships that are to be used in biaxial calculations. However, until more complete information is available on the large biaxial deformation behavior of polymers at high temperatures, constitutive relationships will have to be based solely on uniaxial data.

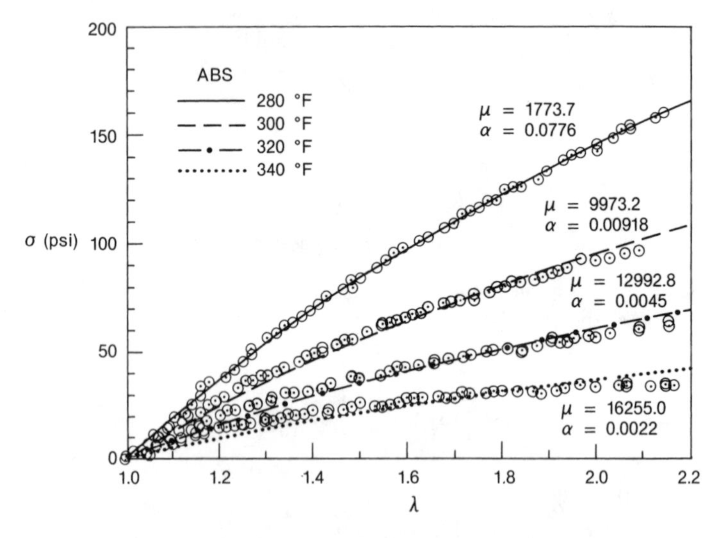

Figure 5.13a Stress vs. stretch data for several polymers. Curve fit with single term Ogden expansion. ABS, displacement rate = 1.05 in/sec (from [43]).

Examples of uniaxial data taken at different temperatures, but only at a single strain rate are shown in Figure 5.13. In this figure, true stress vs. stretch data is shown for Acrylonitrile Butadiene Styrene (ABS) [43], Modified Polyphenylene Oxide (PPO) and an alloy of Polycarbonate and Poly(Butylene Terephthalate) (PC/PBT). The solid line in these figures was generated from a nonlinear least squares curve fit of the experimental data using Eq (72) and a single term Ogden model, i.e., $r = 1$. From these figures it can be seen that a single term Ogden model can represent the data for simple extension reasonably well over a large strain range. In particular, the curve fit to the PPO data, with a single term Ogden expansion, seems to be a remarkably good characterization of the material behavior out to a

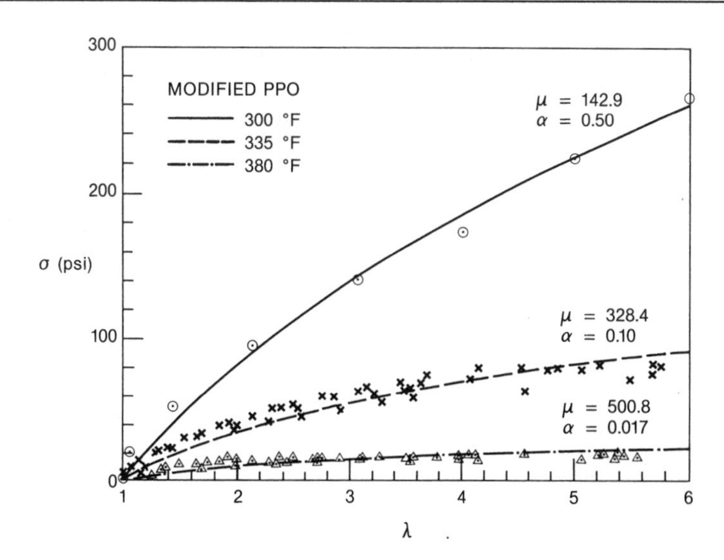

Figure 5.13b Stress vs. stretch data for several polymers. Curve fit with single term Ogden expansion. PPO, displacement rate = 1.0 in/sec.

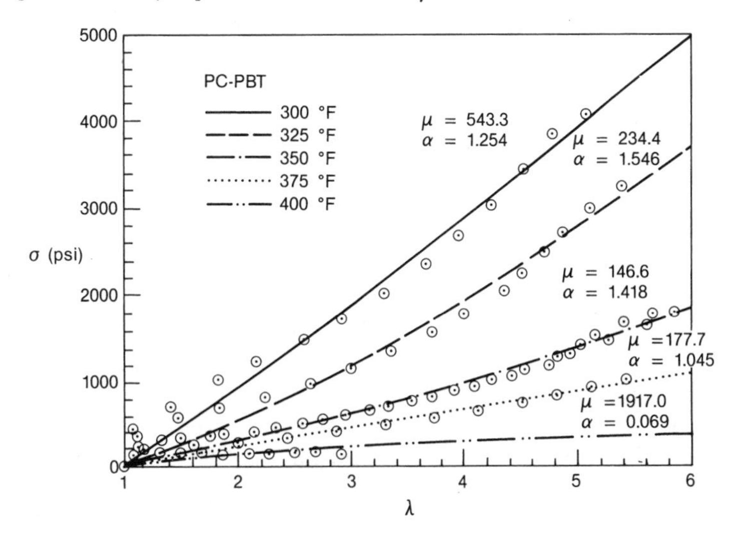

Figure 5.13c Stress vs. stretch data for several polymers. Curve fit with single term Ogden expansion. PC-PBT alloy, displacement rate =1.0 in/sec.

stretch $\lambda = 8$. However, much more uniaxial experimental characterization remains to be done. Once a sufficiently large number of experiments have been conducted at different strain rates and temperatures, it should be possible to obtain an empirical correlation between the material constants and the strain rate and temperature. For example, in the simple one term

Ogden model, the product of $\mu\alpha$ seems to be a smoothly decreasing function of temperature at a given strain rate. If this relationship is known over a wide enough temperature range, it may be possible to "generate" approximate stress vs. stretch curves by interpolation. This can be illustrated by the Ogden constants determined from modified PPO data obtained between the forming temperature range of 280° F and 400° F. Table 5.1 contains Ogden constants determined by a least squares curve fit to uniaxial data for modified PPO at a stretch rate of 1/sec. The product 3/2 $\mu\alpha$ in Table 5.1 corresponds to the elastic modulus for infinitesimal deformations and decreases as a function of increasing temperature. Using a least squares curve fit, the temperature dependency of 3/2 $\mu\alpha$ in Table 5.1 can be empirically correlated with temperature using an exponential expression of the form:

$$\frac{3}{2}\mu\alpha = Ce^{-kT} \tag{76}$$

where $C = e^{12.0}$ psi, $k = 0.0245\ 1/°\,F$.

Equation (76) can be used to estimate the value for one Ogden constant when the other is fixed, for different values of temperature.

Table 5.1 Ogden constants for modified PPO between 280° F and 400° F

Temperature (°F)	μ (psi)	α	$\frac{3}{2}\mu\alpha$ (psi)
280	124.4	0.874	109
300	142.9	0.50	71
335	328.4	0.10	33
350	332.5	0.046	15
380	500.8	0.017	9
400	98.6	0.070	7

What role do material constants play in computer simulation of thermoforming and blow molding? In isothermal calculations, different material properties most directly affect pressure and stability characteristics during inflation. However, different material properties do not seem to have a strong influence on inflation shape (except at very large stretches) and final part thickness. Figure 5.14 illustrates this by comparing the membrane thickness at the pole during free inflation of a circular sheet into a dome, for a three term Ogden model of neoprene rubber at room temperature, a single term fit to the PPO data at 280° F, and a single term fit to the PC/PBT alloy at 400° F. Note that even though the three term Ogden fit for rubber was obtained from uniaxial and biaxial data, the differences in thickness between rubber and the two thermoplastics modeled with a single term Ogden expansion is negligible in this particular example. This does not mean that material properties are unimportant for determining thicknesses in thermoforming and blow molding analysis. On the contrary,

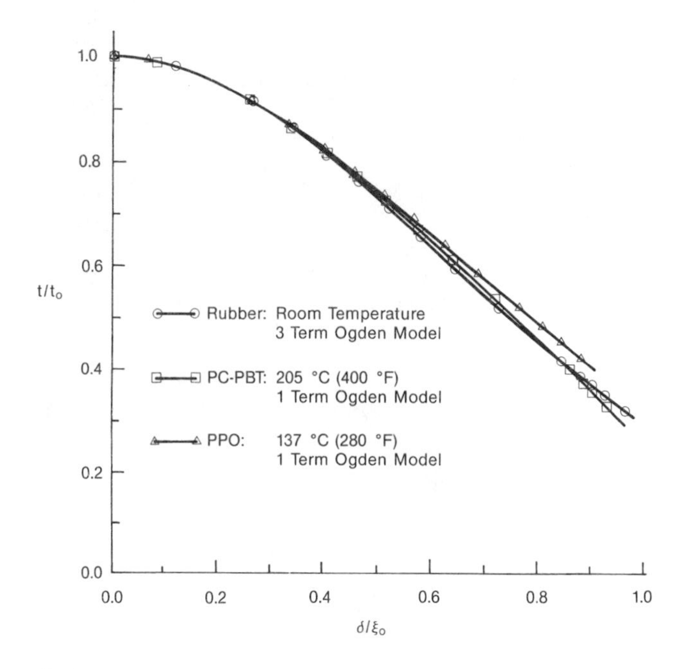

Figure 5.14 Thickness comparison at the pole for different materials during free isothermal inflation of a dome.

dramatic thickness variations result when temperature variations are taken into account. As can be seen in Figures 5.12 - 5.13, slight differences in temperature can greatly change the stress-stretch behavior. When calculations are performed taking temperature gradients into account, the variable "stiffness" of the non-isothermal polymer can cause dramatic thickness differences between non-isothermal and isothermal calculations. This effect is illustrated by an example in Section 5.4.

5.4 THERMOFORMING EXAMPLES

In this section some examples are given to illustrate the application of the finite element method to a variety of thermoforming problems. In the first example, free inflation of a flat membrane is examined for different initial shapes. The results from the finite element calculations are compared with classical analytical solutions. In the remaining examples, finite element solutions are presented for a variety of thermoforming problems, both axisymmetric and general 3-D. For the case of axisymmetric thermoforming into a cylindrical mold cavity, the computed thickness distribution is compared with experimental measurements.

5.4.1 Free Inflation of a Flat Membrane

The classical solution of membrane inflation from an initially flat circular sheet is given in [44]. In the solution presented here, the strain energy function is assumed to have the standard Mooney-Rivlin form given by Eq (52), with the constants A_{10} and A_{01} prescribed by the ratio $A_{01}/A_{10} = 0.005$.

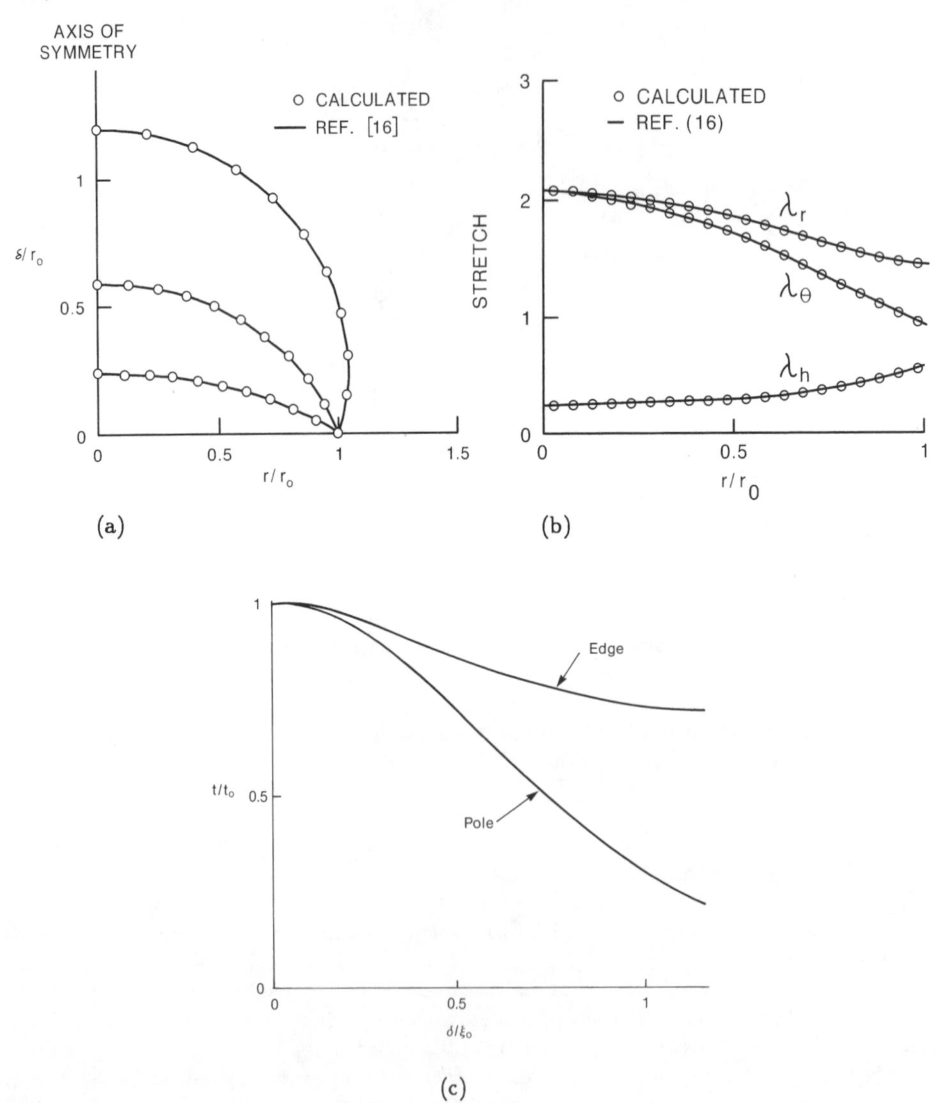

Figure 5.15 Inflation of bubble from flat circular sheet. (a) Bubble profiles at three different inflation stages. (b) Stretches in a bubble as function of initial position. (c) Thickness as function of bubble height.

To verify the results from the finite element model, the solutions to the system of nonlinear equations published in [44] were recomputed using the method of successive approximations, as outlined in [44]. Typical examples of the close comparison between the two methods were reported in [16] and are shown in Figure 5.15 where the normalized bubble profiles and stretches are plotted for a case where the initial thickness was 0.001 of the radius of the sheet. In Figure 5.15a the profile of the bubble is shown at three different pressures. In this figure r is the distance from the axis of symmetry, r_0 is the radius of the initial sheet and δ is the height of the bubble. In Figure 5.15b the radial stretch λ_r, the circumferential stretch λ_θ, and the thickness stretch λ_h are given for the bubble when $\delta/r_0 = 1.2$. Of more interest to thermoforming applications is Figure 5.15c, which shows the change in thickness at the center and at the edge of the sheet as a function of the bubble height. The major effect of the large deformation is clearly seen in the nonuniform thickness variation from the base of the bubble to the pole in Figures 5.15b and 5.15c. It is also evident from these figures that during inflation the bubble assumes a nonspherical shape. The precise shape of the circular sheet during free inflation is dictated by the ratio of the material constants, i.e., A_{01}/A_{10}.

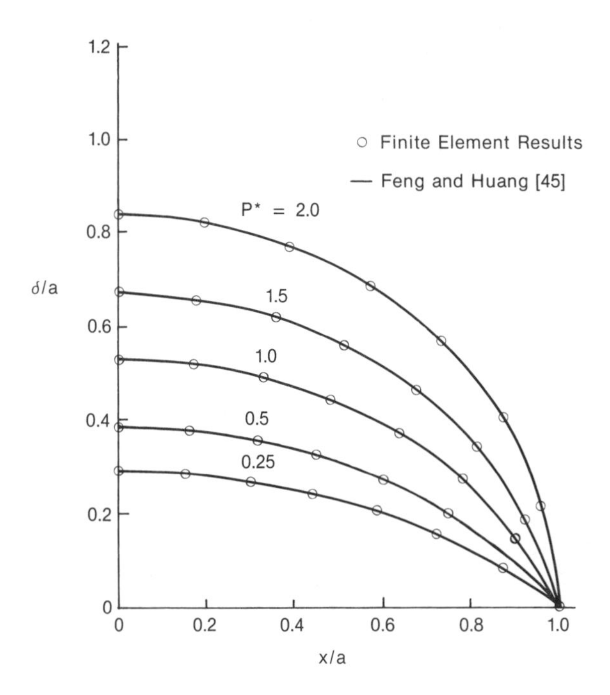

Figure 5.16 Profiles of an inflated square Mooney-Rivlin membrane. Comparison between finite element calculations and results from [45].

While the previous example was modeled with axisymmetric elements, the free inflation of a flat rectangular membrane has to be modeled with general membrane elements. The results for a square membrane are shown in Figure 5.16. In this figure, the cross section profiles of an inflated square Mooney-Rivlin membrane, at different pressures, computed from

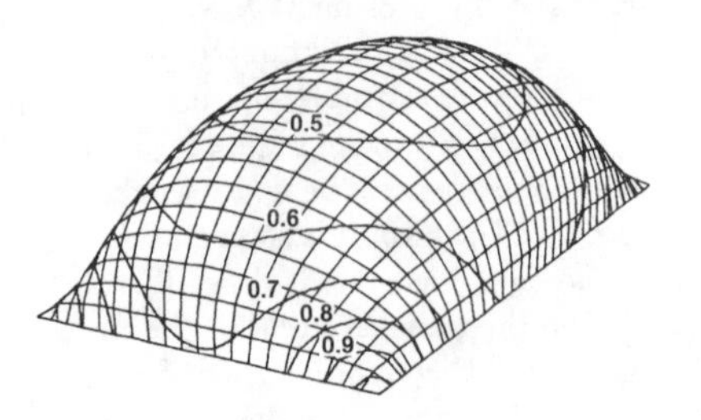

Figure 5.17 Inflated shape for a rectangular membrane.

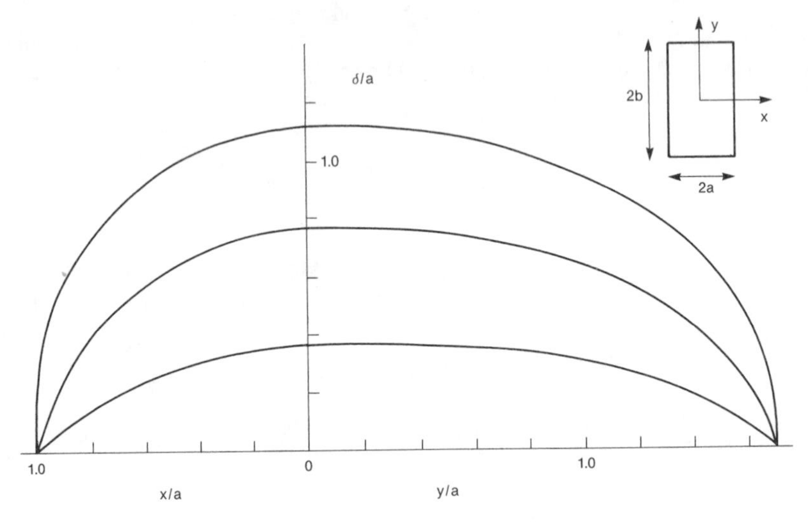

Figure 5.18a Inflation of bubble from a flat rectangular sheet. Bubble profiles in the symmetry cross sections.

the finite element program, are compared with results from [45] that were obtained using a semi-analytical iterative algorithm. In this figure δ is the membrane displacement and the non-dimensional pressure P*, is given by:

$$P^* = \frac{p\ a}{2A_{10}\ t_0} \tag{77}$$

where p is the internal pressure, 2a is the length of a side, and t_0 is the initial sheet thickness. The Mooney-Rivlin constants used in this calculation are given by $A_{01}/A_{10} = 0.2$. There is again excellent agreement between the finite element calculation and numerical results obtained using the alternative formulation [45]. Finally, the inflated shape for a rectangular membrane is shown in Figure 5.17, where the ratio of the sides of the initial rectangular sheet is given by 1:1.7. In this figure the non-dimensional center deflection $\delta/a = 1.1$, where δ is the deflection at the center of the membrane and 2a is the length of the shortest side of the

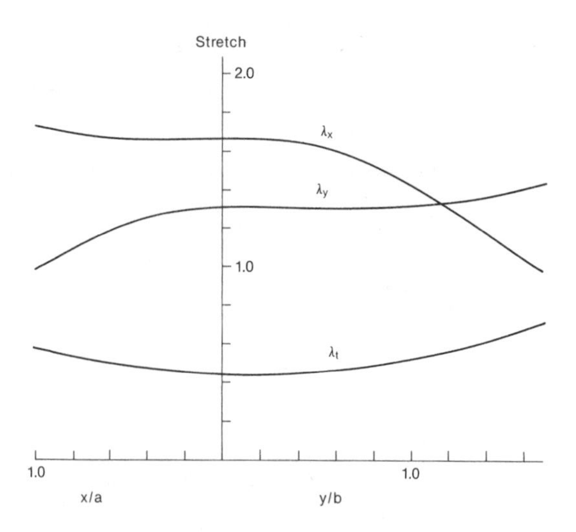

Figure 5.18b Inflation of bubble from a flat rectangular sheet. Stretches in the symmetry cross sections.

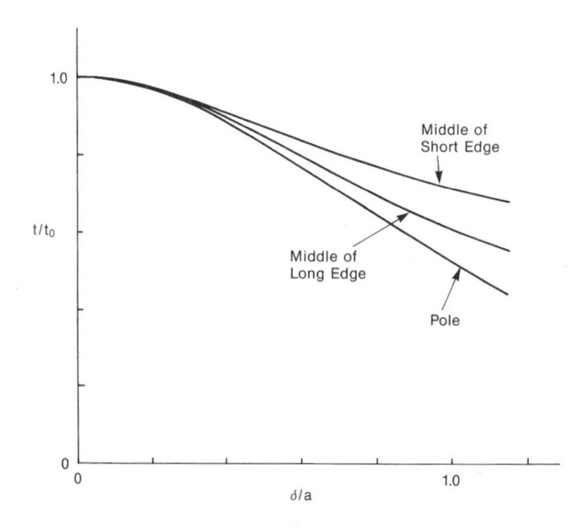

Figure 5.18c Inflation of bubble from a flat rectangular sheet. Thickness of bubble vs. bubble height at three different locations.

rectangle. The computations for this example were performed using the Mooney-Rivlin model with the material constants given by $A_{01}/A_{10} = 0.05$. The contours in Figure 5.17 represent lines of constant thickness stretch for the inflated membrane. Figure 5.18a shows the membrane cross sections in the two symmetry planes at three different inflation heights. The stretches in the symmetry cross sections are shown in Figure 5.18b and the thinning at the top of the membrane and the middle of the two edges is shown in Figure 5.18c as a function of bubble height. It is clear from Figures 5.15c and 5.18c that a polymer sheet will experience large differences in thickness even during the free inflation phase of a thermoforming process.

5.4.2 Comparison with Experimental Results for a Deep Vacuum Formed Cylinder

In this example, an analysis is performed for a vacuum formed, deep conical cylinder, with final dimensions depicted in the inset of Figure 5.19. A comparison of the analytical predictions with experimental data obtained from numerous thermoforming tests described in [46] is discussed in detail. The deep cylinders were thermoformed from high impact

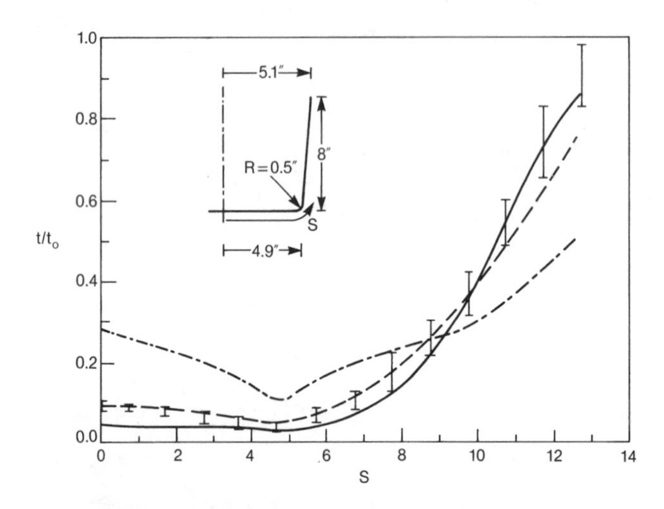

Figure 5.19 Comparison between measured and predicted wall thicknesses in thermoformed dish. (----) uniform temperature model, (—), nonuniform temperature model, (—·—·) conservation of mass calculation from [5], (I) test data: standard deviation from 32 points.

polystyrene (HIPS) and the initial thickness of the polymer sheet was 0.1 in. After thermoforming, the material thickness was measured at approximately 1 in. increments along the side and bottom of the finished part and the standard deviations of these measurements are plotted in Figure 5.19 versus the distance S measured from the axis of symmetry to the part edge (half of the axisymmetric geometry is shown in the inset of Fig. 5.19). In comparison to the experimental measurements, the non-dimensional thickness t/t_0, where t is the final thickness and t_0 is the initial sheet thickness, is plotted in Figure 5.19 for (1) an isothermal finite element analysis, (2) a non-isothermal finite element calculation, and (3) a conservation of mass prediction calculated from [5].

In the isothermal finite element analysis, the material constants for HIPS determined experimentally by Schmidt and Carley [18] were used. The form of the modified Mooney-Rivlin strain energy function they used is given by:

$$W = A_{10}(I_1 - 3) + A_{02}(I_2 - 3)^2 \tag{78}$$

where:

$$A_{10} = 20.7\,\text{lb/in}^3; \quad A_{02} = 3.2 \times 10^{-6}\,\text{lb/in}^3 \tag{79}$$

It should be pointed out that the polymer was constantly cooling during the experiments in [18], therefore Eqs (78) and (79) describe some average material behavior in the temperature range experienced during the tests. Nonetheless, this data is the only high-temperature, large-deformation material data available for HIPS in the thermoforming range.

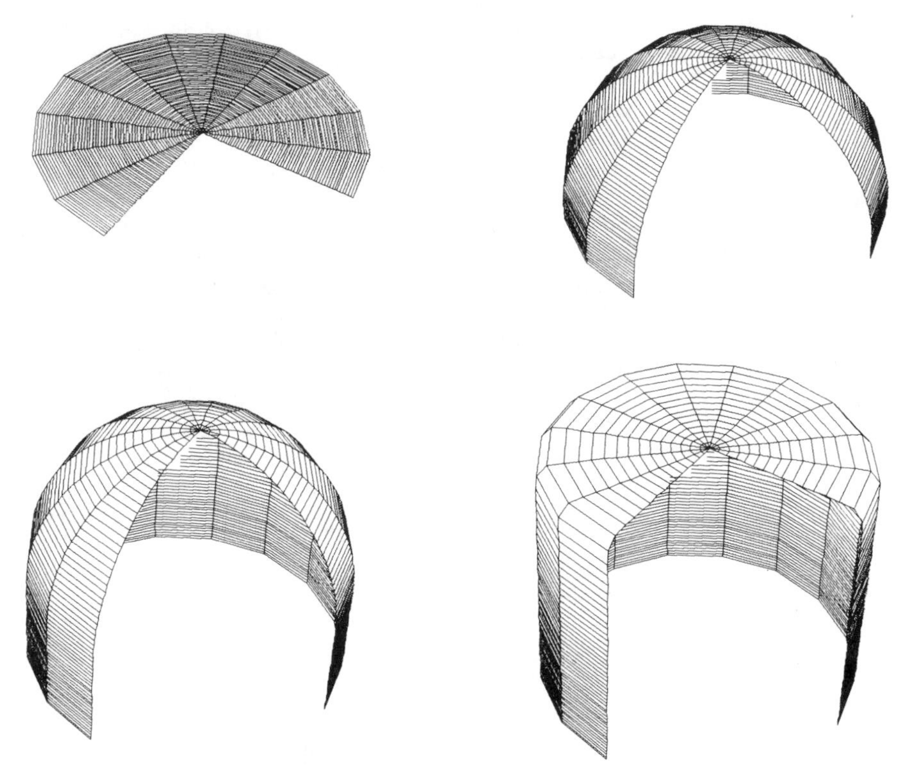

Figure 5.20 Inflation sequence for thermoformed cylinder.

As mentioned previously, in this example as well as in all the other examples in this chapter, it is assumed that the polymer, upon contact with the mold, experiences sudden cooling and/or adhesion with the mold surface, which prevents any further thinning of the polymer after contact. The finite element model used to generate the results shown in Figure 5.19, were obtained using a mesh which consisted of 60 uniformly spaced axisymmetric membrane elements.

Figure 5.20 depicts the inflation sequence as the initially flat sheet is forced into the mold cavity. As mentioned above, the analysis was performed with 2 noded axisymmetric elements, but for display purposes the model has been swept around 270°. Also, for clarity the mold surface is not shown. In Figure 5.20b, the polymer has experienced substantial inflation and has just begun to contact the sides of the mold. After the polymer has contacted the side wall over a significant distance, contact between the polymer and the top of the mold occurs at the point shown in Figure 5.20c. From this point onwards all further deformation, and thus all further thinning, occurs only to the material that is being forced into the corners of the mold. The thickness variation predicted by this isothermal temperature analysis is compared with the experimental data in Figure 5.19 as the uniform temperature model. It is seen that the thickness is underpredicted at the outside edge of the deep cylinder and tends to be a little high at the corner. In general however, the isothermal predictions are in very close agreement with the experimental measurements. It is suspected that the major reason for the discrepancy between the observed and predicted thickness variation at the edge of the sheet is most likely due to the fact that the material temperature is not uniform as assumed in the analysis, but instead decreases radially from the center of the sheet to the outermost edge. For example, in [18] radial temperature gradients of as much as 40° F were measured in similar tests, in spite of an attempt to conduct the tests under relatively isothermal conditions. The thermal gradient in the radial direction causes material at the edge of the cylindrical part to have a greater apparent stiffness than material at the center of the sheet since the material stiffness is sensitive to temperature variations as discussed in Section 5.3. In an attempt to examine the effect that an imposed thermal gradient has on the final part thickness distribution, a "non-isothermal" analysis was performed for the same geometry. The differential thermal softening was simulated by varying the degree of stiffness of the material over the initial sheet, with the softer material located at the center and the stiffer material at the outer edge. This approach assumes that heat transfer during inflation does not have a great influence on final thickness in comparison to the initial temperature distribution. Unfortunately, in the experiments performed in [46], thermal gradients were not measured. In addition, material data at strictly controlled temperature levels in the form reported by Schmidt and Carley in [18], does not exist for HIPS. Thus, in an attempt to examine a hypothetical non-isothermal situation, it was assumed that the thermal softening of HIPS could be described by an equation of the form given by Eq (76), and that a 40° F linear temperature gradient was present. Equation (76) is an empirical relationship which gives the initial slope of the uniaxial stress versus stretch behavior of modified PPO as an exponential function of temperature using an Ogden material model. The initial slope of the uniaxial stress versus stretch behavior for the modified Mooney-Rivlin form given by Eq (78), can be determined by differentiation of Eq (61). Taking the limit of this derivative as $\lambda \to 1$ yields:

$$\lim_{\lambda \to 1} \frac{d\sigma_{11}}{d\lambda} = 6A_{10} \tag{80}$$

Thus the "non-isothermal" model described here is essentially a nonhomogeneous material model in which the constants in Eq (78) are modified for each element according to Eqs (76) and (80), i.e., $6A_{10} = C\exp\{-kT\}$. On an element by element basis, A_{02} is held constant and A_{10} is varied according to the temperature at that particular position. When a 40° F linear radial temperature gradient is assumed, A_{10} increases from 7.77 psi at the center of the sheet to its base value of 20.7 psi on the edge of the sheet. The thickness variation resulting from an analysis based on this approach using 60 uniformly spaced elements is shown in Figure 5.19 as the nonuniform temperature model. This model predicts a greater wall thickness at the edge of the part and thinner corners. The non-isothermal calculation is in agreement with the measured thicknesses at the edge of the part and underpredicts the thickness in the corners and the bottom. It is interesting to note that the measured data is bracketed by the isothermal and non-isothermal solution. This seems to suggest that a calculation using actual temperature distributions (as opposed to a linear thermal gradient) would result in very accurate thickness predictions over the entire part. For comparison purposes, Figure 5.19 contains a plot of the thickness prediction obtained from a conservation of mass model which assumes that during inflation the bubble assumes a spherical shape and the bubble has a uniform thickness prior to contact with the mold surface. The conservation of mass curve was calculated numerically from the equations derived for this model by Rosenzweig et al. [5]. The conservation of mass model greatly underpredicts the vertical wall thickness of the deep cylinder and overpredicts the thickness of the corner and the bottom of the part. The reason for the discrepancy in the conservation of mass model, is that during free inflation of the bubble, there is a significant thickness variation from the base of the bubble to the pole (see Fig. 5.15c), which in turn causes an equivalent thickness variation in the final part.

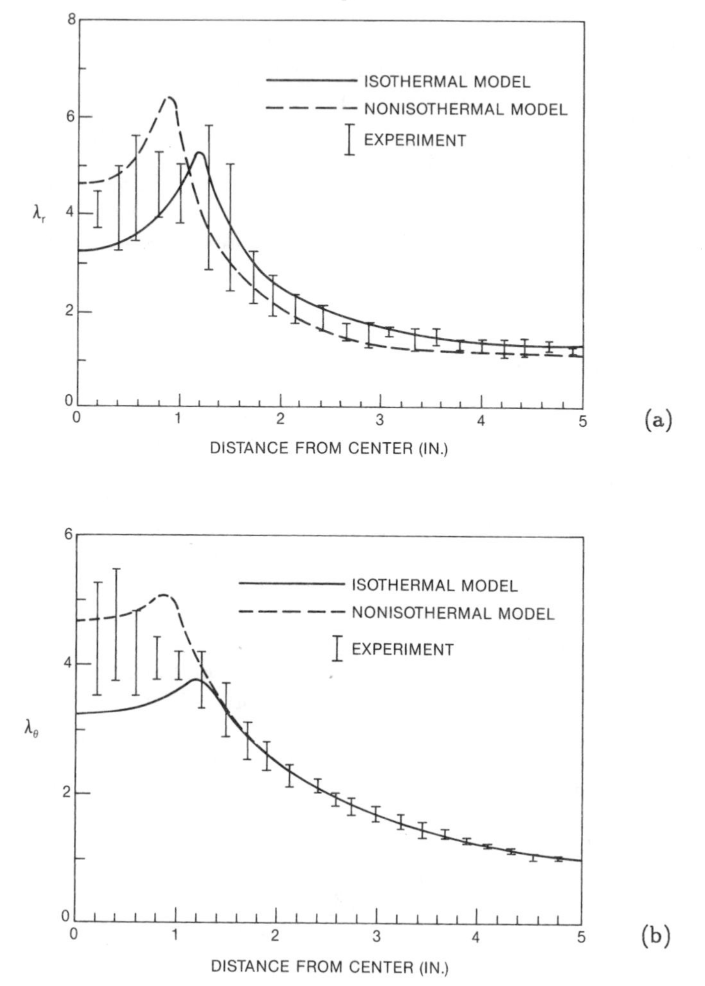

Figure 5.21 (a) Radial stretches in thermoformed cylinder. (b) Circumferential stretches in thermoformed cylinder.

The stretches in the radial and circumferential directions were obtained by measuring the deformation of circles photoetched on the initial polymer sheet. The calculated stretches for both the isothermal and non-isothermal analyses, i.e., uniform and nonuniform material models, are compared to the measured values in Figure 5.21. The measured data is represented by vertical bars which show the entire range of the data, i.e., they are not standard deviation bars. In Figure 5.21, the stretches are plotted versus distance from the center of the original undeformed sheet. It is evident from these plots that the material that ends up in the corner had its origin approximately 1 inch away from the center of the sheet. For the circumferential stretch, both models give excellent agreement with the experimental values on the vertical side of the deep cylinder, while the isothermal model underpredicts the circumferential stretch on the bottom of the cylinder and the non-isothermal model is slightly high at this location. For the radial stretch, the two models also seem to bracket the measured values. The maximum stretch values are in the range of 4 - 6 for both directions.

The agreement of the finite element model with experimental measurement is quite good, especially when it is realized that the thickness predictions are within a few percent of the experimental measurements after the polymer has been subjected to extremely large deformations. The slight discrepancies between the analysis and the test results are most likely due to (a) the absence of temperature information as a function of time at each point in the membrane during inflation, and, more importantly, (b) a lack of knowledge of the actual material behavior and the variation of this behavior with temperature. A better understanding of these two points may permit an even closer correlation with experimental results in the future.

5.4.3 Forming a Deep Cylinder with a Male Mold

This example illustrates the capability of the finite element method to model a more complex thermoforming process such as forming a deep cylinder with a male mold. The inflation sequence predicted with the finite element method is shown in Figure 5.22. The initially flat sheet is inflated to the shape of a large dome, Figure 5.22b, while the male mold is located underneath the plastic. The male mold is then raised up into the dome and the corners of the mold are brought into contact with the plastic. After the plastic has been contacted, the mold is moved further upwards, stretching the vertical sides of the polymer membrane even more (Fig. 5.22c). Finally, the pressure is taken off and the membrane snaps back draping the male mold (Fig. 5.22d). In this example, the ratio of the mold depth h to mold radius r is $h/r = 0.8$.

In Figure 5.23, a comparison is made between the thickness variation calculated for the male mold and that calculated for a female mold with otherwise identical dimensions, i.e. $h/r = 0.8$. Both isothermal calculations were performed with the material properties given in Eqs (78) - (79). It can be seen that the use of a male mold results in thinner sides and a much thicker bottom on the final finished part. If the initial bubble had been inflated to a different size before the male mold was raised, the thickness distribution would have been quite different, which illustrates how computer simulation can be used to quickly examine different techniques for forming a part and thus systematically determine the relevant processing parameters that satisfy minimum thickness criteria.

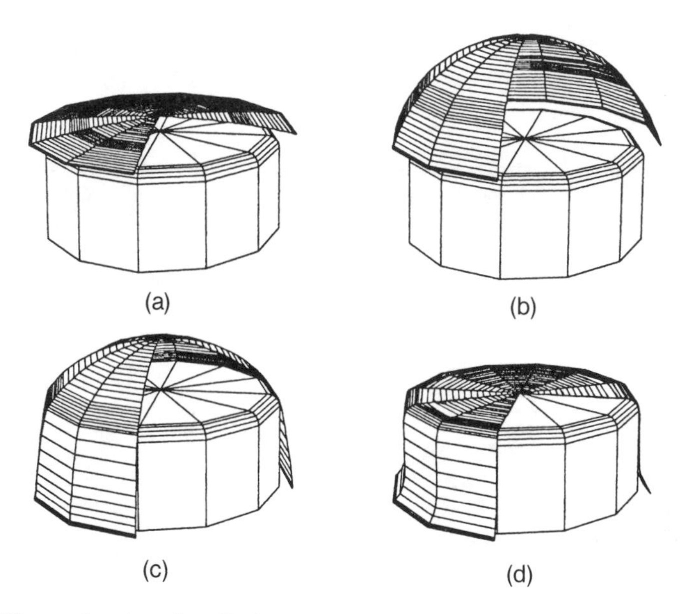

Figure 5.22 Thermoforming of a cylinder on a male mold.

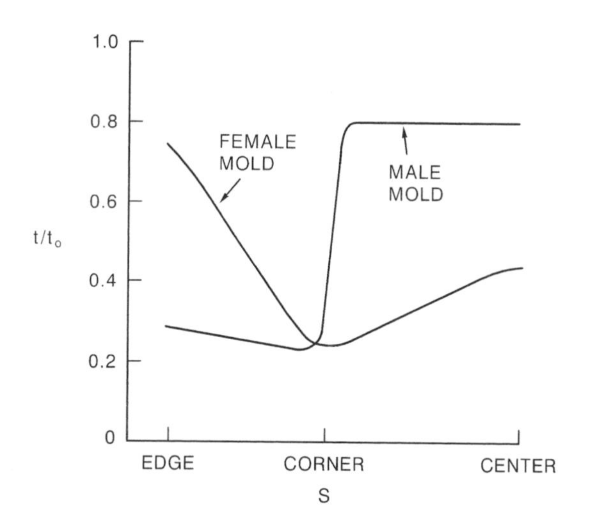

Figure 5.23 Comparison of calculated thicknesses for a cylinder formed in a female mold and with an identical male mold.

5.4.4 Thermoforming a 3-D Box

As a final example of a thermoforming simulation, the thermoforming of a rectangular box is examined. This problem is of particular interest, since the thickness variation that occurs to the polymer as it is "pushed" into a 3-D corner cannot be predicted from a two-dimensional analog. The additional constraint that arises in the 3-D corner can cause severe thinning of the polymer. The dimensions of the box mold in this example are given by a depth-width-length ratio of 0.4:1:2. Figure 5.24 contains a sequence of figures obtained from a finite element analysis. Due to symmetry, only the front half of the inflated sheet is shown as it is being formed into the box mold (mold not shown for clarity). In this particular analysis, an Ogden model for PPO at 300° F (see Fig. 5.13b) was used to simulate the polymer material behavior.

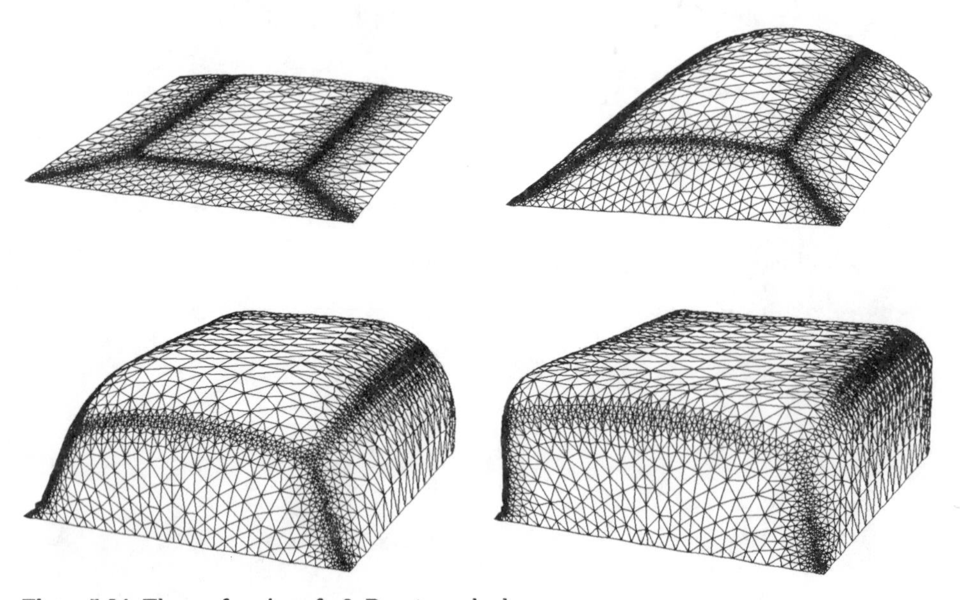

Figure 5.24 Thermoforming of a 3-D rectangular box.

Figure 5.25a contains an enlarged view of the corner of the final thermoformed box structure. Superimposed on this enlarged view of the corner geometry are contour lines which represent the thickness stretch. As can be seen, the polymer thins down in the "crease" regions of the mold cavity and experiences the most severe thinning in the corners. In this particular example, the thickness in the corner is reduced to 22% of the original sheet thickness. As would be expected, the thinning in the corner is not symmetric, with the thickness reduction more pronounced along the long edges that converge at the corner.

Figure 5.25b contains an enlarged view of one of the open corners of the box, i.e., the bottom corner in Figure 5.24d and Figure 5.25a where the plastic sheet is clamped to the mold surface. This very interesting view of the deformed plastic, shows that there is a tendency for the plastic to wrinkle due to compressive stresses in this region of the box. This wrinkling has been observed in practice when thermoforming 3-D box structures.

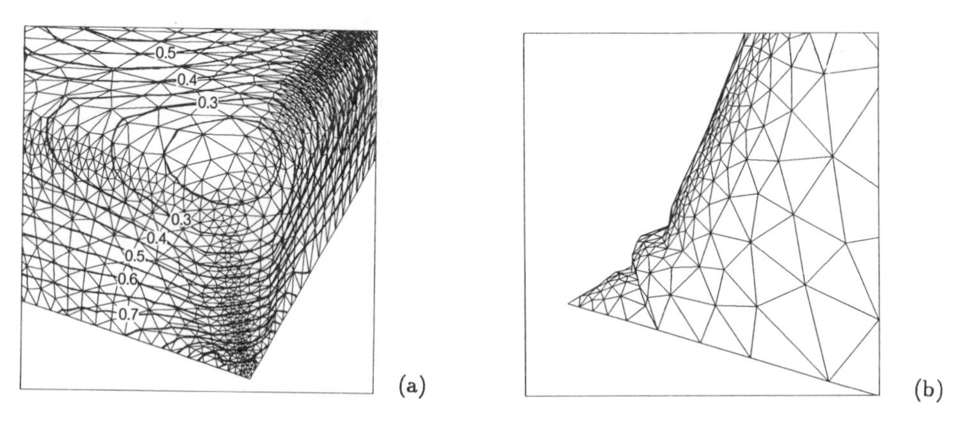

Figure 5.25 (a) Thickness contours at the corner of thermoformed rectangular box, (b) wrinkling at the open corner.

5.5 BLOW MOLDING EXAMPLES

5.5.1 Blow Molded Jar

Figure 5.26 contains a sequence of steps leading to the final wide-mouth jar shown in Figure 5.26c. Unlike the thermoforming examples, which started from an initially flat sheet, the initial configuration in this example is an axisymmetric parison shown in 3/4 view in Figure 5.26a. As the parison inflates it first contacts the bottom of the mold (mold not shown for clarity) and begins to bulge out on the sides (Fig. 5.26b). The polymer will then contact the mold on the middle of the cylindrical section and the corners will be filled out last. Since contact first occurs on the bottom of the jar with very little inflation, the bottom surface ends

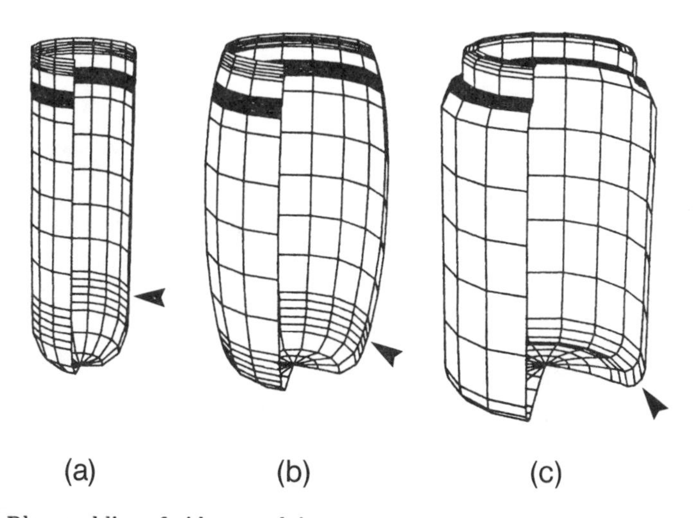

Figure 5.26 Blow molding of wide-mouth jar.

up as the thickest part of the jar. It is interesting to note, that in the final configuration shown in Figure 5.26c, the vertical wall thickness is almost uniform.

One of the most useful features of the finite element method is the ability to track material from the final configuration back to the original parison. Using this analysis tool, it becomes possible to systematically make intelligent changes in the initial parison dimensions utilizing parison programming. In this manner one can design a process and/or mold for fabrication of a specific part with specified thickness dimensions without resorting to the trial and error development process. The arrows in Figure 5.26 illustrate this material tracking capability and point to the location on the parison that was the origination site for material that ended up in the corner of the jar.

5.5.2 Blow Molding of a Rectangular Box

Figure 5.27 shows the inflation sequence for a cylindrical parison blow molded into a

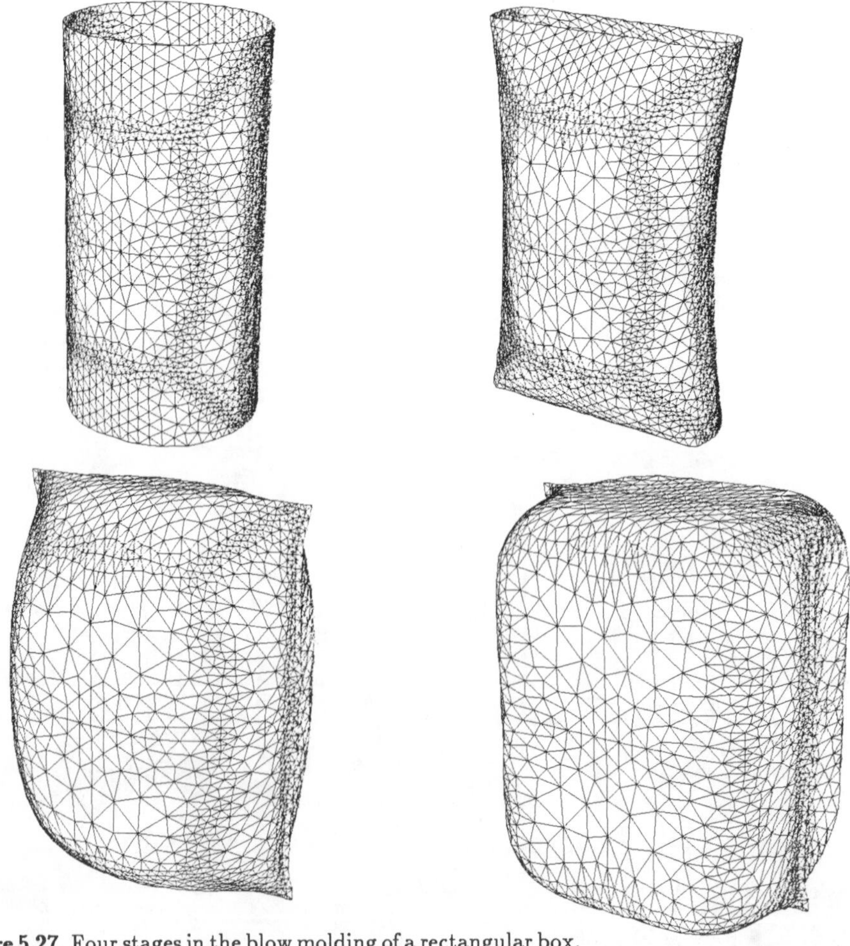

Figure 5.27 Four stages in the blow molding of a rectangular box.

rectangular box. The overall dimensions of the box are 24 x 24 x 12 in. (610 x 610 x 305 mm). The diameter of the parison is 12 in. (305 mm), and it was modeled with a 2 term Ogden material with the constants $\mu_1 = 143$, $\mu_2 = 1.4$, $\alpha_1 = 0.5$, and $\alpha_2 = 4.0$. The cylindrical parison is shown in Figure 5.27a and it is noted that the finite element mesh seems highly irregular. The mesh was constructed this way so that the material would penetrate all the way out into the corners of the box. In Figure 5.27b the mold is starting to close and it is starting to pinch off the parison at the top and bottom. During this stage a slight pressure is applied on the inside of the parison to prevent it from collapsing. In Figure 5.27c the mold has closed and the parison is starting to inflate. Both in this picture and in the finished box, Figure 5.27d, the flash on the sides of the box arising from the closing of the mold is noticeable. The thickness variation in the finished box is shown in the plot of the thickness contours in Figure 5.28. It is seen that the corners of the box are less than 30% of the original parison thickness.

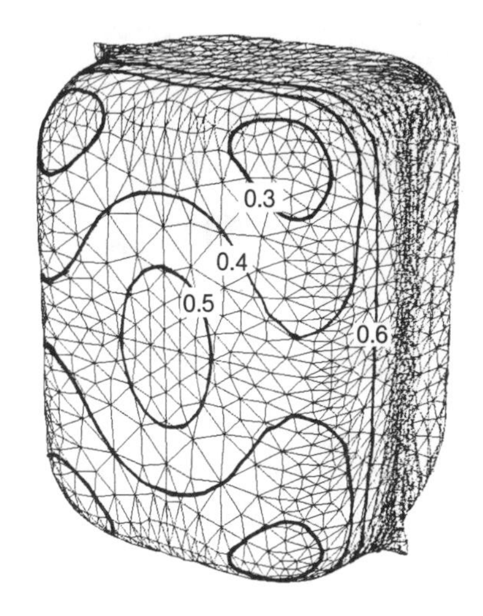

Figure 5.28 Thickness contours on finished box.

5.6 DESIGN METHODOLOGY

5.6.1 Design Philosophy

The greatest benefit of a computer simulation is obtained when it is used early in the design stage, i.e., before a part is fully designed and before the design of the mold has begun. If used at this early stage, the economic impact can be substantial. As mentioned in the introduction, the blow molding and thermoforming processes offer a number of process controls that make it possible to start with a variety of initial conditions. However, presently it is only possible to use these process controls through trial and error procedures after the design has been finalized. By changing the process conditions in a simulation program, the effects on the thickness distribution in the final part can be examined. If the thicknesses are not satisfactory, the analysis can be repeated with different tool designs and/or process parameters altered, until acceptable thicknesses are obtained. The final part thicknesses may not necessarily match the predicted part thicknesses exactly due to approximations in the analysis and uncontrollable conditions in the process. However, with help of such computer simulations, sufficiently many cases can be analyzed so that final part thicknesses will fall within acceptable tolerances. The immediate payback from performing simulations at this early stage is that critical changes can be made early in the design process resulting in savings in time and tool development costs.

Moreover, if the part is to be used in a highly stressed structural application, the wall thicknesses predicted by the simulation can be used as input to a structural analysis program. The part can then be analyzed for the expected loads and the working stresses can be predicted. If the stresses are not within the desirable range (either too high and the part will fail, or too low and material is wasted) the tooling or process conditions can be changed in the blow molding simulation. A new thickness prediction can then be made and a new stress calculation performed. Since the first design rarely will be optimal, several iterations are frequently needed before the design is finalized. The resulting part will have a thickness distribution which does not overstress the part nor waste material.

Even though the savings in tool development time and cost can be significant, the major economic payback from a computer simulation is that the part can be designed with the minimum amount of material necessary for proper structural performance. The minimum material design obtained this way could generate substantial savings in material cost over the lifetime of the mold.

5.6.2 Example of Iterative Design Procedure

In an effort to illustrate how computer simulation can be beneficially used to design blow molded containers, an axisymmetric example from injection stretch-blow molding is presented. A more detailed account may be found in [47]. The main objective in this particular example is to determine the initial preform geometry which will result in an essentially uniform thickness container. Obviously, other objectives such as, minimization of deflection due to internal or external pressure, maximum overall crushing strength, or optimal material utilization, might equally well be used as design objectives. The iterative procedure demonstrated in this particular example applies to these other design objectives as well.

Stretch-blow molding is a multistep forming process which begins with the injection molding of an initial "preform." Since the preform is formed in a confined mold, the initial overall geometry as well as the initial thickness variations can be accurately controlled. Once the preform has been molded, the forming process usually proceeds as a two stage sequence, i.e. a stretching stage and an inflation stage as illustrated schematically in Figure 5.29. In Figure 5.29a the initial preform shape is shown clamped at the mouth of the mold just prior to the stretch phase. The stretch phase, Figures 5.29b-c, is performed with the assistance of a "plug" which can take a variety of shapes. The shape of the plug and the distance of plug travel can have a dramatic effect on the final container thickness. In this illustrative example, the plug is pushed all the way to the bottom of the mold. At this point the blow molding inflation phase begins, Figure 5.29d, which results in the final container shape shown in Figure 5.29e.

Current design practice usually starts with a uniform thickness preform. This, however, results in very nonuniform container wall thicknesses, with excessive thinning in the corners. At this point, a trial and error process development begins with the aim of determining a preform thickness which will result in acceptable final wall thicknesses at the container corners. Eventually, preform and plug shapes will be developed, which will permit formation of the container, but this will, in some cases, take months of process development. It is highly unlikely, though, that this design will be optimal, in the sense that far more polymer will be used than is actually needed for a structurally sound part.

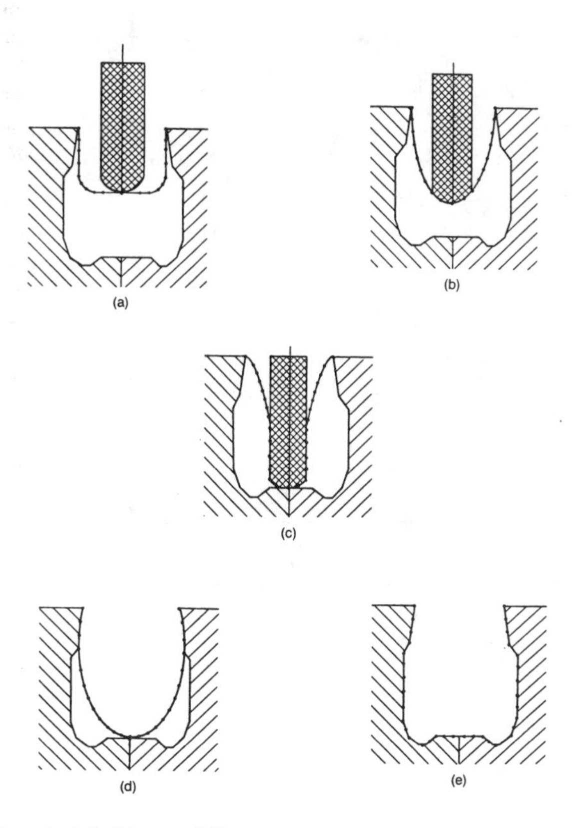

Figure 5.29 Injection stretch-blow molding sequence.

Design optimization using computer simulation proceeds in a similar manner, with the exception that successive design iterations are performed without costly process developments. A simple starting point for the design of this particular container would be the uniform thickness preform. In this example, the preform thickness is taken to be 0.200 in. (5.1 mm) and the total volume of material will be held constant at 4.3 cu. in. (1303 cm^3). Figure 5.30 shows dimensions of the preform, the plug, and the final container. For simplicity, it is assumed that all the material has a uniform temperature and that the temperature remains constant throughout the process. With the uniform thickness preform as the starting point, an analysis can be performed to predict the wall thickness in the finished container. Subsequent design iterations can now be performed by observing where excessively thinned material in the finished container came from on the original preform. The preform can then be made thicker in these locations and a new iteration started. The actual number of iterations needed to meet design objectives will depend on the complexity of the part and the process. In this example, the iterative process was stopped after 6 iterations and Figure 5.31 shows the dimensions of the final preform. The thicknesses vary from 0.093 in. (2.4 mm) at the center of

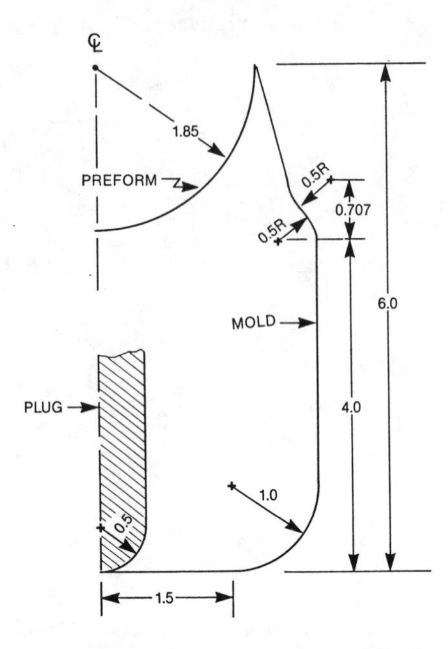

Figure 5.30 Preform, plug and final part dimensions (all dimensions in inches, 1 in. = 25.4 mm).

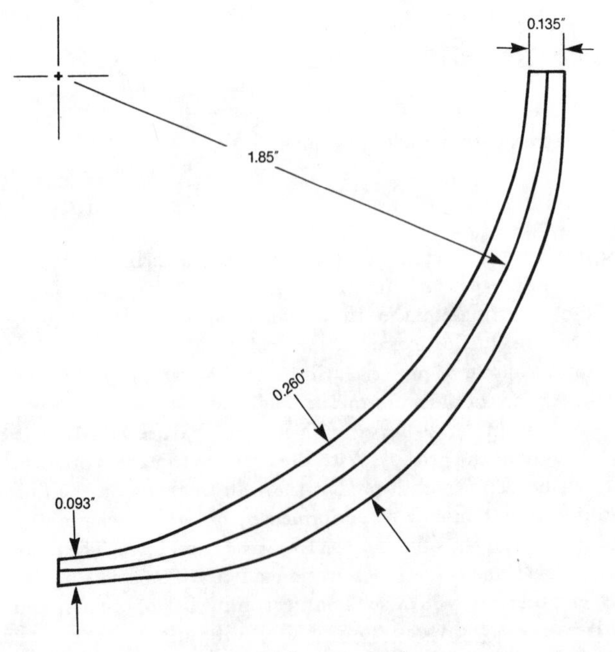

Figure 5.31 Preform shape for uniform thickness container.

the preform to a maximum thickness of 0.260 in. (6.6 mm) on the preform arc and then back down to a thickness of 0.135 in. (3.4 mm) where the plastic preform is clamped over the mold mouth. A comparison of the thickness distributions arising from the initial and final preforms at two stages in the process is shown in Figures 5.32 - 5.33. Figure 5.32 shows the thickness variation at the end of the stretch phase when the plug touches the bottom of the mold. In Figure 5.33, it can be seen that the final container wall has a more or less uniform thickness averaging 0.040 in. (1 mm) except at the bottom center of the container (location 12). The

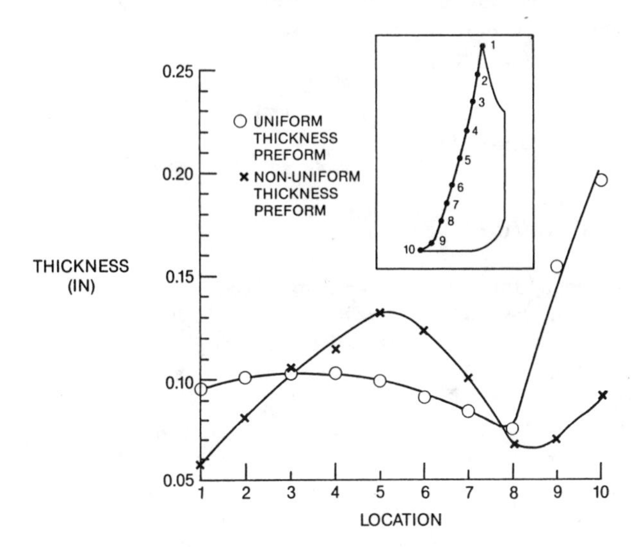

Figure 5.32 Thickness vs. location after stretch part of injection stretch-blow molding process.

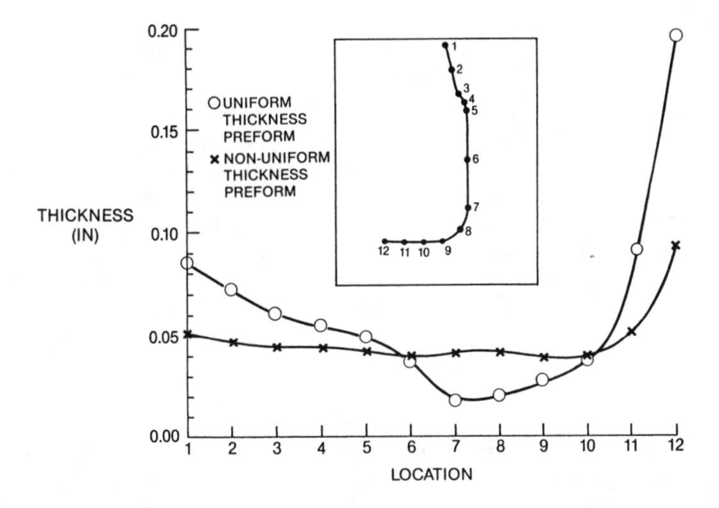

Figure 5.33 Final thicknesses vs. location on stretch-blow molded container.

bottom is thicker due to the fact that the plug is pushed all the way to the bottom of the mold and the polymer is assumed not to slide over the plug. Making the very bottom of the container thinner would require modification of the process, by either changing the plug geometry or distance of travel. In contrast, the wall thickness using the initial preform is very nonuniform with a minimum thickness of 0.017 in. (0.4 mm) at location 7, well below an acceptable level. It must be emphasized that this example utilized the assumption of a uniform temperature distribution whereas in practice this may not be the case. The incorporation of nonuniform temperatures is likely to yield different final preform dimensions.

Though this simple example did not examine other process variables such as, nonuniform temperatures and variable plug shapes, such information could be incorporated into a computer simulation as well. For example, if the plastic's stretching behavior is accurately known as a function of temperature, calculations with nonuniform thickness and temperatures could be performed simultaneously.

5.7 SOME UNRESOLVED ISSUES

It is readily acknowledged that our understanding of the "process physics," which governs both thermoforming and blow molding, is at best incomplete. Extensive experimental studies remain to be conducted to accurately characterize these processes and verify the efficacy of the assumptions used to construct the computer simulation model outlined in this chapter. For example, one of the fundamental simplifications made in this chapter is the assumption that the polymer can be treated as a nonlinear rubberlike material. Because of the relatively high speeds with which these processes are performed, it is probably a good approximation in many cases. In processes where the speed is relatively low, viscous effects should be taken into account. This is, for example, the case in blow molding during the period of time between parison extrusion and inflation, while the parison is hanging under its own weight. If time-dependent effects are to be taken into account, the polymer must be modeled as a viscoelastic material. For small strains, the mechanical behavior of many polymers can be characterized by so-called linear viscoelasticity. Unfortunately, for large deformations, the relationship between time-dependent stress and strain deviates significantly from this simple linear relationship. Therefore, it is not possible to use traditional linear viscoelastic material descriptions as outlined in [48-50]. Many nonlinear viscoelastic models have been developed. These models have been based on either (1) case specific empirical relations, (2) a continuum mechanics approach, or (3) a molecular theory approach [27]. At present, the continuum mechanics approach seems to offer the best framework for the development of constitutive relationships which capture the most important features of multiaxial polymer deformation. However, a rigorous multiaxial formulation based on nonlinear viscoelastic constitutive equations, which does not assume that the strain and time dependence of the response are separable, results in a multiple integral representation which would require measurements from an exceedingly large number of biaxial experiments [27,50]. The magnitude of the effort required to rigorously characterize even a single material, makes such a formulation unsuitable for practical use. Simpler single integral viscoelastic formulations, such as the BKZ-model [50], offer the potential of incorporating some, but not all, viscoelastic effects into thermoforming and blow molding simulation. However, at present there is insufficient experimental data to permit the application of even these constitutive models to a wide class of materials and forming processes.

In the applications given in this chapter, it has been assumed that the polymer sticks to the mold surface as soon as it touches. This assumption seems to agree with most observations.

As mentioned in the introduction, when the polymer contacts the mold, it starts to cool down, and since the stiffness of the polymer is highly temperature sensitive in the range where the forming takes place, the cooler polymer is stiffer than the hot polymer which has not yet contacted the mold. It is hypothesized that this cooling helps to prevent any further thinning of the polymer that has contacted the mold. In addition, polymers have a certain tackiness at elevated temperatures which tends to make the polymer stick to the mold. Nevertheless, in some applications it has been observed that the polymer can slide along a mold wall in situations where there is highly localized stretching. For example, in plug assist forming, the polymer membrane has been observed to slide around the corners of the plug when the stretch is sufficiently large. A natural extension of the present formulation would be to allow this sliding to occur when the forces attain a certain threshold. However, such a formulation would require knowledge about the friction coefficient between the hot polymer and the cooler mold surface. This information is presently not available and the tests which would be required to obtain this information would be difficult to perform. Without this experimental knowledge, the friction behavior could only be included in a qualitative way. However, it is expected that including sliding will only make a difference in membrane thickness in a limited number of applications, and then only in locally highly stretched areas.

The finite deformation of nonlinear elastic membranes leads to instabilities which arise if the loading parameter is continuously increased. This can be seen even in simple tension where the load P, for a single term Ogden expansion, is given in terms of stretch λ by:

$$P = A_0 \mu \left[\lambda^{\alpha-1} - \lambda^{-\left[\frac{\alpha}{2}+1\right]} \right] \tag{81}$$

where A_0 is the original cross-sectionaal area. Equation (81) shows that it is possible to have extremum values in the P vs. λ relationship. The extremum values of P can be obtained by setting the derivative of Eq (81) with respect to λ equal to zero. Rearranging terms we find the critical λ at which an extremum point occurs given by:

$$\lambda = \left[\frac{\left[\frac{\alpha}{2} + 1\right]}{(1-\alpha)} \right]^{\frac{2}{3\alpha}} \tag{82}$$

A similar relationship can be obtained for equibiaxial stretching:

$$\lambda = \left[\frac{(2\alpha + 1)}{(1-\alpha)} \right]^{\frac{1}{3\alpha}} \tag{83}$$

It is interesting to note that for these special cases, and for positive values of α, the applied force will have a maximum at a finite value of λ for $0 < \alpha < 1$. This implies that for these cases an instability will occur, i.e. at $\lambda_{critical}$ a small increase in load P will result in a large increase in λ. For values of $\alpha > 1$ the character of the solution changes in these examples and instead of a maximum, the slope $dP/d\lambda$ remains positive for all values of stretch. It should be noted that these observations cannot be generalized to other loading situations. For example, Ogden [31] shows that for the pressure inflation of a sphere and for a single term strain energy function, the relationship between pressure P and stretch λ is given by:

$$\frac{P\, r_0}{2h_0} = \mu \left[\lambda^{\alpha-3} - \lambda^{-(2\alpha+3)} \right] \tag{84}$$

where r_0 and h_0 are the dimensions of the undeformed radius and initial membrane thickness, respectively. Figure 5.34 contains a plot of this pressure vs. stretch relationship for $\mu = 7$ psi and $\alpha = 2$ (for a sphere $\lambda = r/r_0$). Equating the derivative of Eq (84) to zero yields:

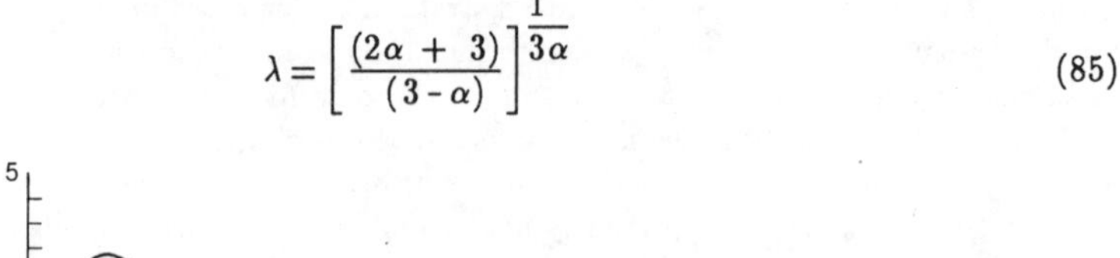

$$\lambda = \left[\frac{(2\alpha + 3)}{(3 - \alpha)} \right]^{\frac{1}{3\alpha}} \tag{85}$$

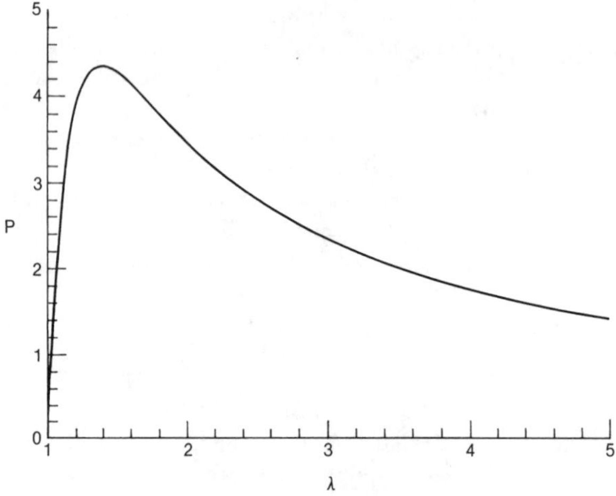

Figure 5.34 Normalized pressure vs. radial expansion for a spherical bubble modeled with a single term Ogden expansion. $\mu = 7$ psi, $\alpha = 2$.

Thus, it can be seen from Eq (85), that for the inflation of a spherical membrane, a maximum for the pressure exits for values of α within the range $-3/2 < \alpha < 3$. Therefore, in this case, stable solutions can only be assured for values of $\alpha > 3$. However, such large values of α seem to be physically unrealistic. Experimental observations on the inflation behavior of neoprene rubber balloons [41] indicate that as a balloon is inflated, the applied pressure increases to a maximum as the balloon initially expands and that further inflation is achieved at a reduced pressure. Eventually, at large stretches a minimum pressure is reached and then the pressure increases until the balloon bursts. This suggests that for rubber, no single-term strain energy function of the Ogden type is sufficient to completely describe the sinusoidal pressure vs. radial expansion curve observed experimentally for the inflation of a sphere. The important point to be made, is that a pressure peak is observed during inflation of a membrane and if the volume increase is sufficiently great, a pressure drop takes place while the membrane is expanding. Accurate pressure measurements have confirmed that such a pressure drop also occurs during thermoforming of plastics. This pressure drop during inflation can cause difficulty with numerical algorithms that rely on incremental increases in pressure to drive the numerical simulation, since large increases in λ will suddenly occur for a small pressure increase in the neighborhood of the pressure peak (Fig. 5.34). Alternative numerical algorithms have been developed, based on increasing volume in a controlled manner, to obtain stable solutions for these situations.

In addition to the instabilities that arise due to large increases in deformation with small increases in pressure, other types of instabilities can occur in nonlinear elasticity when the solution suddenly bifurcates during inflation. This behavior is analogous to buckling phenomena. The simplest example of this kind of physical instability is the inflation of an elongated balloon. At first the balloon is very hard to blow up, but suddenly one part of it starts to inflate creating two distinct regions, one with a localized bubble. The balloon is now much easier to blow up further, and during this process the localized bubble propagates axially until it extends along the length of the balloon. A theoretical study of this phenomenon is given in [51-52]. The geometries that result from such physical instabilities are not necessarily symmetric, even when the initial geometry is symmetric. For example, it has been observed that a spherical balloon can bifurcate to a nonsymmetrical shape [41]. Similar instability phenomena can also occur in blow molding during the inflation phase and may have been observed experimentally in [8-10] where sudden bulges appear on the middle of a parison right before it is expanded all the way out to the mold. Since the physical instabilities just described are associated with relatively large shape changes for small variations in inflation pressure, these types of instabilities invariably lead to numerical instabilities in a finite element analysis. Methods which have been used successfully for buckling of structures [53-59] may be used for solving some classes of these instability problems, but it is expected that the class of instabilities which may occur in polymer inflation problems is actually much larger than can be solved with present numerical techniques and that new numerical methods might have to be developed.

While the above instability phenomena are related to geometric instabilities, there is another class of problems which behave similarly, but are related to the non-uniqueness of the material response during large deformations. It has been shown that nonlinear elastic materials can have a multitude of deformation solutions for the same stress state [60-63]. For example, a unit cube exposed to a certain loading may have up to 7 different displacement solutions. Which of these multiple solutions are obtained in an analysis depends on the loading sequence [62-63]. Polymers for large strain rates and elevated temperature behave much like rubber and it is therefore likely that can have multiple deformation states for a specific loading. If these multiple solutions come into play in a thermoforming or blow molding process, it is possible that just small changes in the process parameters may lead to large differences in polymer thickness distributions. In the numerical treatment, a whole host of new numerical problems could arise. However, it is still too early to tell whether these effects will become major problems.

Another subject that has not been fully addressed in this chapter is the influence of temperature gradients along the polymer membrane. Except for the deep dish thermoforming example 5.4.2, it has been tacitly assumed that all polymers have been at a uniform temperature throughout the process. The difference in material stiffness for even small changes in temperature can be drastic as seen from the discussions in the materials section. This has the effect that any temperature gradients along the surface of the polymer can have a major effect on the deformation of the polymer and, therefore, on the final thickness distribution. In fact, the temperature gradient is probably the single most important effect to include in an expanded finite element formulation. However, the temperature distribution is rarely, if ever, known. If, however, the temperature distribution in a particular process were known, it could be utilized in an analysis by using this information as input. A more challenging approach would be to perform a transient thermal analysis simultaneously with the inflation analysis. This would, however, require a better knowledge of the conductive and radiative heat transfer for the hot polymer during the entire process.

From a user viewpoint, one of the major obstacles in performing a finite element analysis is the effort required for generating a finite element mesh. Several commercial mesh generation packages are available for this work, but in a thermoforming or blow molding analysis it is generally impossible to know a priori how a mesh should be constructed so that it is refined in the appropriate locations. Mesh refinement should be performed where the polymer gets most highly deformed. Typically, this is where a mold has geometric complexities such as edges and corners. It is, however, impossible to predict which part of the polymer ends up in these locations and, therefore, the mesh refinement cannot be done with precision before an analysis has been performed.

Presently, two different approaches may be used for generating meshes. In one approach an initial analysis is performed with a relatively coarse mesh. This analysis supplies the information about where different parts of the polymer ends up contacting the mold. A second mesh is then constructed with the necessary mesh refinement in the appropriate places. In the second approach, the finite element mesh is constructed in such a manner that it automatically meets the mesh refinement requirements everywhere. The first approach requires the construction of two finite element meshes and is, therefore, time-consuming for the analyst. The second approach results in very large finite element models and might exhaust the computing resources of all but the largest computers. It is, therefore, desirable to couple an adaptive mesh refinement scheme to the analysis so that the mesh can be continually refined where necessary as the analysis progresses. Adaptive meshing is presently an active area of development for 2-D and to a limited extent also for 3-D analyses. The application of adaptive meshing to continuously deforming curved surfaces in space is, however, a field where little work has been done. The immediate application of adaptive meshing to thermoforming and blow molding analyses, therefore, is highly unlikely. A somewhat less ambitious approach, which will reduce the applied time for the analyst and not overly tax computer resources, is possible. In this approach, the analyst would start with a coarse finite element mesh. The calculations would proceed until some predefined criterion signals that the mesh needs to be refined at a certain location. At this point the analysis would automatically be stopped, the original starting mesh would be refined in this particular area, and the analysis would be restarted from the beginning. This process can be repeated several times, each time starting a new analysis from the beginning. Even though this method uses more computer time than the first method described above, it would free the analyst from the work of generating a second refined finite element model.

5.8 CONCLUSIONS

Finite element simulations of thermoforming and blow molding processes have only been applied over the last few years, and as shown in the present chapter, then only to relatively simple geometries. The method has, however, the potential to have a major impact on the way new parts and processes are designed and could lead to drastic savings in material costs through design optimization. The next few years should see a flurry of activity in this area with several special purpose finite element programs emerging and some of the problems discussed in Section 5.7 on unresolved issues will undoubtedly be solved. The progress along this path should be a challenge to any numerical analyst and should greatly benefit the thermoforming and blow molding industry.

ACKNOWLEDGEMENT

The authors gratefully acknowledge the support provided by General Electric's Plastic Business Group, Pittsfield, MA and Major Appliance Business, Louisville, KY. We would like to note the contribution of members of the staff of General Electric's Plastic Application Center, Louisville, KY, who performed key experiments, in an effort to accurately characterize thermoforming behavior. Thanks are also due to Mr. L.P. Inzinna at General Electric's Corporate Research and Development Center for performing numerous experimental measurements on polymers at high temperatures. Finally, the authors would like to especially acknowledge the contributions of Mr. C.A. Taylor, GE's Plastic Business Group, Pittsfield, MA, who assisted in the development and coding of the finite element program and who generated the solutions for many of the example problems.

REFERENCES

1. "Modern Plastics Encyclopedia," 1984-85, McGraw-Hill Inc., New York, NY (1984).
2. Throne, J.L., "Thermoforming," Hanser Publishers, New York (1986).
3. Sheryshev, M.A., Zhogolev, I.V., Salazkin, K.A., *Soviet Plast.*, No. 11, 30 (1969).
4. Tadmor, Z., Gogos, C.G., "Principles of Polymer Processing," John Wiley & Sons, New York (1979).
5. Rosenzweig, N., Narkis, M., Tadmor, Z., *Polym. Eng. Sci.*, *19*, 946 (1979).
6. Crawford, R.J., Lui, S.K.L., *Europ. Polymer J.*, *18*, 699 (1982).
7. Williams, J.G., *J. Strain Anal.*, *5*, 49 (1970).
8. Ryan, M.E., Dutta, A., *Polym. Eng. Sci.*, *22*, 1075 (1982).
9. Dutta, A., Ryan, M.E., *Polym. Eng. Sci.*, *24*, 1232 (1984).
10. Ryan, M.E., Dutta, A., *Polym. Eng. Sci.*, *22*, 569 (1982).
11. Coleman, B.D., *Proc. R. Soc. London*, *A306*, 449 (1968).
12. Wineman, A.S., *J. Non-Newtonian Fluid Mech.*, *4*, 249 (1978).
13. Allard, R., Charrier, J.-M., Ghosh, A., Marangou, M., Ryan, M.E., Shrivastava, S., Wu, R., *J. Polym. Eng.*, *6*, 363 (1986).
14. Charrier, J.-M., Shrivastava, S., Wu, R., *J. Strain Anal.*, *22*, 115 (1987).
15. Nied, H.F., deLorenzi, H.G., *SPE Tech. Papers*, *33*, 418 (1987).
16. deLorenzi, H.G., Nied, H.F., *Computers Struct.*, *26*, 197 (1987).
17. Schmidt, L.R., Carley, J.F., *Int. J. Eng. Sci.*, *13*, 563 (1975).
18. Schmidt, L.R., Carley, J.F., *Polym. Eng. Sci.*, *15*, 51 (1975).
19. DeVries, A.J., Bonnebat, C., Beautemps, J., *J. Poly. Sci.: Polym. Symp.*, *58*, 109 (1977).
20. Burnett, D.S., "Finite Element Analysis: From Concepts to Applications," Addison-Wesley, Reading, Massachusetts (1987).
21. Zienkiewicz, O.C., "The Finite Element Method," McGraw-Hill, London (1977).
22. Bathe, K.J., "Finite Element Procedures in Engineering Analysis," Prentice-Hall, Englewood Cliffs, NJ (1982).
23. Oden, J.T., "Finite Elements of Nonlinear Continua," McGraw-Hill, New York (1972).
24. Truesdell, C., Toupin, R., "The Classical Field Theories," in Handbuch der Physik (Edited by S. Flugge), Springer, Berlin, (1960).
25. Eringen, A.C., "Nonlinear Theory of Continuous Media," McGraw-Hill, New York (1962).

26. Green, A.D., Adkins, J.E., "Large Elastic Deformations," Oxford University Press (1960).
27. Ward, I.M., "Mechanical Properties of Solid Polymers," 2nd Ed., John Wiley & Sons, New York (1983).
28. Green, A.E., Zerna, W., "Theoretical Elasticity," Oxford Press, (1954).
29. Treloar, L.R.G., *Proc. R. Soc. London, A351*, 301 (1976).
30. Stokes, V.K., Nied, H.F., "Constitutive Modeling for Nontraditional Materials," AMD-Vol. 85, The American Society of Mechanical Engineers (1987).
31. Ogden, R.W., *Proc. R. Soc. London, A326*, 565 (1972).
32. Blatz, P.L., Ko, W.L., *Trans. Soc. Rheol.*, *6*, 223 (1962).
33. Meissner, J., *Polym. Eng. Sci.*, *27*, 537 (1987).
34. Meissner, J., Raible, T., Stephenson, S.E., *J. Rheol.*, *25*, 1 (1981).
35. Munstedt, H., Middleman, S., *J. Rheol.*, *25*, 29 (1981).
36. Hoover, K.C., Tock, R.W., *Polym. Eng. Sci.*, *16*, 82, (1976).
37. Denson, C.D., Hylton, D.C., *Polym. Eng. Sci.*, *20*, 535 (1980).
38. Rhi-Sausi, J., Dealy, J.M., *Polym. Eng. Sci.*, *21*, 227 (1981).
39. Chung, S. C.-K., Stevenson, J.F., *Rheol. Acta*, *14*, 832 (1975).
40. DeVries, A.J., Bonnebat, C., *Polym. Eng. Sci.*, *16*, 93 (1976).
41. Alexander, H., *Int. J. Eng. Sci.*, *9*, 151 (1971).
42. Nied, H.F., Stokes, V.K., Ysseldyke, D.A., *Polym. Eng. Sci.*, *27*, 101 (1987).
43. Goldsmith, J., "High Temperature/High Speed Tensile Testing of Amorphous and Crystalline Polymers," Masters Thesis, University of Louisville (1987).
44. Adkins, J.E., Rivlin, R.S., *Phil. Trans.*, *A244*, 505 (1952).
45. Feng, W.W., Huang, P., *J. Applied Mech.*, 767 (1974).
46. Wooldridge, J.M., "Private Communication of Experimental Results," GE Louisville, January, 1986.
47. deLorenzi, H.G., Nied, H.F., Taylor, C.A., *SPE Tech. Papers*, *34*, 797 (1988).
48. Findley, W.N., Lai, J.S., Onaran, K., "Creep and Relaxation of Nonlinear Viscoelastic Materials," North-Holland, Amsterdam (1976).
49. Christensen, R.M., "Theory of Viscoelasticity," Academic Press, New York (1982).
50. Lockett, F.J., "Nonlinear Viscoelastic Solids," Academic Press, London (1972).
51. Chater, E., Hutchinson, J.W., "Phase Transformations and Material Instabilities in Solids," Academic Press (1984).
52. Yin, W.-L., *J. Elasticity*, *7*, 265 (1977).
53. Riks, E., *J. Appl. Mech.*, Trans. ASME, 1060, December, 1972.
54. Riks, E., *J. Pressure Technol.*, *109*, 33 (1987).
55. Crisfield, M.A., *Computers Struct.*, *13*, 55 (1981).
56. Crisfield, M.A., Int. *J. Numerical Methods Eng.*, *19*, 1269 (1983).
57. Bellini, P.X., Chulya, A., *Computers Struct.*, *26*, 99 (1987).
58. Bathe, K.J., Dvorkin, E.N., *Computers Struct.*, *17*, 871 (1983).
59. Duffett, G.A., Reddy, B.D., *Computer Methods in Applied Mechanics and Engineering*, *59*, 179 (1986).
60. Rivlin, R.S., *Quart. Appl. Math.*, 265 (1974).
61. Gurtin, M.E., "Topics in Finite Elasticity," Chapter 14, Soc. Ind. Appl. Mathematics (1981).
62. Tabaddor, F., *Computers Struct.*, *26*, 33 (1987).
63. Tabaddor, F., *Rubber Chem. Technol.*, *60*, 957 (1987).

REFERENCES FOR FURTHER READING

Nied, H.F., Taylor, C.A., deLorenzi, H.G., "Mechanics of Plastics and Composites," AMD-Vol. 104, Stokes, U.K., ed., ASME, 257 (1989).

Zamani, N.G., Watt, D.F., Esteghamatian, M., *Int. J. Numerical Methods Eng.*, *28*, 2681 (1989).

Taylor, C.A., "Structural Plastics '90", Proceed. 18th Annual Confer., Soc. Plast. Ind., 41 (1990).

CHAPTER 6

SIMULATION OF THE PACKING AND COOLING PHASES OF THERMOPLASTICS INJECTION MOLDING

by D. Huilier and W.I. Patterson

Ecole d'Application des Hauts Polymeres/Institut Charles Sadron
Universite Louis Pasteur
4,rue Boussingault
67000 Strasbourg
FRANCE

ABSTRACT
6.1 INTRODUCTION
6.2 MODELING THE INJECTION MOLDING PROCESS
6.2.1 Filling Modules in CAD/CAM/CAE Systems
6.2.2 Packing-Cooling in CAD/CAM/CAE Systems
6.2.3 Literature Review of Packing-Cooling Modeling
6.2.4 Problem Formulation
6.2.5 Definition of the Packing and Cooling Phases
6.3 A DIFFERENT MODELING APPROACH: PACK1 AND PACK2
6.3.1 Pack1
6.3.2 Computational Results: Pack1
6.3.3 Pack2: A Modification of Pack1
6.3.4 Computational Results: Pack2
6.3.5 Summary
6.4 EXTENSIONS AND DEVELOPMENTS
6.4.1 Physics and Thermodynamics
6.4.2 Modeling and Industrial Practice
6.5 CONCLUSION
ACKNOWLEDGEMENTS
NOMENCLATURE
REFERENCES
APPENDIX

ABSTRACT

The injection molding process can be divided into three stages: filling, packing and solidification. In this chapter only the latter two stages are discussed. These are of prime importance as far as performance of the finished product are concerned; the packing and solidification determines the final dimensions (i.e. tolerances), as well as the morphology and anisotropy of the material. The developed modeling procedure is based on the finite difference method with the staggered mesh. The mass, momentum and energy equations are solved assuming, on the mesh element scale, a quasi-steady state incompressible Hele-Shaw creeping flow, controlled by the pressure computed from the thermodynamic equation of state (Tait model). The heat losses are assumed to occur via thermal diffusivity and heat conduction in the thickness direction. Onset of the solifification takes place once the polymer temperature reaches either the glass or the crystallization temperature. When the pressure reaches the atmospheric the local variation of density (i.e. shrinkage) can be computed from the equation of state. For simple mold geometries the simulation of the pressure, velocity and temperature distribution during packing and solidification stages agreed quite well with the experimental data.

6.1 INTRODUCTION

The molding of thermoplastics consists of injecting a hot polymer whose temperature, T_{melt}, is much higher than its glass transition temperature T_g (or the crystallization temperature T_c), into the cavity of a mold regulated at a temperature T_{mold} much lower than T_g or T_c. The injected melt progressively solidifies during the molding cycle due to heat transfer.

Historically, the injection molding process has been divided into three phases: filling, packing and solidification. The first, or filling, phase occurs at a pressure only sufficient to cause the viscous melt to completely occupy the mold cavity. It is then necessary to force additional material at high pressure into the cavity to compensate the significant thermal contraction of the polymer melt ($\alpha = -1/V_s\,(\partial V_s/\partial T)_p \approx -10^{-4}\ ^\circ C^{-1}$). This introduction of extra polymer constitutes the second, or packing, phase and is enabled by the low thermal conductivity of the polymer melt. The definition of the transition from packing to the third phase has not been unanimous. Usually, once the gate is sealed, one considers that the last, cooling and/or solidification phase has begun. This extends until the whole polymer mass is at temperature lower than T_g or T_c (approximately at the mold temperature) and the molding cycle is complete.

6.2 MODELING THE INJECTION MOLDING PROCESS

The first modeling efforts concentrated on developing a description of the filling phase. This was useful since incomplete filling (short-shots) or overfilling (flashing) could be predicted. The early models only treated flow in the cavity itself and, although the non-Newtonian behavior of the melt was recognized, the assumption of isothermal flow was usual. Guided by experimental tests, the models have evolved to include simultaneous flow and heat transfer and predict flow in the sprue and delivery channels for multi-cavity molds. Some models allow for viscoelastic as well as non-Newtonian descriptions of the melt. Thus, the initial analytical studies of injection molding have evolved into complex physical models primarily concerned with the filling phase. The use of these models requires sophisticated numerical methods and is made possible by the evolution of computer hardware and software technologies.

6.2.1 Filling Modules in CAD/CAM/CAE Systems

The filling phase has been extensively studied on a mathematical basis [1]. Filling is usually modeled as a transient, isothermal or non-isothermal, laminar flow of a non-Newtonian incompressible fluid, with a moving free surface. Finite difference (e.g. marker and cell) or finite element (Galerkin, penalty) methods are generally used and several flow analysis (filling) programs for molten polymers are now commercially available (e.g. Moldflow, C-Flow, Cadmold, Timon, Simuflow, Procop, Polyflow...). Most of them exist as part of integrated CAD/CAM/CAE packages [2] and they are primarily intended to facilitate and automate mold design. Although each program has certain merits and inconveniences due to its modeling approach, all perform a fundamental analysis of the filling process. They predict essential behavior such as:

- optimal number and position of injection points
- melt front advances and weld line positions
- balanced filling of multi-cavity molds
- fill time
- order of magnitude of required clamping force
- prediction of a solidifying layer during filling

A number of them also provide the pressure, velocity and temperature distributions during the filling phase. The usefulness of these programs is obvious because the predictions are directly related to machine operating parameters such as mold and melt temperatures and filling flow rate or ram velocity.

6.2.2 Packing-Cooling in CAD/CAM/CAE Systems

A well filled cavity does not however imply that the desired final part properties will necessarily be obtained. These properties, which may include mechanical, optical, geometrical or surface qualities, are closely associated with the concepts of: (i) tolerances (allowable accuracies of the molded part and of the mold cavity), (ii) morphology (orientation of molecules or fibers and degree of crystallization), and (iii) anisotropy (frozen stresses caused by simultaneous flow and thermal exchange). These factors affect final part characteristics since they directly determine residual stresses, warpage and shrinkage. They can only be controlled by optimizing the process variables related both to the filling and to the post-filling stages. Typically, these parameters are not only mold and melt temperature but also packing pressure, holding and cooling time.

There are two possible means to solve this problem. The first approach, currently used industrially, combines a simple filling simulation with a rudimentary expert system. The expert module relies upon measurements (such as shrinkage or spiral flow tests obtained from standard molds) which are correlated with the operating conditions. Together with pressure-volume-temperature (PVT) information, this semi-empirical approach can guide the choice of the molding conditions and predict peak and holding pressures, local part shrinkage, final part weight and the required clamping force. Although this provides an immediately useful tool for the molder, it has the disadvantage of requiring experimentation for each new resin or particular application.

The second approach is to develop theoretically based simulations of the packing and cooling/solidification phases. This, although more difficult, provides insight and an understanding of these phases of the cycle that will ultimately give a more efficient molding process. It is this second approach that we wish to examine in this chapter and to describe our efforts of simulating the packing and cooling/solidification phases. The perspective of this is obtained by a brief review of prior modeling approaches which is given in the following paragraphs.

6.2.3 Literature Review of Packing-Cooling Modeling

Initially, the pressure in the cavity was the most accessible measurement that provided information about the molding cycle. It is therefore not surprising that the models elaborated in the 1970s relied on a division of the injection molding cycle into three phases, ordered by a typical cavity pressure curve as shown in Figure 6.1. These phases are distinguished as (1) cavity filling, (2) packing, and (3) solidification/cooling.

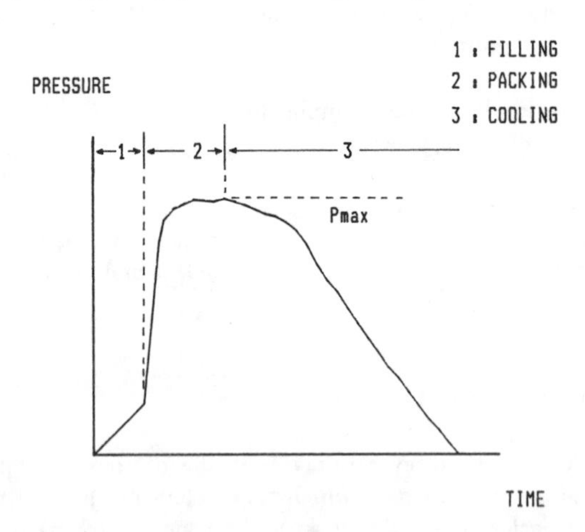

Figure 6.1 Classical decomposition of an injection molding cycle into filling, packing and cooling phases.

Packing, the rapid increase of pressure to its maximum value P_{max} and the corresponding density increase of the molten polymer, were treated as an isothermal dynamic process (phase 2). The resolution of the momentum equations was inspired by the studies employed for the filling models and the compressible character of the fluid mechanics problem was taken into account by an equation of state for the molten polymer.

The general technique utilized was to substitute simplified momentum equations (e.g., the Hele-Shaw creeping flow approximation) and the Spencer-Gilmore equation of state (a Van der Waals equation type having a linear relation between pressure and density for constant temperature) into the continuity equation. The result was a nonlinear partial differential equation for pressure (or density). This equation was solved analytically with further simplifying assumptions or numerically by finite difference or similar techniques [3-9].

The cooling phase (3), typified by a slow pressure decrease, was treated as a non-isothermal process with zero flow. The classic heat conduction equation, plus the equation of state provided the solution. An alternative approach of the post-filling phases, advocated in [10], allows for compressibility effects and assumes the packing phase to be also isothermal, leading to a diffusion equation for pressure. The isothermal restriction is removed for the cooling phase calculations.

It is evident from these studies that the transition from packing to cooling/solidification is not as clear and unambiguous as it is for the filling to packing phases. One cannot be certain that no further flow is occurring when pressure reaches a local peak value as seen by a pressure transducer. Indeed, the core of the molded product cools slowly due to the low thermal conductivity of the polymer, and molten polymer continues entering the mold as long as a pressure gradient remains at the entrance to the cavity and gate sealing is not complete.

Local viscous heating of the melt, due to the high shear rate, occurs simultaneously with appreciable heat exchange between the hot polymer and the mold during the filling phase. Thus, non-isotropic and nonuniform temperature variations are present already during filling, particularly in the form of high thermal gradients near the mold wall and must be considered during the packing stage. They influence in part the final microstructural properties and render the pressure based model subdivision of the molding cycle untenable.

The demarcation between packing and cooling cannot be based solely on pressure since flow and thermal effects have to be taken into account.

6.2.4 Problem Formulation

One can see that the description of packing, compared to the filling stage, has a higher degree of complexity. This is caused by the coupling of the governing equations and the associated moving boundary conditions. The system of equations for amorphous polymers consists of the continuity, energy and momentum equations (nonlinear partial differential equations) coupled through an equation of state and rheological constitutive laws. The unknowns are the velocity components, the temperature, the symmetrical stress tensor (6 components), density and pressure. For crystalline polymers, the crystallinity represents an additional complication, and a further equation describing the kinetics of crystallization has to be known.

The cooling/solidification phase is considerably simpler since only conductive heat transfer and the equation of state (and possibly crystallization) need to be considered.

The first attempts to model, in extended form, the packing and cooling phases under non-isothermal conditions were developed simultaneously by Deterre [11], Lafleur [12] and Titomanlio et al. [13] in the early 1980s. Since Deterre's work, the modeling of packing has been active in France and recent workers [14-16] have since developed two successive models, Pack1 and Pack2. These models are based on the separation of packing from cooling by examining the velocity field locally in the cavity. The definition of these phases is given below and the development and solution techniques of these models is described in subsequent sections.

6.2.5 Definition of the Packing and Cooling Phases

The beginning of the packing phase is taken as the instant that the cavity is completely filled at low pressure. It is manifested by the discontinuity in the cavity pressure-time profile when the cavity pressure begins to increase rapidly. Packing then continues until there is no more flow of polymer as evidenced by the velocity field and pressure gradient becoming null throughout the cavity. The nullity of the velocity field marks the transition from the packing to the cooling/solidification phase, which then continues until the part temperature becomes that of the mold.

From a fluid mechanical viewpoint, packing is the transient, non-isothermal and weakly compressible laminar flow of a non-Newtonian fluid into a mold cavity which is already filled and bounded by metallic walls maintained at a low temperature. There may be a solid polymer layer adjacent to the mold wall. The flow is caused by the presence of pressure gradients and happens concurrently with a progressive liquid to solid phase transformation of the injected material. The thermal effects are essentially induced by heat conduction, from the hot polymer melt towards the cold cavity walls.

6.3 A DIFFERENT MODELING APPROACH: PACK1 AND PACK2

Packing is a weakly compressible (compressibility $\beta = -1/V_S\,(\partial V_S/\partial P)_T \approx 10^{-9}\,\mathrm{Pa}^{-1}$) laminar flow ($Re \ll 1$), thus the methods of trans/supersonic compressible aerodynamic flows are not applicable. Both Pack1 and Pack2 use a combination of finite difference and finite volume methodologies as described for each model. The unknowns are assigned to a basic mesh or to a shifted mesh. The use of staggered mesh systems, in contrast to conventional grids, provides certain numerical advantages as described elsewhere [17,18]. The shifted meshes and the associated variables of Pack2 differ from those used for Pack1 as described later.

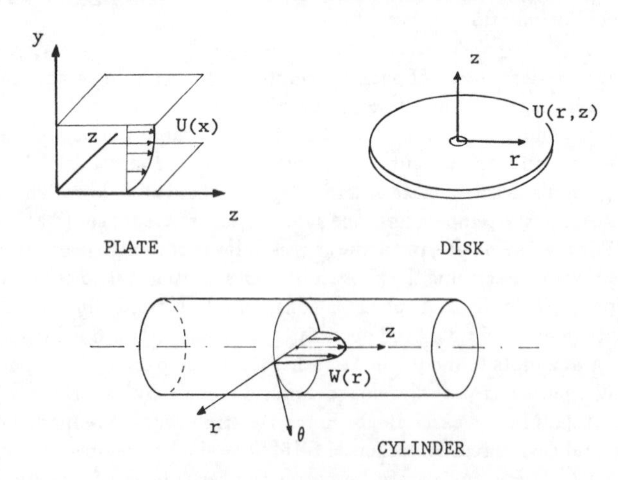

Figure 6.2 Simplified elementary geometries occuring after lay-flat of a complex mold cavity.

We formulate the models using the equations and limits given earlier but, due to the complexity of the problem, restrict our solutions to simple 1D-flow models of elementary geometries (Fig. 6.2) such as thin plates and disks, or cylinders in which the flow can be assumed to be uni-dimensional. This is inspired by the first filling models. These models rely on the lay flat assumption and the thin-cavity approximation to treat complex three dimensional mold cavities as combinations of these various simple two dimensional geometries [19]. The methodology will be illustrated using a particularly simple example: the case of an amorphous polymer injected into a rectangular plate cavity as illustrated in Figure 6.3, of a length L and width w; L and w are taken much larger than the thickness, e.

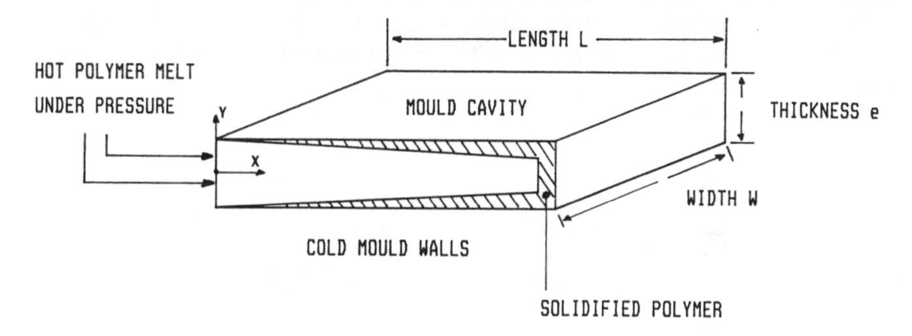

Figure 6.3 Thin rectangular mold cavity used for simulation.

Assumptions: The thin approximation condition $e \ll w,L$ permits the assumption that, to a first approximation, the pressure is constant throughout the thickness and the velocity field is 1-dimensional, given in the cartesian coordinate system by:

$$\vec{V} = (U(x,y,t),0,0)$$

The no-slip condition (velocity $U = 0$) is taken at the liquid-solid interface ($y = h(x,t)$). The solid at the interface may be metallic or polymeric. The position of the solid layer, h, is discrete and it coincides with the mesh nodes. The polymeric solid is assumed non-deformable and thus has zero velocity.

The criterion chosen to determine nil velocity at a node is the value of the viscosity $\eta = \eta(T, \partial U(y)/\partial y, P)$ at this point. If the viscosity is calculated to be > 1000 times the value of η at 220° C (after correction of the pressure effect), then the velocity is taken as zero and the material at that node is considered solid.

The transient problem is of course solved in discrete time, so the main algorithm is iterated. Packing typically takes about one second in a real process while the simultaneous cooling phase continues for ten seconds or more. It is therefore reasonable to assume that during the packing phase the process is thermally quasi-steady. The thermal and fluid-mechanical calculations can thus be separated.

The rheological equations used represent inelastic fluids. The Pack models therefore treat mass density, flow velocity, pressure and temperature evolution (as well as crystallinity in case of crystalline polymers) but do not predict properties such as frozen stresses.

The initial conditions required for the packing stage simulation are: pressure, temperature, and velocity fields at the end of filling. For this work, these conditions are the final results calculated by the mold filling simulation program "Fill" developed by Philipon et al. [20].

6.3.1 Pack1

The development of the model Pack1 commenced in 1982. The numerical methodology utilized to solve the system of equations is essentially based on finite difference procedures with the use of staggered meshes. The finite difference grid is applied to the elementary plate geometry (Fig. 6.4). The temperature, T, and velocity, U, are calculated at the meshnodes i,j and the pressures, P, are evaluated at the staggered meshes i+1/2, j+1/2 and averaged throughout the thickness to give $P_{i+1/2}$. The densities are also computed at the staggered nodes.

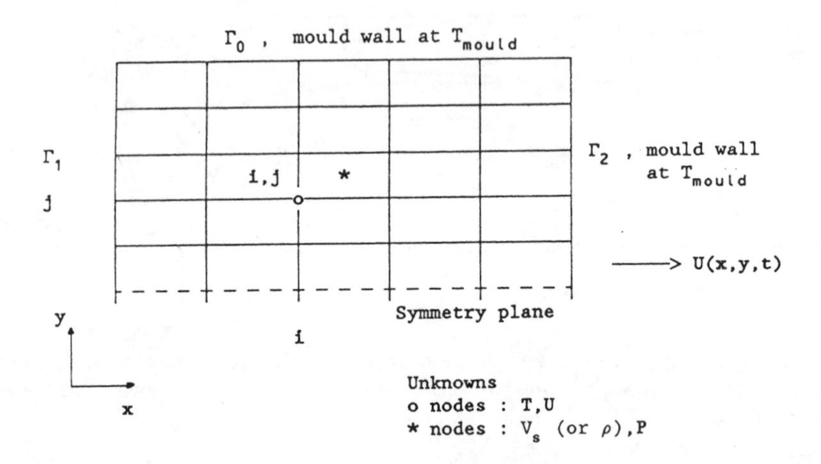

Figure 6.4 Staggered meshes used for the Pack1 model.

Main Algorithm

The mass, momentum and energy balances and the equation of state are necessarily coupled. The main question is how to solve the unsteady flow equations. Two principal ideas govern the simplification of our problem:

1. The unsteady-state compressible flow is divided into a succession of locally (order of mesh size) quasi-steady state incompressible (during a time step) Hele-Shaw creeping flows. Corrections to the melt density are made at each time step by integrating the conservation form of the continuity equation.
2. To establish the pressure profile with an equation of state suitable for polymers, the densities and the temperatures being respectively evaluated by the continuity and energy equations.

The global flowchart is given in Figure 6.5. The principal loop is for time and the first step begins after filling time, t_f, with the appropriate initial conditions. The elements of the algorithm are explained below.

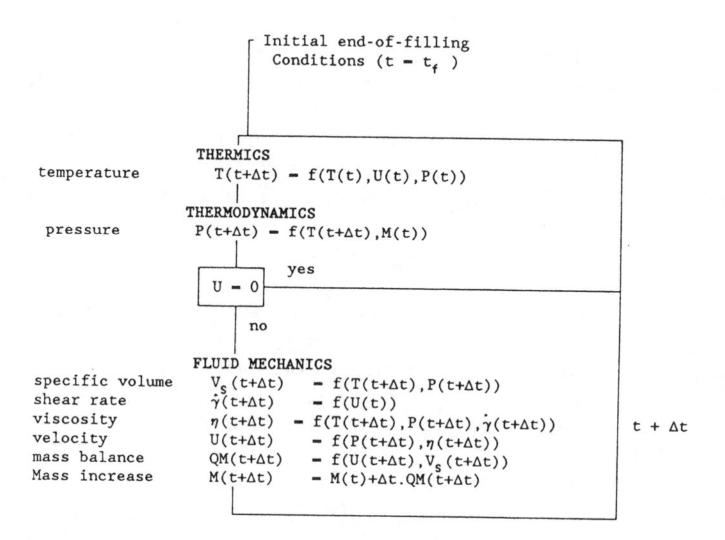

Figure 6.5 Global flowchart of the algorithm of Pack1.

Heat transfer

First one proceeds to calculate temperature in the time step $t+\Delta t$, in order to establish the temperature field $T_{ij}(t+\Delta t)$. The principal assumptions are: (i) the thermal diffusivity, $a_T = k/\rho c_p$), is constant, and (ii) thermal conduction occurs only in the thickness direction. A constant wall temperature, T_{mold}, and a constant temperature of the molten polymer, T_{melt}, at the cavity entry constitute the boundary conditions:

- at the entry,on Γ_1 $\quad T = T_{melt}$
- at the wall,on Γ_0 $\quad T = T_{mold}$
- at the end of the mold,on Γ_2 $\quad T = T_{mold}$
- at the symmetry plane $\quad \partial T/\partial y = 0$

The following Eq (1) applies to the planar heat exchange noting that thermal convection and viscous dissipation are not neglected:

$$\rho c_p \left(\partial T/\partial t + U\,\partial T/\partial x\right) = k\,\partial^2 T/\partial y^2 + \dot{W}(\text{shear rate}) \tag{1}$$

where $\dot{W}(\text{shear rate}) = \eta\dot{\gamma}^2$ is the viscous heating and c_p and k are respectively the specific heat and thermal conductivity of the polymer. This heat equation is easily discretized by the classic Crank-Nicholson scheme and the resulting linear system is directly solved [21].

Flow Analysis

The velocity field U_{ij} is established by neglecting the volume forces, mass and inertia. The packing pressure at the entrance to the cavity, P_{pack}, is for simplification assumed to be directly determined by the machine hydraulic pressure and the downstream condition is zero rate of flow of polymer at the solid boundaries. As mentioned above in the Main Algorithm section, it is assumed that the molten polymer within a mesh cell behaves like an incompressible material during the time step. The local melt motion is governed by a pressure gradient dP/dx balancing the viscous laminar Hele-Shaw flow.

The rheological model chosen is the inelastic generalized Newtonian fluid (GNF) law:

$$\eta' = \eta'(T,\dot{\gamma}) \tag{2a}$$

where $\dot{\gamma}$ is the local shear rate. Equation (2a) was not reduced to a functional form by fitting the data to one of the usual relationships. Instead, the viscosity η' is calculated by linear interpolation of flow curves obtained from capillary rheometer measurements. It is clear that, due to the level of pressures occurring during packing, pressure effects on viscosity are not negligible. They are taken into account by an empirical exponential relation, as given by Glasstone et al. [22]:

$$\eta = \eta' \exp\{P/P_0\} \tag{2b}$$

where P_0 is a reference pressure. The rheological laws are then given by the GNF relations:

$$\tilde{\tilde{\sigma}} = 2\eta\tilde{\tilde{\Delta}} - P\tilde{\tilde{\delta}} = 2\eta'(T,\dot{\gamma})\exp\{P/P_0\}\,\tilde{\tilde{\Delta}} - P\,\tilde{\tilde{\delta}} \tag{3}$$

where $\tilde{\tilde{\sigma}}$, $\tilde{\tilde{\Delta}}$ and $\tilde{\tilde{\delta}}$ are respectively the stress tensor, the rate of deformation tensor and the unit tensor.

The thin plate (Fig. 6.3) of constant thickness reduces the velocity vector $\vec{V}$ to one component and, for a given (x,t), we have $\vec{V}(U(y),0,0,)$. From the momentum equations we obtain, for a given station x:

$$dU/dy = (dP/dx)(y/\eta(y)) \tag{4}$$

The nonlinear differential equation is solved numerically by the Euler finite-difference scheme to yield U_{ij}, an approximation at the meshnode ij of:

$$U(y) = dP/dx \int_h^y \xi/\eta(\xi)\, d\xi \tag{5}$$

It is essential to recall that the momentum equations are solved separately from the continuity equation. The errors introduced by such a procedure are restricted to the orders of the mesh size and time step. The increasing density of the molten compressed polymer is corrected by a mass flow rate balance in each mesh. This is done by integrating the conservation form of the continuity equation over the finite mesh volumes and over a time step, utilizing the new velocity distribution.

Thermodynamics of Amorphous Polymers

The local density or specific volume can be easily deduced from the mass contained in each mesh volume, V. Temperature is given by the solution of the energy equation. Pressure, P, in a mesh is obtained from an equation of state (or P-V-T diagram). In fact, pressure is averaged throughout the thickness (over j) to yield $P_{i+1/2}$ since it is assumed that $\partial P/\partial y = 0$.

Recent work [11] has shown that Tait equation of state, given by:

$$V_s(P,T) = f(T)\{1 - c_t \ln[1 + P/g(T)]\} \tag{6}$$

is a good representation of the thermodynamic behavior of amorphous polymers (Fig. 6.A1). This empirical equation is simple and describes, for amorphous polymers, changes of specific volume in the liquid state with good accuracy. It is also satisfactory for the glassy state if the thermal history is ignored. We take $f(T) = aT + b$ and $g(T) = ce^{-dT}$ for this computation. V_s is the specific volume and c_t is Tait's constant. The set of coefficients a,b,c,d is different for the liquid and the glassy states, determined according to:

$$T_g = T_{go} + 0.25\ P \tag{7}$$

where P is the pressure in MPa and T_{go} is the glass transition temperature at P_{atm} (see Appendix).

Cooling

After the velocity field has vanished, the pressure, temperature and density distributions during the solidification phase are found from the thermal and thermodynamic relations, and the fluid mechanics branch is skipped (Fig. 6.5). The heat conduction equation and the equation of state are solved remembering that the total mass of injected polymer is constant (mass conservation). When pressure is greater than atmospheric, the temperature field is first obtained at each time step and these values used, via the inverted Tait equation, to obtain the pressure field. Once the pressure has reached atmospheric the density variations describe the volume loss of the molding and the local final shrinkages can be estimated.

6.3.2 Computational Results: Pack1

Model Validation

Figure 6.6 compares experimental and simulation results for the reference conditions given in the Appendix. The pressure was measured at the mid-point of the cavity by a transducer placed at a distance of 20% of the length ($x/L = 0.2$).

The model predicts with good accuracy the rapid increase of the pressure and the magnitude of its peak. It calculates a high pressure maintained over a longer period than was found experimentally. This difference is likely due to machine operating procedures which produced actual boundary conditions at variance with those assumed for the model. There does not appear to be a major defect in the model since it predicts the overall shape of the packing and cooling phases including pressure fall and total cycle duration reasonably well.

Pack1 was subjected to tests for the effects of heat transfer mode and sensitivity to material property and changes in operating conditions. The results are presented below. The calculated prediction shown in Figure 6.6 is used as a reference for purposes of comparison.

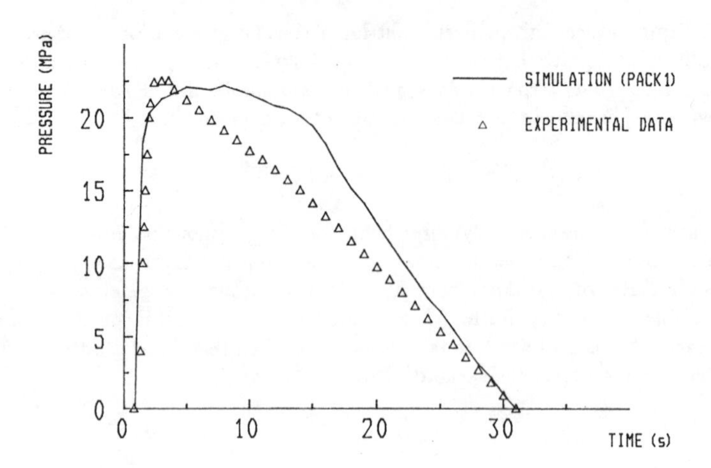

Figure 6.6 Comparison between experimental and simulated pressure profiles for the reference conditions.

Thermal Effects

Figure 6.7 shows the influence of the different modes of heat transfer - conduction, convection and viscous heating - which have been successively eliminated. The elimination of the heat conduction term makes packing a nearly isothermal process. In this case, the injected polymer remains close to the melt temperature and the holding pressure fully propagates into the cavity. This unrealistic simulation does not predict the correct pressure evolution, especially the pressure peak. The presence of convection allows more hot polymer to be packed into the

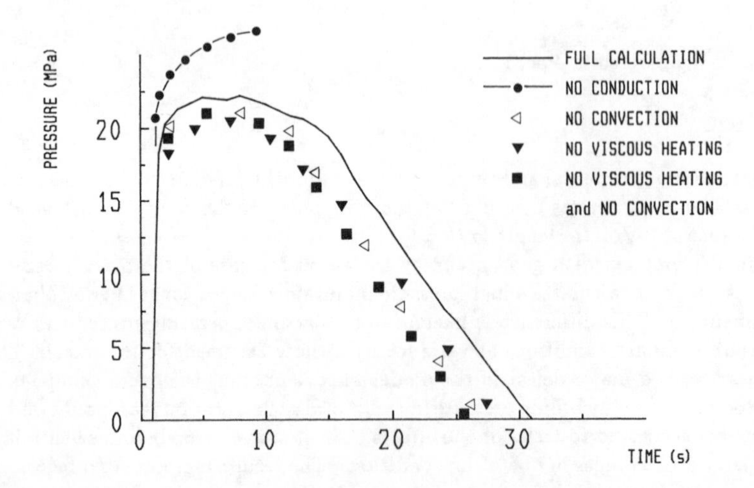

Figure 6.7 Influence of the various heat transfer terms of the energy equation on pressure.

cavity aiding the packing phase. Without the viscous heating term (which increases local temperature near the walls), the polymer solidification rate is somewhat increased and less matter is injected into the cavity. Perforce, packing pressure is less well transmitted and the resulting pressure curve lies beneath the reference one.

Influence of Molding Conditions

The influence of packing pressure is evident and needs no comment (Fig. 6.8). Mold and melt temperatures directly influence the cooling rates. It is clear that an increase in melt or mold temperature furthers the propagation of the holding pressure (and results in packing additional polymer) and delays somewhat the solidification and pressure decrease. The mold temperature change of 10° C affects the pressure response less than a corresponding melt temperature change. This is because the melt temperature has a larger effect on the viscosity of the injected material, hence on the development of packing.

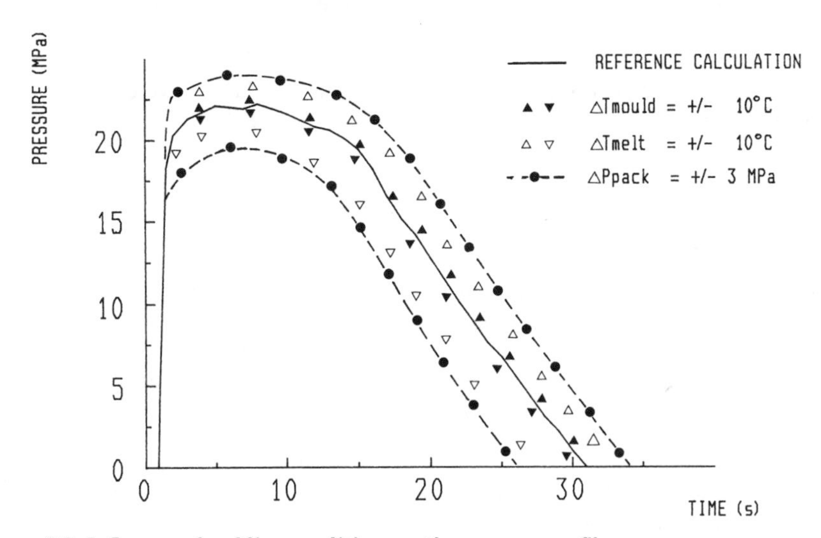

Figure 6.8 Influence of molding conditions on the pressure profile.

Influence of Physical Properties

Variations in the thermodynamic properties (glass temperature, compressibility, thermal expansion) have little effect as seen from Figure 6.9. Thermal diffusivity $a_T = k/\rho c_p$ is the parameter controlling the heat exchange rate between the hot injected mass and the cold walls. Thus, although the pressure evolution is somewhat changed during the packing phase, the rate of pressure decrease during the solidification stage is accelerated for higher values of a_T. This is clearly confirmed by the simulations. The model is very sensitive to viscosity changes. Viscosity variations greatly change the pressure transmission and evolution. High viscosities lower the mass flow rates and impede the transfer of polymer during packing so that the holding pressure cannot achieve its original task.

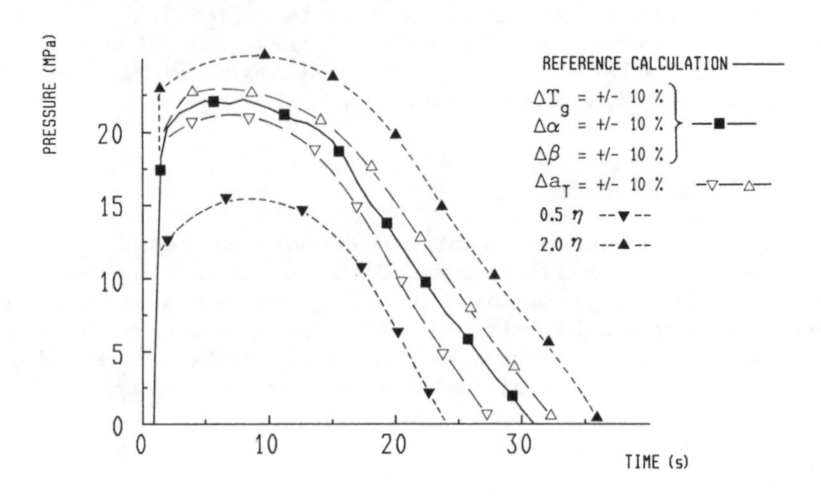

Figure 6.9 Influence of some material properties on the calculated pressure profile (note $T_g = 100°C$).

Limitations of Pack1

While Pack1 correctly shows the effects of property variations and molding conditions, it requires a long calculation time. For the preceding rectangular cavity divided into 100 meshpoints (10 in the flow direction, x, and 10 in the thickness, y) Pack1 needs respectively 10 hours on a micro-computer (IBM AT), 1 hour on a DEC MicroVAX and 7 minutes on a mainframe (IBM 3081 D). This problem is inherent in the explicit character of the algorithm utilized in Pack1, which requires a very small time step to insure stability. The way in which compressibility is treated and the decoupling and the successive solution of the equations do not allow large time steps. The time step is restricted to values between a millisecond and a microsecond or less, especially for the packing phase.

6.3.3 Pack2: A Modification of Pack1

Mathematical basis

Recently, the calculation time has been reduced by using a more efficient procedure. A different technique to solve the compressible flow equations is utilized. The method relies entirely on the finite difference/finite control-volume integration approach [23,24] and is a special case of the Galerkin method or weighted residuals. This technique has been used with success in studies of unsteady compressible and two-phase flows and nuclear systems [17].

The principles of the technique are the integration of the conservation forms of the governing partial differential equations over finite volumes; the main advantage is that the physical meanings of the differential equations are retained when working on their conservation form.

Staggered Mesh Scheme

Staggered meshes are used in a manner similar to Pack1. The benefit of this is an improved numerical stability and better conditioning of the matrix associated with the resulting linear system [25].

Figure 6.10 shows the staggered grids used for the rectangular plate and explains how the different groups of unknowns are assigned to the meshnodes. This assignment is one of the key differences between Packl and Pack2. The temperature T is calculated at the meshnodes ij, the pressure is assumed constant throughout the thickness and calculated at the sections of abscissa i. The local densities are determined from the equation of state $\rho = \rho(P,T)$ and correspond, to a first approximation, to the densities of the meshes centered on i,j. The mass flow rates $G(=\rho U)$ are calculated at the staggered nodes (i+1/2,j).

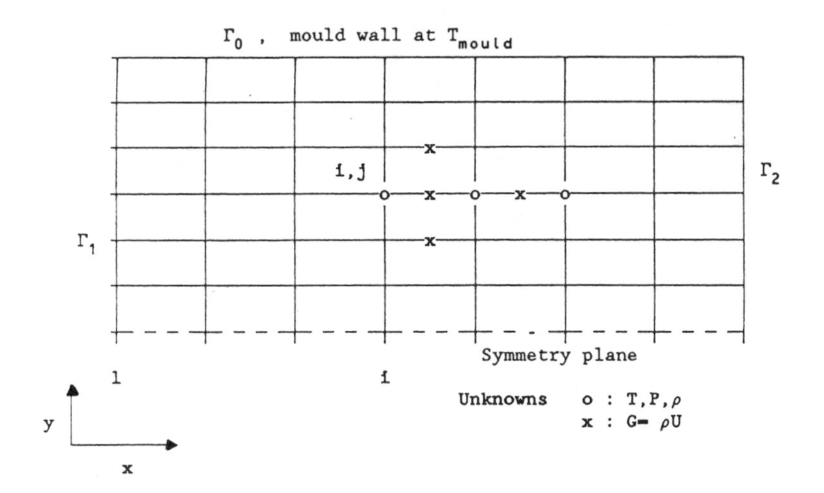

Figure 6.10 Staggered meshes used for the Pack2 control-volume model.

General Algorithm

The post-filling stage is divided into two phases as was done for Pack1. The first packing phase (with non-trivial velocity field) is followed by the second no-flow cooling/solidification phase. The corresponding algorithmic treatment is given in the flowchart of Figure 6.11.

Packing (Compressible Flow) Phase

The equation of energy (in T) is first solved at each time step followed by the equations of mechanics (in P,G) and then the equation of state establishes the new density field $\rho(T,P)$. The choice of the time step is related to the mass flow over the finite volumes.

Formulation of heat transfer in packing: During packing, which has a non-zero velocity field ($U \neq 0$), the heat equation contains not only the convective term but also the

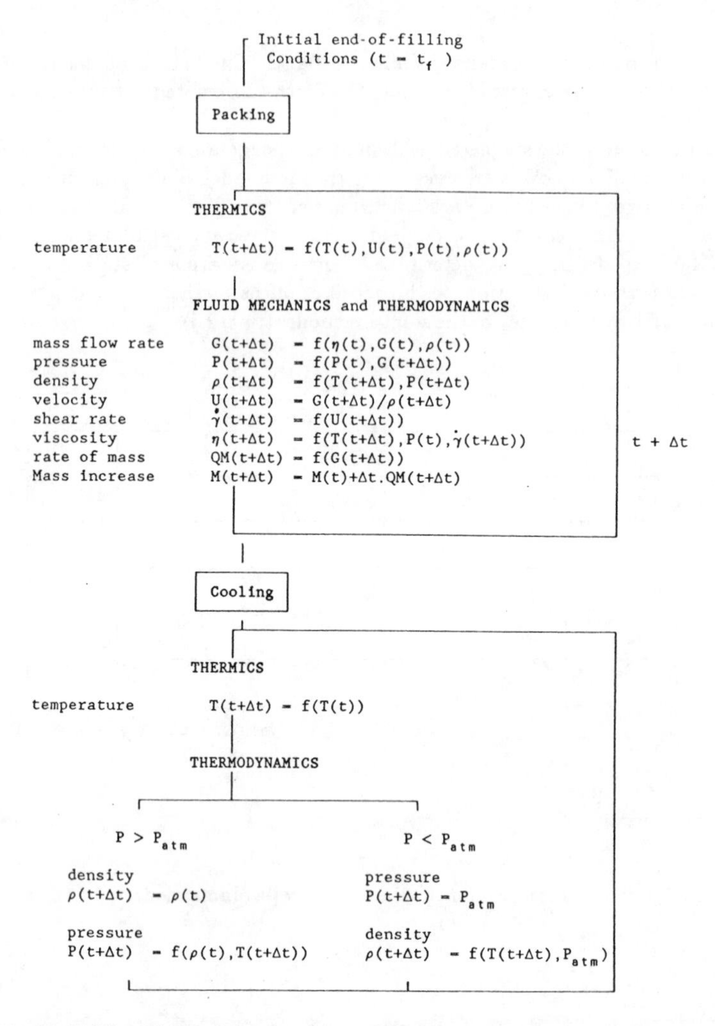

Figure 6.11 Global flowchart of the algorithm of Pack2.

viscous heating term and the conduction is assumed to occur only in the thickness direction as previously:

$$\partial(\rho c_p T)/\partial t + \partial(\rho c_p U T)/\partial x - \partial(k \partial T/\partial y)/\partial y - \dot{W} = 0 \tag{8}$$

Equation (8) is integrated over Δt and over the finite control-volumes centered at the corresponding temperature nodes, taking into account the same thermal boundary conditions used for Pack1.

The discretization of the convective terms uses an implicit and spatial backward scheme and the temperature variables T are treated in an implicit manner. The resulting linear system in $T(t + \Delta t)$ is solved by an algorithm suitable for tridiagonal matrices.

The boundary conditions are those of Pack1 except that they are expressed in terms of mass flow rates G and not in terms of velocity U:

- No flow at the end of the cavity, $G = 0$ at Γ_2
- $G = 0$ at the wall (no-slip condition on boundary Γ_0)
- $\partial G/\partial y = 0$ at the symmetry plane
- $P = P_{pack}$ at Γ_1.

The fluid mechanics problem is treated in the following way (see Figs. 6.10 and 6.11). The equation of mass conservation reduces to the one dimensional flow equation:

$$\partial\rho/\partial t + \partial G/\partial x = 0 \quad ; \quad G = \rho U \tag{9}$$

This equation is integrated over the time step Δt and over the control finite volume ΔV centered on a section i:

$$\int_{\Delta t} \iiint_{\Delta V} (\partial\rho/\partial t + \partial G/\partial x)\, dV\, dt = 0 \tag{10}$$

Using the equation of state, $\rho = \rho(P,T)$, and assuming the process quasi-steady state thermally, leads to the following system relating the implicit flow rates G to the implicit and explicit pressure field:

$$P(i,t + \Delta t) + \Delta t\,[A]\,[G(i + 1/2, t + \Delta t) - G(i - 1/2, t + \Delta t)] = P(i,t) \tag{11}$$

The conservation of momentum is determined by Eqs (12) and (13):

$$\begin{cases} P = P(x,t) & (12) \\ \rho\partial U/\partial t + \rho U\,\partial U/\partial x = -\,\partial P/\partial x + \partial(\eta\partial U/\partial y)/\partial y & (13) \end{cases}$$

The effects of gravity are ignored, but not the inertial terms which are not negligible compared to the convective term or to the pressure gradient. The momentum Eq (13) is integrated in terms of mass flow rate, G, (conservation form) over a time step and a control-volume centered on a G-node. The system of equations, relating the implicit pressure gradient to implicit mass flow rates, is then obtained:

$$[B]\,[G(i+1/2, t+\Delta t)] + [C]\,[G(i-1/2, t+\Delta t)] + [D]\,[P(i+1) - P(i)]_{t+\Delta t} = [E] \tag{14}$$

A judicious combination of the two systems (11) and (14) for continuity and momentum allows the elimination of the implicit pressure terms. The result, including the appropriate boundary conditions, is a linear matrix system for the remaining implicit unknowns G (the flow rate terms):

$$[F]\,[G(t + \Delta t)] = [H] \tag{15}$$

where F is a band matrix and this system is solved by a special algorithm adapted to this form of matrix [26].

Once the mass flow rates $G(t+\Delta t)$ are calculated, the equations of mass conservation, Eq (11), are used to establish the new implicit pressure profile $P(t+\Delta t)$. The remainder of the calculations for the time increment are made in a straightforward manner.

Solidification (Zero Flow) Phase

There is no flow (velocity field = 0) in this phase. The only equations required for the second phase are the conventional one-dimensional transient conduction equation and the equation of state.

The heat equation, without the convective and viscous dissipation terms, reduces to the one-dimensional transient conduction equation:

$$\partial(\rho c_p T)/\partial t - \partial(k\,\partial T/\partial y)\,\partial y = 0 \tag{16}$$

This equation is solved in a nearly implicit way by the same control-volume technique used for Eq (8). The thermodynamic calculations, for each section i, depend on the values of P_i compared with the atmospheric pressure, P_{atm}. Two different branches have to be considered during cooling:

a) $P > P_{atm}$ - the constant density calculation

The local pressure, not yet at the atmospheric value, requires the density to be constant within the corresponding finite volume. The decreasing evolution of the local pressure is calculated from the simplified continuity equation:

$$\partial\rho/\partial t = 0 \tag{17}$$

which, related to the equation of state, gives:

$$\partial P/\partial t = -(\partial\rho/\partial T)_P\,(\partial T/\partial t)\;/\;(\partial\rho/\partial P)_T = -\alpha/\beta\,(\partial T/\partial t) \tag{18}$$

where $(\partial\rho/\partial T)_P$ and $(\partial\rho/\partial P)_T$ are calculated from Tait's equation of state Eq (6), and the cooling rate, $\partial T/\partial t$, is given by the solution of the heat Equation (16).

b) $P = P_{atm}$ - the constant pressure calculation

Once the local pressure has reached the atmospheric level, considering further temperature decreases at $P = P_{atm}$ needs only the equation of state to calculate the local variations in densities. The increases in density characterize the shrinkage evolution in the different sections i.

6.3.4 Computational Results: Pack2

Calculation time

The simulation runs using Pack2 (reported in the next paragraphs) indicate that this model is an order of magnitude faster than Pack1, using time steps between 10^{-4}s and 10^{-2}s. This improvement is inherent in the implicit algorithm used to solve the compressible flow problem, as well as the finite volume approach itself. The implicit character of the scheme and the choice of the staggered meshes leads to a numerically stable solution of the coupled continuity and momentum equations.

Recent numerical tests [27] have shown that for the plate simulation, the inertial terms (in Eq (13)) are negligible. This allows the systems of equations to be simplified with an attendant decrease of time required for their solution.

The results reported below are for a polystyrene (Gedex 1541) injected at 250° C into a 2 mm thick rectangular cavity held at 50° C. The operating parameters used in the simulations are given in Table 6.A2 of the Appendix.

Cavity Pressure Evolution

Figure 6.12 shows the pressure-time evolution at different points within the plate cavity during a typical injection cycle. At the end of filling, the pressure rises within half a second to almost its highest value and holds over more than a second. This constitutes the compressible packing phase. Thereafter, the pressure progressively decreases, this constitutes the cooling phase (solidification). Particularly, it is noted that the pressure first reaches the atmospheric level at the far end of the mold, then progressively at stations closer to the entrance. The particular conditions and simple geometry of this simulation cause the maximum pressure to coincide with the packing/solidification transition. This point is discussed in a later section.

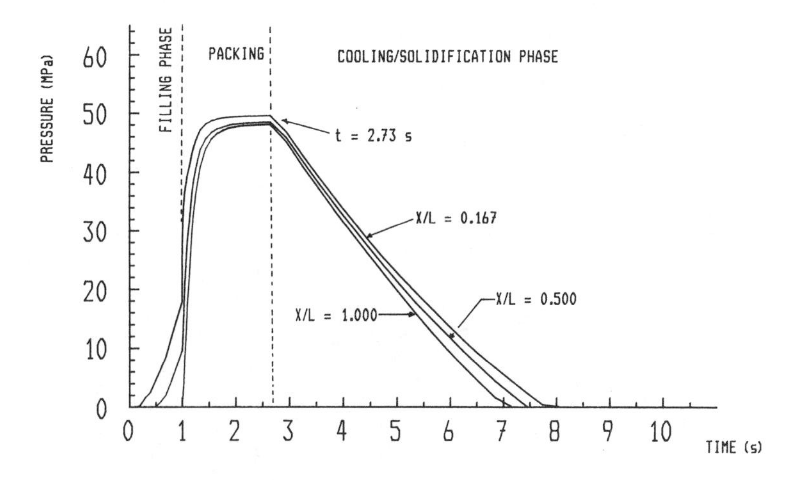

Figure 6.12 Cavity pressure simulation during a typical injection cycle.

The pressure space distribution at various times is given in Figures 6.13a and 6.13b. Figure 6.13a shows how the pressure distribution evolves with time. The pressure is initially, at the end of the filling phase (t = 1 second), almost linearly decreasing in the flow-direction, x. Then packing pressure propagates, spreading out rapidly downstream in less than half a

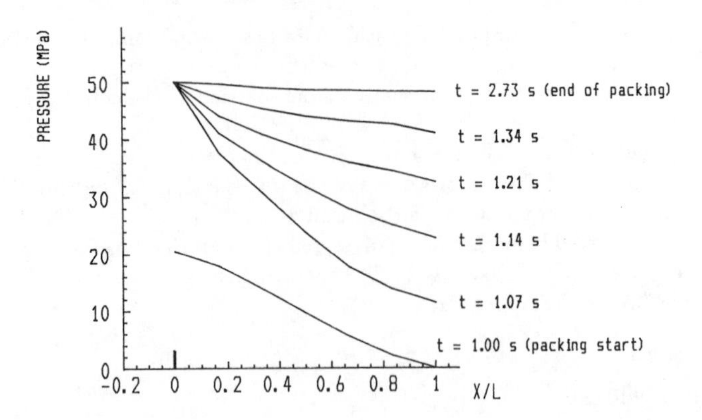

Figure 6.13a Pressure distribution during the packing phase.

second, and cavity pressure becomes almost uniform. The pressure is maintained close to its maximum for a second or more. This constitutes the packing phase. These results are comparable to those reported by Lafleur [12] for a geometry of the same type. Subsequently (Fig. 6.13b), the pressure decreases due to cooling and falls to the atmospheric pressure first at the end of the cavity, then progressively closer to the cavity entrance. The rate of pressure decrease is determined by the cooling rate.

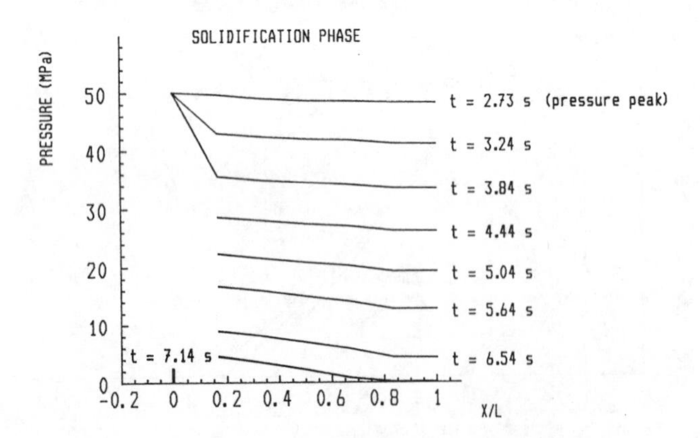

Figure 6.13b Pressure distribution during the cooling/solidification phase.

<u>Velocity Profiles</u>

Figures 6.14a-c show the velocity distributions at various moments during packing. Initially and immediately after the end of filling ($t = 1.001$ s) the large pressure gradient at the entrance

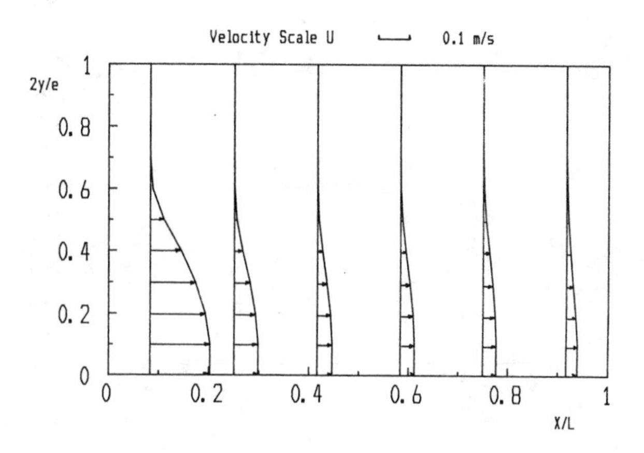

Figure 6.14a Velocity profile just after filling (t = 1.001 s).

of the cavity induces an acceleration of the melt; consequently, the flow rate for small x/L values is much more important than elsewhere (Fig. 6.14a).

This large magnitude transient velocity rapidly disappears (within less than 10^{-1}s; at t = 1.05 s on Fig. 6.14b) and, due to pressure propagation, the pressure gradient becomes more homogeneous along x as seen in Figure 6.13a.

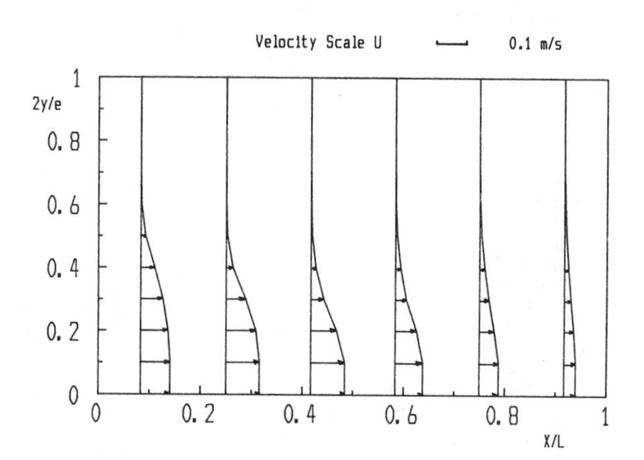

Figure 6.14b Velocity profile during packing (t = 1.05 s).

The cooling effects increase the local viscosity and reduce the liquid melt zone and the pressure gradient tends rapidly to zero. Consequently, the velocity field vanishes (Fig. 6.14c) and the packing phase ends.

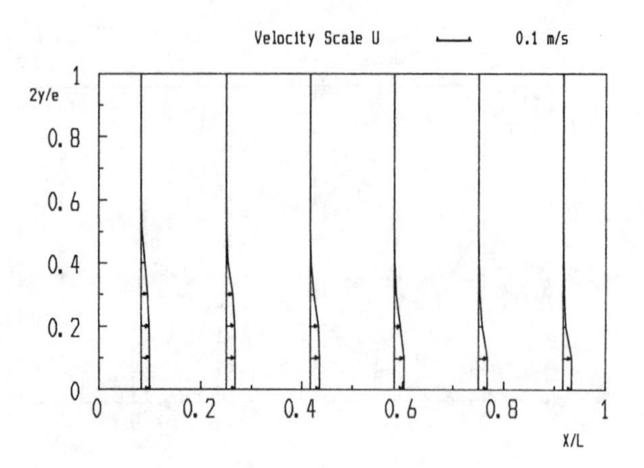

Figure 6.14c Velocity profile at time t = 1.25 s.

Temperature Profiles at the Cavity Middle (x = 0.5 L)

Figure 6.15 shows the temperature versus time evolution in the middle of the cavity (x/L = 0.5) at different stations, y/e, from the mold walls. The high pressure gradients at the beginning of packing cause sufficient shearing that viscous heating is observed as an increase in temperature, especially near the entrance of the cavity. To a first approximation, it is expected that the temperature will decrease exponentially: this is clearly indicated by the calculated temperature curves.

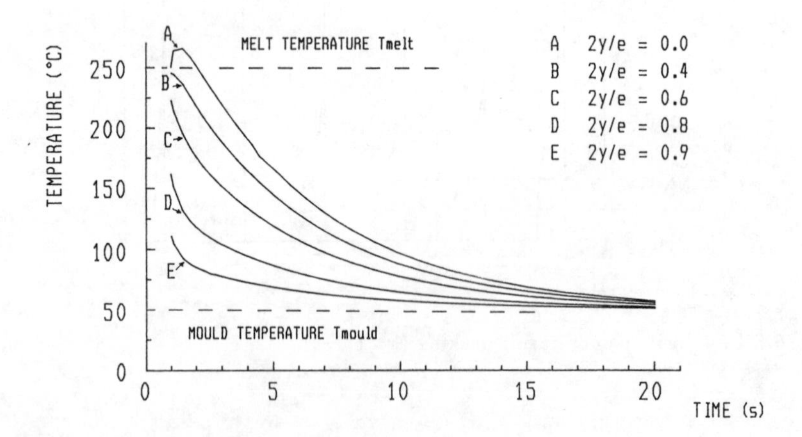

Figure 6.15 Temperature evolution profiles in the middle of the cavity (x = 0.5 L).

Shrinkage Development

Once the pressure has reached the atmospheric value P_{atm}, the densities increase and shrinkage develops from the far end towards the injection point (Fig. 6.16). The final shrinkage is of course larger for high x/L stations because the corresponding material reaches P_{atm} first, and this happens at higher temperatures. Shrinkage can be reduced by additional compression of the polymer melt by higher packing pressures. Simulations were done with different packing pressures and the principal results are reported in Table 6.1.

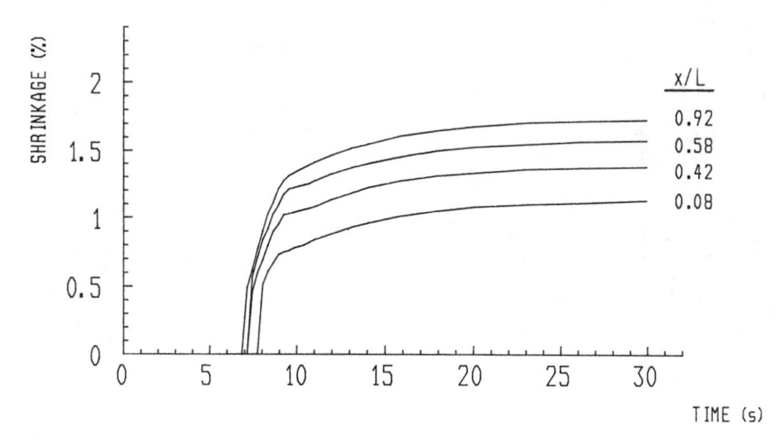

Figure 6.16 Shrinkage development during cooling. Ultimate shrinkages (T=50° C) are: x/L=0.92 1.73%, x/L=0.58 1.58%, x/L=0.42 1.38%, x/L=0.08 1.13%.

Table 6.1 Simulation of shrinkage at different pressures.

Packing Pressure P_{pack} (MPa)	30	40	50	60	70
t_0 (s)	5.3	6.5	7.7	9.6	15.4
T_0 (° C)	158.3	140.0	124.0	104.2	68.9
$R_{0 \cdot 5}$ (%)	2.35	1.89	1.48	1.00	0.52

- t_0 is the time at which the mid pressure returns to P_{atm}.
- T_0 is the temperature at the middle of the plate ($x/L = 0.5$, $y = 0$).
- $R_{0 \cdot 5}$ is the final shrinkage magnitude at $x/L = 0.5$.

It is evident that all three parameters are strongly affected by the packing pressure and vary in the manner expected. Pack2 also predicted that the slope of the pressure-time profile and the time of the packing to cooling transition to be essentially independent of packing pressure as shown in Figure 6.17. Overpacking, when the pressure never returns to P_{atm}, was also successfully simulated.

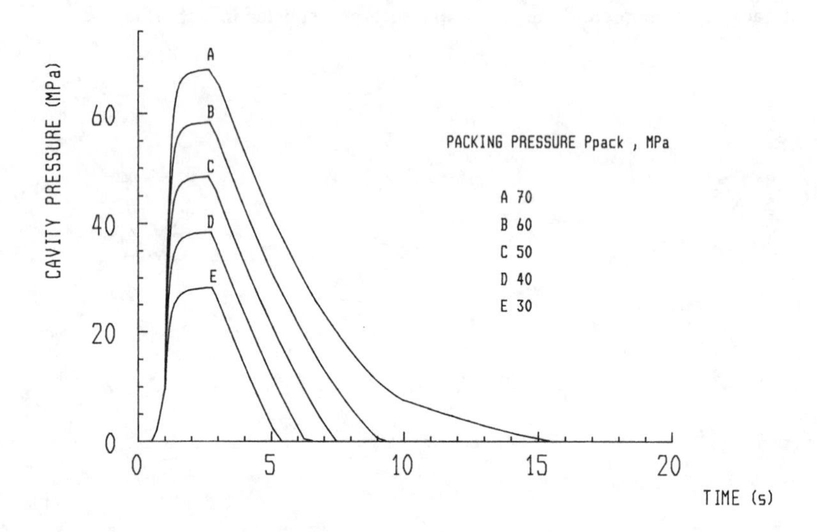

Figure 6.17 Influence of the upstream packing pressure on the cavity pressure at x = 0.5 L.

6.3.5 Summary

Methods have been developed for simulating the interaction between flow, heat transfer and thermodynamic behavior of polymers during the post-filling stage in injection molding of thermoplastics. Their originality comes from the numerical techniques utilized to solve the general governing nonlinear system (simultaneously the continuity, momentum and energy equations coupled to an equation of state). The inelastic generalized Newtonian fluid (GNF) law including pressure effects was used as the rheological model during packing and Tait's equation of state applied to amorphous polymers.

The simulations calculate the pressure, velocity and temperature distributions, and density variations in simple mold cavity geometries during the packing and cooling phases. The final shrinkage field of the resulting molded parts is also estimated. The model predictions agree with the pressure evolution recorded on an experimental mold of simple geometry. It is evident that thermal effects are important during the packing phase affecting the flow via the P-V-T relationship and through changes in viscosity. The flow cannot be treated as if it were isothermal. Results obtained from varying the physical properties and molding operating conditions are consistent and sensible but need experimental confirmation.

6.4 EXTENSIONS AND DEVELOPMENT

The Pack models show some difficulties, in common with other efforts, to simulate the packing and cooling phases, although the means of eliminating the problems are not necessarily the same. All models have currently been tested for only simple geometries and physical property relationships. The numerical difficulties that may be encountered as the algorithms are developed are largely unknown.

This requirement of enhanced model capability must be balanced against the need to reduce the amount of calculation time, e.g. by decreasing the number of mesh points and iterations. A combination of modeling effort and exploitation of numerical technology guided by experimental results seems to be the most promising avenue to achieve this.

6.4.1 Physics and Thermodynamics

There is a need for the judicious incorporation of improved relationships. Dietz [28], for example, reports thermal diffusivity variations on the order of 10-15% over the range of typical molding conditions. The inaccuracies of the Tait (or other) relationships for P-V-T have a direct and important bearing on the quality of the prediction. Much work is needed both at theoretical and experimental levels to gain a better understanding of the phenomena and the process.

There is a less urgent requirement for knowledge about the compressibility, thermal expansion and glass transition phenomena as they are affected by operating conditions. This is, however, still subject to experimental verification.

Crystalline Polymers

The Pack models can be applied to crystalline or partially crystalline polymers if the kinetics of crystallization are known for temperature far from the melting point. The extension to crystalline polymers has to take into account (by an additional term in the heat equation) the rate of release of latent heat:

$$\rho c_p (\partial T/\partial t + W \partial T/\partial z) = k \partial^2 T/\partial y^2 + \eta \dot{\gamma}^2 + \dot{W}(\text{enthalpy}) \tag{19}$$

where $\dot{W}(\text{enthalpy}) = (dX_c/dt)\ \Delta H$ is the enthalpy release rate, ΔH is the heat of crystallization and dX_c/dt is the rate of crystallization. The energy equation must be coupled with an equation for the crystallization kinetics.

In the Pack models, the amorphous fraction of polymer is treated separately (with Tait's equation) and the crystalline fraction is defined by $X_c = (V_a - V_s)/(V_a - V_c)$, where V_a, V_c and V_s are the specific volumes of amorphous, crystalline and real phases and obey locally isothermal Avrami laws during the short time intervals, Δt. The whole non-isothermal crystallization process is divided into a sequence of isothermal and isopressure crystallizations. The specific volume V_c of the pure crystal is assumed to be temperature and pressure independent.

The cooling rates in injection molding are extremely high during the filling and packing stages: $\partial T/\partial t$ of 10 to 100°C/s are not uncommon (Fig. 6.14). The further development of crystallization kinetics will probably be achieved either by using existing theory [29] or by explicitly modeling the non-isothermal transformation kinetics.

Rheology

Modeling of the filling phase using elastic fluids has been reported by several research groups over the past few years [30,31]. They also signal the need for realistic viscoelastic laws to be utilized in packing-cooling models to analyze and quantify shear and normal stress development. In this way, orientation, thermo-mechanical history, and thermal stresses could be predicted, leading to the calculation of the final part properties.

However, even for purely inelastic filling and packing models, which are very viscosity dependent, the technology of viscometric apparatus has to be improved. The lack of precision of viscosity measurements made by the usual techniques (capillary rheometry for instance) is a problem. Packing also requires studies of the pressure dependence on viscosity (Utracki [32], So and Klaus [33]), and of the pressure influence on the glass transition temperature T_g or crystalline melting point T_C. Viscosity measurements at temperatures near T_C or T_g with pressure effects will allow the specification of realistic no-flow criteria based on viscosity, temperature and crystallinity levels.

6.4.2 Modeling and Industrial Practice

There are a number of differences between current modeling capabilities and industrial practice. We briefly enumerate, in this section, some of the more important areas which require further work. It is assumed in the Pack and other models that the packing pressure rises immediately to a value which is held constant throughout the packing and cooling phases. An injection molding machine cannot, of course, make an instantaneous transition from a given mass flow rate (filling) to an imposed pressure (packing). Further, the polymer in the runner and sprue system undergoes packing and, in some cases, solidification. The effects of cooling on this polymer and compressibility both in the delivery channels and mold cavity must be taken into account.

The currently reported modeling efforts use very simple boundary conditions to predict the behavior of the polymer in the cavity. The industrial molding of products uses molds that are much more complex with respect to both pressure and temperature boundary conditions: the real conditions typically are quite removed from the uniform ones used in modeling. These non-uniformities also cause dynamic deformations of the cavity. The simulation of this effect has only been approached in a very rudimentary way.

The "lay-flat" technique has been very successful in predicting the advance of the melt front and the weld-line positions for complex molds. It is not clear however that pressure transmission from one plane to another, around corners, can be treated in a simple isotropic manner when compressible flow is occurring in the presence of solidifying polymer.

Finally, there are the questions about the ability to predict structure development for composites and at the molecular level for crystalline polymers. This capability is one of the ultimate objectives since it leads directly to the final part properties. Throughout all future efforts and improvements, we must always strive to validate the models using experimental data of the highest quality.

6.5 CONCLUSION

There is no doubt that progress in the understanding and modeling of the injection molding process cannot ignore the experimental and theoretical analysis of the packing and cooling phases. Post-filling simulation models will in the future find their place alongside the developing expert systems within CAD/CAM/CAE packages.

ACKNOWLEDGEMENTS

Many persons have contributed to the work and development presented here and the authors wish to recognize their aid. A special thank-you is due to M.E. De la Lande who worked on much of the code of Pack2.

NOMENCLATURE

A: Matrix of coefficients, Eq (11)
a: Fitting parameter in Tait's equation, $m^3\,kg^{-1}\,K^{-1}$
a_T: Thermal diffusivity, $m^2\,s^{-1}$
B: Matrix of coefficients, Eq (14)
b: Fitting parameter in Tait equation, $m^3\,kg^{-1}$
C: Matrix of coefficients, Eq (14)
c: Fitting parameter in Tait's equation, Pa
c_p: Specific heat at constant pressure, $J\,mol^{-1}\,{}^\circ C^{-1}$
c_t: Constant in Tait's equation, dimensionless
D: Matrix of coefficients, Eq (14)
d: Fitting parameter in Tait's equation, K
d_c: Characteristic dimension for Re, m
E: Matrix of coefficients, Eq (14)
e: Thickness of cavity (Fig. 6.3), m
F: Matrix of coefficients, Eq (15)
$f(T)$: Function in Tait's equation, K
G: Mass flow rate $(= \rho U)$, $kg\,s^{-1}$
$g(T)$: Function in Tait's equation, K
H: Heat of crystallization, $J\,mol^{-1}$
h: Position of the solid layer, m
i, j: Indices identifying mesh nodes
k: Thermal conductivity, $J\,s^{-1}\,m^{-1}\,{}^\circ C^{-1}$
L: Length of cavity (Fig. 6.3), m
M: Mass, kg
P: Pressure, Pa
P_{atm}: Atmospheric pressure, Pa
P_c: Clamping force, N
P_{melt}: Magnitude of the peak of the pressure curve, Pa
P_o: Reference pressure for Eq 2b, Pa
P_{pack}: Packing pressure, Pa
Re: Reynolds number $= d_c U \rho / \eta$, dimensionless
$R_{0.5}$: Volumetric shrinkage, % , dimensionless
T: Temperature, ${}^\circ$C or K
T_c: Crystallization temperature, ${}^\circ$C or K
T_{ij}: Local temperature at meshnode ij, ${}^\circ$C
T_g: Glass temperature, ${}^\circ$C or K

T_{go}: Glass temperature at $P = P_{atm}$, °C or K
T_{melt}: Melt temperature, °C
T_{mold}: Wall temperature, °C
T_0: Temperature at the middle of the cavity (Table 6.1), °C
$t, \Delta t$: Time, time increment, s
t_f: Time at the end of filling phase, s
t_0: Time at which $P = P_0$, s
U: x velocity component, m s^{-1}
$V, \Delta V$: Volume, volume element, m^3
V_a: Specific volume of the amorphous material, $m^3\,kg^{-1}$
V_c: Specific volume of the pure crystal, $m^3\,kg^{-1}$
V_S: Specific volume, $m^3\,kg^{-1}$
$\vec{V}$: Velocity vector, cartesian coordinates
$\dot{W}$: Rate of viscous dissipation
w: Width of cavity (Fig. 6.3), m
X_c: Degree of crystallinity, dimensionless
x, y, z: Cartesian coordinates (Fig. 6.3)
α: Coefficient of thermal contraction, $°C^{-1}$ or K^{-1}
β: Compressibilty
$\dot{\gamma}$: Shear rate, s^{-1}
$\tilde{\tilde{\delta}}$: Unit tensor
$\tilde{\tilde{\Delta}}$: Rate of strain tensor, s^{-1}
η', η: Viscosity, Pa s
ρ: Mass density, kg m^{-3}
$\tilde{\tilde{\sigma}}$: Stress tensor, N m^{-2}
ξ: Variable of integration
$\Gamma_0, \Gamma_1, \Gamma_2$: Boundaries of the rectangular cavity.

REFERENCES

1. Mavridis, H., Hrymak, A.N., Vlachopoulos, J., *Adv. Polym. Techn.*, *6*, 457 (1987).
2. Manzione, L.T. (Ed.), "Applications of Computer Aided Engineering in Injection Molding," Hanser Publishers, Munich (1987).
3. Kamal, M.R., Kenig, S., *Polym. Eng. Sci.*, *12*, 294 (1972).
4. Kamal, M.R., Kuo, Y., Doan, H., *Polym. Eng. Sci.*, *15*, 863 (1975).
5. Kuo, Y., Kamal, M.R., *AIChE J.*, *22*, 661 (1976).
6. Ryan, M.E., Chung, T.S., *Polym. Eng. Sci.*, *20*, 642 (1980).
7. Chung, T.S., Ryan, M.E., *Polym. Eng. Sci.*, *21*, 271 (1981).
8. Chung, T.S., Ide, Y., *J. Appl. Polym. Sci.*, *28*, 2999 (1983).
9. Chung, T.S., *Polym. Eng. Sci.*, *25*, 772 (1985).
10. Hieber, C.A., in "Injection and Compression Molding Fundamentals," Isayev A.I., Ed., Marcel Dekker, New York (1987).
11. Deterre, R., D. Eng. thesis, Louis Pasteur University, Strasbourg, France (1984).
12. Lafleur, P., PhD thesis, McGill University, Montreal (1983).
13. Titomanlio, G., Piccarolo, S., Levati G., *J. Appl. Polym. Sci.*, *35*, 1483 (1988).
14. Huilier, D., "Modeling the packing-cooling stage," 2nd Annual Meeting of the Polymer Processing Society, Montreal, April 1-4 (1986).
15. Huilier, D., Lenfant, C., Terrisse, J., Deterre, R., *Polym. Eng. Sci.*, *28*, 1637 (1988).

16. Huilier, D., De la Lande, M.-E., "An improved simulation for the packing-cooling stage in injection molding of thermoplastics," North American meeting of the Polymer Processing Society, Buffalo, Sept. 27-30 (1987).
17. Nakamura, S., "Computational Methods in Engineering Science," 2nd ed., R.E. Krieger Publishing Co., Malabar, FA (1986).
18. Patankar, S.V., "Numerical Heat Transfer and Fluid Flow," McGraw Hill, New York (1980).
19. Menges, G., Schmidt, L., Kemper, W., *Kunststoffe*, *72*, 446 (1982).
20. Philipon, S., Agassant, J.-F., Vincent, M., *Rev. Gen. Therm. Fr.*, *279*, 291 (1985).
21. Carnahan, B., Luther, H.A., Wilkes, J.O., "Applied Numerical Methods," John Wiley & Sons, New York (1969).
22. Glasstone, S., Laidler, K.J., Eyring, H., "The Theory of Rate Processes," McGraw Hill, New York (1941).
23. Ames, W.F., "Numerical Methods for Partial Differential Equations," Thomas Nelson, London (1969).
24. Roache, J., "Computational Fluid Dynamics," Hermosa Publishers, Albuquerque, NM (1972).
25. Latrobe, A., "Fast finite difference methods for solving 1D two-phase flows," OCDE Specialists' Meeting on Transient Two-phase Flows, San Franscisco, (1981).
26. Martin, R.S., Wilkinson, J.H., *Numerische Math.*, *9*, 279 (1967).
27. Maillot, I., De la Lande, M.-E., Latrobe, A., private communication (1987).
28. Dietz W., D. Eng. thesis, Universite Pierre et Marie Curie, Paris (1976).
29. Nakamura, K., Watanabe, T., Katayama, K., Amano, J., *J. Appl. Polym. Sci.*, *16*, 1077 (1972).
30. Mavridis, H., Hrymak, A.N., Vlachopoulos, J., *Polym. Eng. Sci.*, *26*, 449 (1986).
31. Kamal, M.R., Chu, E., Lafleur, G., Ryan, M.E., *Polym. Eng. Sci.*, *26*, 190 (1986).
32. Utracki L.A., *Polym. Eng. Sci.*, *25*, 655 (1985).
33. So B.Y., Klaus E.E., *ASLE Trans.*, *23*, 409 (1980).

APPENDIX: Operating conditions and material properties.

The data presented here concern the molding conditions and characteristics of an amorphous polymer. The polymer used is a polystyrene (Gedex 1541, CdF-Chimie). Its material properties, used for computation, are given in Table 6.A1 and PVT data are presented in Figure 6.A1. The shear dependent viscosity has been characterized by measurements on a capillary rheometer at different temperatures with Bagley and Rabinovitch corrections applied (Figure 6.A2). Experiments were conducted on a 90 ton BILLION screw plasticating injection molding machine equipped with a Visumat control system. The mold itself contained piezo-electric pressure transducers and the outputs were recorded by a data acquisition system. The reference molding conditions utilized in the computations of Pack1 and Pack2 are given in Table 6.A2.

Table 6.A1 Material properties (polystyrene Gedex 1541-Cdf Chimie)

Glass temperature		$T_{go} = 100°C$
Specific heat		$c_p = 2.10^{+3}$ J/kg °C
Thermal diffusivity		$a_T = 8.10^{-8}$ m²/s
Coefficients of Tait's equation		
$c_t = 0.0894$	$T > T_g$	$T < T_g$
a ($m^3 kg^{-1} K^{-1}$)	62.10^{-8}	12.10^{-8}
b ($m^3 kg^{-1}$)	74.10^{-5}	93.10^{-5}
c (Pa)	74.10^{+6}	64.10^{+7}
d (K^{-1})	46.10^{-4}	89.10^{-4}

Table 6.A2 Geometric parameters and molding conditions used in the simulation

	Pack 1	Pack 2
Lenght of cavity, L	0.150 m	0.215 m
Thickness, e	0.005 m	0.002 m
Width of the cavity, w	0.010 m	0.070 m
Mold wall temperature, T_{mold}	55° C	50° C
Melt temperature, T_{melt}	200° C	250° C
Packing pressure, P_{pack}	30 MPa	50 MPa

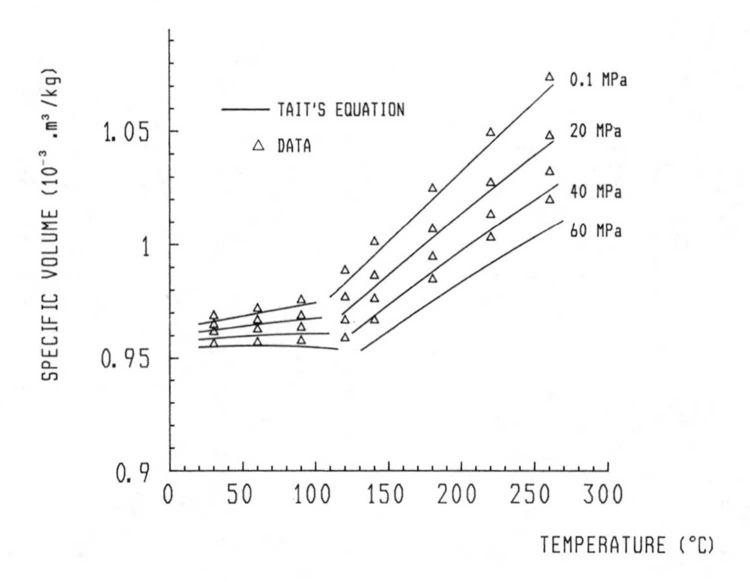

Figure 6.A1 Fitting of Tait equation to P-V-T data for polystyrene (from "Thermodynamik" vol. 1, p. 258, VDMA-Carl Hanser Verlag, Munich (1979)).

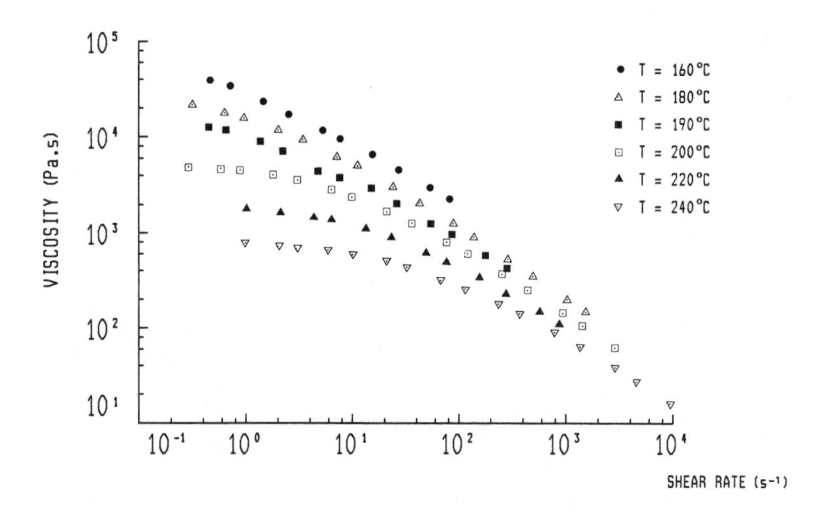

Figure 6.A2 Shear rate and temperature dependence of viscosity for polystyrene 1541 Gedex.

CHAPTER 7

SIMULATION OF INJECTION MOLDING OF RUBBER COMPOUNDS

By M. Sobhanie and A.I. Isayev

Institute of Polymer Engineering
The University of Akron
Akron, Ohio 44325
U.S.A.

ABSTRACT
7.1 INTRODUCTION
7.2 THEORETICAL
 7.2.1 Governing Equations
 7.2.2 Viscous Heating
 7.2.3 Curing Kinetic Model
7.3 NUMERICAL SOLUTION METHOD
 7.3.1 Pressure Profile
 7.3.2 Temperature Profile
 7.3.3 Avdancement of the Melt Front
 7.3.4 Treatment of the Melt Front
 7.3.5 Solution Algorithm
 7.3.6 Comparison of the Techniques
7.4 NUMERICAL EXAMPLE
 7.4.1 Processing Conditions and Cavity Dimensions
 7.4.2 Physical Properties of Rubber
 7.4.3 Rheological Constants
 7.4.4 Curing Kinetic Constants
 7.4.5 Predictions for Filling Stage
 7.4.6 Predictions for Post-Filling Stage
7.5 CONCLUDING REMARKS
ACKNOWLEDGEMENT
NOMENCLATURE
REFERENCES
APPENDIX

ABSTRACT

This chapter presents a general, finite element and finite difference model for simulation of the injection molding of rubber compounds. The program makes use of a control volume finite element method for solution of the continuity and momentum equations, and a finite difference method for solution of the energy equation. The model is general enough to accommodate any rheological and vulcanization model. Both, the filling and post-filling stages are considered. In the filling stage, flow through runners and mold filling are described. The pressure, velocity, temperature, induction time, melt front, and weld lines are computed. In the post-filling stage the kinetics of vulcanization leads to a map of the degree of cure within the molded part.

7.1 INTRODUCTION

Injection molding is defined as an automatic cyclic process in which the strip or granular form of the rubber compounds is plasticated and heated to a temperature just below the vulcanization temperature. Then the melt is injected through a nozzle, sprue, runner, and gate into a hot mold having high enough temperature to initiate vulcanization, and subsequently vulcanize the rubber inside the cavity and eject the molded part. The advantages of the injection molding process over conventional compression molding are now obvious, and perhaps the most important feature is that it introduces a large degree of automation into the molding operations of the rubber industry. Nevertheless, this advantage has to be set against the high cost of injection equipment and mold.

Despite the increasing popularity of injection molding in the polymer industry, the injection molding of rubbers has received much less academic interest in the past as compared to thermoplastics. This is partly due to the difficulties associated with the rheological and thermal characterization of rubber compounds. These difficulties could be due to (i) the presence of filler, oil, and reactive agents which have an influence on flow dynamics of the rubber compounds; (ii) modeling of the cross-linking reaction which has significant effect on flow and heat transfer; (iii) presence of a viscoelastic behavior of rubber compounds during flow even in the uncured state. Modeling of the injection molding process of rubber compounds involves consideration of the unsteady, non-isothermal flow of a non-Newtonian medium with viscous heat generation. This includes simulation of the flow in the screw extruder, nozzle, sprue, runner, gate, and cavity until the cavity is filled. Furthermore, a consideration of heat generation due to vulcanization reaction and calculation of the state of cure is required [1].

The vulcanization process in a mold is a transient heat transfer problem with internal heat generation due to exothermic reaction of vulcanization. It is postulated that during vulcanization under an isothermal condition, the heat of reaction at any particular time is directly proportional to the degree of cure [2]. Thus, development of a curing kinetic model is a prerequisite for a successful modeling of the injection molding process. Various isothermal reaction kinetic models are reported in the literature [3-5]. In order to apply the kinetic model based on an isothermal formulation to the non-isothermal process, temperature history must be taken into account [6-9]. For rubber compounds, the vulcanization starts after time t_i has been reached. This period is called the induction time [10]. From a kinetic study under isothermal conditions, for example, by means of differential scanning calorimeter (DSC), the induction time can easily be found experimentally. It is a material function dependent on reaction temperature. It has been proposed that this function has an Arrhenius-type temperature dependence [10]. Recently, prediction of the induction time for a non-isothermal process was developed [11].

In general, the viscosity of rubber compounds depends on the shear rate, pressure, temperature, volume fraction of the fillers, and the degree of cure. However, in order to avoid scorch, an optimized processing design requires that filling time be equal to the non-isothermal induction time. Therefore, one can always find an optimized processing condition leading to the start of the vulcanization only after the cavity is completely filled. In this case, viscosity may be assumed to be independent of the degree of cure during the filling stage.

Solution of the governing equations (momentum, continuity, and energy) with an appropriate rheological and curing kinetic model eventually provides a method to optimize the injection molding cycle aimed at uniform cure without the scorching problem. During the last decade or so, there have been major successful efforts to develop computer codes for the injection molding of thermoplastics [1,12,13]. An extension of these developments to the simulation of injection molding of rubber compounds is needed.

In the early development of a numerical simulation for thermoplastics, the governing equations were solved for one-dimensional flow for a simple geometry. Examples in this category are a rectangular cavity [14-16], a center-gated or half-disk [17-25], a circular cross-section tube [26,27], and a non-circular channel [28]. More complicated geometries have been decomposed into a series of one-dimensional flow paths [29-32]. Each flow path may consist of a network of basic units of simple geometry including a strip, disk, or tube. A one-dimensional flow analysis is utilized for each flow path. These solutions are coupled by requiring that the total pressure drop be the same along each flow path subject to the constraint that the total flow rate be satisfied [32]. For most mold networks, this decomposition is straightforward although it is not unique. The main complication arises in the presence of the weld-line, where two different flow paths are joined together. In this case, the position of the weld-line has to be guessed, and its actual position be found by a trial-and-error procedure [29]. The above method is called branching flow analysis in the literature. A true two-dimensional analysis was presented by Richardson [33], Kamal et al. [34,35], Hieber et al. [36,37], Tadmor et al. [38-41], and White [42]. Very often, these simulations utilized a finite-element representation for the planar geometry and a finite-difference mesh in the gap-wise direction. The melt flow is treated as generalized Hele-Shaw flow [43] with the fluid being inelastic but non-Newtonian under non-isothermal conditions. The melt front is treated as flat and is advanced according to the average gapwise velocity.

In particular, in Flow Analysis Network [38-41] (FAN), the flow domain is divided into a series of equal square elements. It is assumed that the fluid in each element is concentrated at the center, or node, of the square. The nodes of adjacent squares are interconnected by links and, as a result, the total flow field in the cavity can be represented by a network of links and nodes. For computational purposes, five kinds of nodes are defined including gate node, full node, border node, front node, and empty node. By satisfying the conservation of mass for a fully filled node together with the associated force balance, one is able to calculate the pressure profile, velocity, etc. Since the underlying square elements cannot fit exactly into an arbitrary mold geometry, special treatment is required at the boundary of the cavity.

The Galerkin finite-element method was utilized for simulation of cavity filling by Hieber and Shen [36]. In this method, the melt domain is divided into a series of triangular elements. Numerical solutions of the momentum and the continuity equations, using the finite element method, provide quantities such as pressure, shear stresses, velocities, etc.

Finite-difference methods are utilized for the evaluation of temperature in the gapwise direction.

Upon the advancement of the melt front, the user should intervene and create new meshes for the next time step. This in turn requires preparing a new input data. Special attention should be given to the selection of time increments. For a large time step, one may lose the accuracy of the solution whereas, for a small time increment, triangular elements of poor aspect ratio may result. For curved-type boundaries, some portion of the melt might move outside of the cavity during the advancement of the melt front. This portion of the melt may be allocated inside the cavity by considering the melt front as orthogonal to the cavity boundary. Later on, these shortcomings were improved in [37] by employing a control-volume finite-element method. Although they have not presented details of mathematical formulation, their procedure is somewhat similar to one proposed by Tadmor et al. [38-41]. The geometry of the cavity is divided into a series of triangular elements. For each vertex node an arbitrary volume is assigned. Each node could be filled, empty, or partially filled. By satisfying the conservation of mass for each fully-filled node together with the associated force balance, the pressure-related quantities are obtained.

There have been numerous studies in the non-isothermal cavity-filling simulation of thermosets. Recently, Kamal and Ryan [2] have extensively reviewed this subject. In dealing with the cavity filling of thermosets, it is customary to assume that the shear-rate dependence of the viscosity is not affected by curing due to the short filling time in comparison with long curing time [2]. On the other hand, in reactive injection-molding studies carried out by Castro et al. [44,45] and Manzione [46], the fluid is considered to be Newtonian but with the viscosity dependent on the extent of reaction. Furthermore, in the simulation of cavity filling of reactive injection molding reported by Domine and Gogos [47], the Carreau model for a shear-rate dependent viscosity influenced by state of cure has been incorporated. In these simulations, the induction time concept has not been considered in the curing stage.

Several attempts have already been made to model the injection molding process of rubber compounds. In particular, Bowers [48] has considered the one-dimensional cavity-filling process. In his simulation, the heat of vulcanization has been neglected and the induction time is assumed to be constant. The state-of-cure calculations are based upon the Claxton model [10] by defining an equivalent cure time. Hsich and Ambrose [49] used their own cure kinetic model [5] which is related to a physical property of the rubber compound. The degree of cure is defined based upon the difference between the maximum and minimum viscosity during one cycle. Similar to Bower's approach, the heat-source term was omitted from the energy equation and the induction time assumed to be constant. In addition, the solution of the momentum and the continuity equations was based upon a finite-difference method which is appropriate only for simple geometries.

In our previous paper [50], the Galerkin finite-element method was used for numerical formulation of the injection molding of rubber compounds. The shortcoming which was observed in [36] was improved by introducing a control volume concept to the Galerkin method, as developed in [38-41] and later adopted in [37]. In this procedure, the mesh is first laid out over the entire geometry of the cavity. Then the cavity-filling simulation is done without creation of any new meshes. In the present paper, a control-volume finite-element and finite-difference computer program has been developed for filling a thin cavity with an arbitrary three-dimensional shape. In this program, the runner systems are modeled with two-node circular-tube elements and the cavity domain is divided into a set of three-node triangular elements. The location of the nodal points of these elements are represented by x, y, and z coordinates. In other words, the present approach is not restricted to a flat geometry in the x-y plane. That is, the cavity and the runner could be located in three-dimensional space.

However, the flow movement in the cavity is planar, i.e., the cavity is assumed to be thin. Flow in the runner is considered to be axisymmetric and one-dimensional.

The Galerkin and the control-volume finite-element methods are suitable techniques for simulation of an injection molding process. The documentation of these techniques are given for the Galerkin finite-element method in [36,50] and the final form of the field equations for the control-volume method in [37]. However, a comparative numerical study of these techniques and the rate of their convergence with respect to the number of elements and mesh size especially in the presence of a "singularity" have not yet been discussed in the literature. Therefore, we attempt here to present a comprehensive review and evaluation of these techniques for the solution of the governing equations related to the injection molding process. Furthermore, a comparison of the solutions from these techniques is given.

A detailed example of this program is presented for the filling of a quarter disk cavity fed by two gates. In this simulation, a modified Cross model [51] is used as a viscosity function together with a curing kinetics formulation based upon our non-isothermal vulcanization model with incorporated non-isothermal induction time [11]. This program consists of the filling and post-filling stages. During the filling stage, pressure, velocity and temperature profiles, location of the melt fronts and weld-lines, and induction-time distributions are determined from the numerical solution of the momentum, continuity and energy equations together with the rheological and curing model. The evolution of the induction time is determined during the cavity-filling process. After the cavity is filled, the evolution of the state-of-cure distribution in the rubber molding is predicted from the solution of the energy equation with heat conduction and heat generation due to curing and with the initial temperature distribution and induction times predicted from the filling stage at the instant of fill.

7.2 THEORETICAL

7.2.1 Governing Equations

The governing equations given below are for the filling of a thin cavity with a generalized Newtonian inelastic fluid under non-isothermal conditions. The kinematics of the flow in the melt front (fountain flow) started to attract attention in the literature related to the simulation of the injection molding of thermoplastics [52–55]. Its importance in the injection molding of thermoplastics is due to the fact that highly-oriented fluid at the melt front is deposited at the cold wall and instantly solidifies leading to the highly-oriented frozen-surface layer [56]. In contrast, in injection molding of rubber compounds, a melt is cold while the wall is hot, and vulcanization starts after the non-isothermal induction time has been reached. Thus, detailed knowledge about temperature and velocity distribution in the melt front is needed in order to predict non-isothermal induction time accurately. Rigorous numerical simulation of the flow for a strip or tube geometry has been done [53–55] by treating the gapwise-curved melt front nodes as additional unknowns. In order to expand these techniques to an arbitrary planar geometry, one has to develop a full three-dimensional finite-element algorithm for determination of the location of the curved melt front. In the absence of a regrèssive and expensive numerical analysis, an approximate method [56] should be utilized for treatment of the kinematics of the melt front which appears to hold reasonably well compared to a rigorous numerical solution [54]. In order to avoid a cumbersome numerical calculation, an approximate method (explained in Section 7.3.4) is used here.

In general, the governing equations for a fluid-mechanics problem are expressed by the conservation of mass, momentum, and energy. The continuity equation for an incompressible flow is:

$$\frac{\partial u}{\partial x}+\frac{\partial v}{\partial y}+\frac{\partial w}{\partial z}=0 \tag{1}$$

where u, v, w are velocity components in the x, y, and z directions, respectively. Integration of Eq (1) with respect to the gapwise coordinate z and using the boundary condition $w = 0$ at the center line and at the wall, leads to expression of the continuity equation in terms of the gapwise-averaged velocity components:

$$\frac{\partial(b\bar{u})}{\partial x}+\frac{\partial(b\bar{v})}{\partial y}=0 \tag{2}$$

where 2b is the cavity thickness, and $\bar{u}$ and $\bar{v}$ are defined as:

$$\bar{u}=\frac{1}{b}\int_{o}^{b} u\, dz \tag{3}$$

$$\bar{v}=\frac{1}{b}\int_{o}^{b} v\, dz \tag{4}$$

It is customary to utilize the classical Hele-Shaw approximation for simulation of thin cavity filling. In particular, the momentum equation is formulated for shear-dominated flow in which the gapwise variation in shear stress is balanced by pressure variations in the x-y plane with the pressure gradient assumed to be zero in the gapwise direction. The x- and y-components of the momentum equation for an inelastic non-Newtonian fluid in the absence of body and inertial forces are:

$$\frac{\partial}{\partial z}(\eta\frac{\partial u}{\partial z})-\frac{\partial P}{\partial x}=0 \tag{5}$$

$$\frac{\partial}{\partial z}(\eta\frac{\partial v}{\partial z})-\frac{\partial P}{\partial y}=0 \tag{6}$$

where P is pressure and η is viscosity.

In a non-isothermal problem, the momentum and the continuity equations are coupled with the energy equation. Since the transverse dimension in a thin cavity is much smaller than the x-y dimensions, the thermal conductivity in the x-y directions can be ignored compared to that in the gapwise direction. For the filling of a thin cavity, the energy equation can then be written as follows:

$$\rho C_p(\frac{\partial T}{\partial t}+u\frac{\partial T}{\partial x}+v\frac{\partial T}{\partial y})=k_{th}\frac{\partial^2 T}{\partial z^2}+\Phi+\dot{Q} \tag{7}$$

where ρ, C_p, k_{th}, Φ, and $\dot{Q}$ are density, specific heat, thermal conductivity, viscous heating and the rate of heat generated by reaction of vulcanization. This latter quantity is defined according to a curing kinetic model given below.

In the energy equation, for convenience, the convective heat term $w(\partial T/\partial z)$ has been omitted although this is not justified on an order-of-magnitude basis. This has been done in most other cavity-filling simulation without any obvious loss in accuracy. Also, without considering fountain flow in the melt front region, the inclusion of this term does not seem justified.

Similarly, the governing equations for one-dimensional flow with circular geometry are:

$$\frac{\partial (R^2 \bar{u})}{\partial \ell} = 0 \tag{8}$$

$$\frac{1}{r}\frac{\partial}{\partial r}\left[\eta r \frac{\partial u}{\partial r}\right] - \frac{\partial P}{\partial \ell} = 0 \tag{9}$$

where Eqs (8) and (9) are the continuity and momentum equations, respectively, R is radius of the tube cross-section, u is the velocity in the length direction ℓ, and $\bar{u}$ is the average velocity:

$$\bar{u} = \frac{2}{R^2}\int_0^R urdr \tag{10}$$

The energy equation is:

$$\rho C_p \left[\frac{\partial T}{\partial t} + u\frac{\partial T}{\partial \ell}\right] = \frac{k_{th}}{r}\frac{\partial}{\partial r}\left[r\frac{\partial T}{\partial r}\right] + \Phi + \dot{Q} \tag{11}$$

Further, it should be noted that, during the post-filling stage, the energy Eq (7) or (11) degenerates to a simple transient heat-conduction equation with heat generation term due to vulcanization.

7.2.2 Viscous Heating

The viscous heating term, Φ, in the energy equation is defined as:

$$\Phi = \eta \dot{\gamma}^2 \tag{12}$$

where $\dot{\gamma}$ is the shear rate and η is the viscosity function which is given by a rheological model. The shear rate $\dot{\gamma}$ for two-dimensional cavity flow is:

$$\dot{\gamma} = \sqrt{\left[\frac{\partial u}{\partial z}\right]^2 + \left[\frac{\partial v}{\partial z}\right]^2} \tag{13}$$

which, for one-dimensional tubular flow, reduces to:

$$\dot{\gamma} = \left|\frac{\partial u}{\partial r}\right| \tag{14}$$

For many polymer melts, the power-law model could be employed for description of the flow behavior at high shear rates. However, this model overpredicts the viscosity behavior at low shear rates. In order to approximate the viscosity dependence on shear rate over a wide range of shear rates, the following rheological equation is adopted in our simulation:

$$\eta = \frac{\eta_0}{1 + \left[\frac{\eta_0 \dot{\gamma}}{\tau} \right]^{1-n}} \tag{15}$$

where

$$\eta_0 (T) = A \exp \{T_b/T\} \tag{16}$$

is the temperature dependent zero-shear-rate viscosity which can be applied over a wide range of shear rates. This equation is the so-called modified-Cross model [51]. It includes four material constants independent of temperature, namely τ, n, A and T_b, and has extensively been used in the simulation of injection molding of thermoplastics [36,37,57].

7.2.3 Curing Kinetic Model

The heat source term, $\dot{Q}$, included in the energy equation, represents heat generated from the vulcanization of rubber. The empirical model for the vulcanization reaction, developed earlier [11], is used in order to perform this calculation.

Let $\alpha = Q/Q_\infty$ denote the state or degree of cure, where Q is the heat released up to time t and Q_∞ is the total heat of reaction. Then an empirical model for isothermal curing kinetics is:

$$\frac{d\alpha}{dt} = \frac{\bar{n}}{k} t^{-(1+\bar{n})} \alpha^2 \tag{17}$$

where t is measured relative to induction time. This equation follows from the following equation [4]:

$$\alpha = \frac{k t^{\bar{n}}}{1 + k t^{\bar{n}}} \tag{18}$$

by taking a derivative of α with respect to time, t. We assume that k is based on an Arrhenius-type temperature dependence, i.e.,

$$k = k_0 \exp\{- E/RT\} \tag{19}$$

where k_0, E, $\bar{n}$ are kinetic constants independent of temperature. It should be noted that the ordinary derivative in Eq (17) has to be replaced by a substantial derivative if vulcanization occurs during the cavity-filling stage.

It is generally accepted that for most rubber compounds there are three periods during vulcanization, namely, induction period, curing stage and post-curing stage. During the induction period [10], chemical reaction does not take place. Accordingly, $\dot{Q} = 0$ for $t \leq t_i$, where t_i is the induction time. So, the question of how to predict the induction time for a non-isothermal process, such as injection molding, becomes relatively important. From a kinetic study under isothermal conditions (for example, by means of DSC) the induction time can easily be found experimentally. It is a material function dependent on reaction temperature. It has been proposed that this function has an Arrhenius-type temperature dependence [10]:

$$t_i = t_0 \exp\{T_0/T\} \tag{20}$$

where t_0 and T_0 are material constants independent of temperature.

Under isothermal conditions, the state of cure, α, is only a function of reaction time for any given sample. However, it becomes a function of both temperature and time during non-isothermal processes [2,6-11], i.e. $\alpha = \alpha(T,t)$. Total differential of α is:

$$d\alpha = \left(\frac{\partial \alpha}{\partial T}\right)_t dT + \left(\frac{\partial \alpha}{\partial t}\right)_T dt \tag{21}$$

By integration of this equation, one can get a numerical expression for the cumulative state of cure as follows:

$$\alpha_i = \alpha_{i-1} + \int_{T_{i-1}}^{T_i} \left(\frac{\partial \alpha}{\partial T}\right)_t dT + \int_{t_{i-1}}^{t_i} \left(\frac{\partial \alpha}{\partial t}\right)_T dt \tag{22}$$

Consequently, for non-isothermal conditions one can get the rate of vulcanization reaction based on the cumulative state of cure as:

$$\left(\frac{d\alpha}{dt}\right)_{\text{non-isothermal}} = \frac{(\alpha_i - \alpha_{i-1})}{(t_i - t_{i-1})}_{\text{cumulative}} \tag{23}$$

if a treatment of non-isothermal induction time is available.

During the non-isothermal process, it was proposed [11] to treat induction time in integral form. For this purpose, dimensionless time, $\bar{t}$, was introduced such that:

$$\bar{t} = \int_0^t \frac{dt}{t_i\,(T)} \tag{24}$$

where $t_i\,(T)$ denotes the dependence of induction time on the temperature during isothermal vulcanization as shown by Eq (20). When the value of the dimensionless time $\bar{t}$ reaches unity, the corresponding upper limit of the integral in Eq (24) is considered to be the induction time for the non-isothermal process. In the above approach, it is assumed that the non-isothermal process during the induction period is composed of many infinitesimal isothermal steps. Therefore, the onset of curing can now be easily identified during a simulation of the cavity-filling and post-filling stages of rubber compounds by employing this induction-time function. It is evident that induction time varies with position within the rubber bulk during curing under the same initial and boundary conditions. This also means that the onset of curing is different from point to point within the rubber bulk, which leads to a more accurate prediction of the temperature profile and state of cure distribution during injection molding.

This approach for evaluation of the induction time is still deficient in one respect. Fluid particles are in continuous motion during cavity filling. Thus, a Lagrangian approach or an Eulerian method with inclusion of a convective term would be more appropriate for treatment of the induction time during cavity-filling process [58]. However, here for the sake of simplicity, the Eq (24) is used.

7.3 NUMERICAL SOLUTION METHOD

7.3.1 Pressure Profile

In the previous section, the governing equations were presented for the filling of a thin cavity of arbitrary shape under non-isothermal conditions for a non-Newtonian inelastic fluid. Solution of these equations with a set of appropriate boundary conditions gives the velocity, pressure, and temperature profiles for a particular problem. The boundary conditions for two-dimensional flow are:

$$u = v = 0 \text{ and } T = T_W \quad \text{at} \quad z = b \tag{25}$$

$$\frac{\partial u}{\partial z} = \frac{\partial v}{\partial z} = \frac{\partial T}{\partial z} = 0 \quad \text{at} \quad z = 0 \tag{26}$$

where T_W represents the wall temperature. For one-dimensional tubular flow, the appropriate boundary conditions are:

$$u = 0 \quad T = T_W \quad \text{at} \quad r = R \tag{27}$$

$$\frac{\partial u}{\partial r} = \frac{\partial T}{\partial r} = 0 \quad \text{at} \quad r = 0 \tag{28}$$

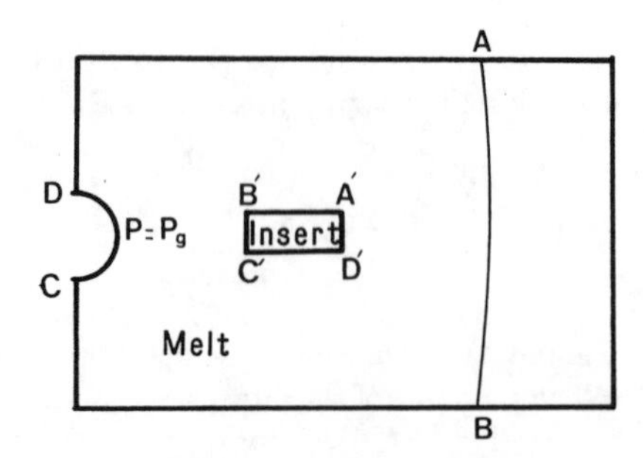

AB – Melt Front, R_m

CD – Gate, R_g

BC, AD, & A′B′C′D′ – Impermeable Boundary, R_i

Figure 7.1 Boundary of a cavity during the filling process.

In the case where the velocity and temperature profiles are not symmetric with respect to $z = 0$, Eq (26) is omitted and Eq (25) is considered for both cavity walls. The symmetry boundary conditions hold always for the tubular cross section.

In addition, the temperature at the inlet is assumed to be uniform, and equal to the melt temperature. Temperature in the melt front is assumed to be non-uniform, and evaluated based on the previous melt front temperature. This will be further elaborated in Section 7.3.4. The boundary of a cavity during the filling process consists of a melt front, impermeable boundaries, and the gate (Fig. 7.1).

In the melt-front region, R_m, melt is in contact with the atmosphere. In this region, we assume atmospheric pressure $P = 0$, and use this as a reference pressure. In the impermeable boundary region, R_i, melt is in contact with the outer boundary of the mold or an insert. The normal velocity components vanish on this type of boundary, thus $(\partial P/\partial n) = 0$. The pressure P_g at the gate, R_g, depends on the delivery system. This pressure is specified, or it could be related to the flow rate Q.

There are some similarities between the governing equations for one and two-dimensional flow. Thus, numerical formulations will be derived for two-dimensional flow, and then the results will be summarized for one-dimensional flow.

The velocity gradients are determined using a simple integration of Eqs (5) and (6) with the boundary conditions Eq (26) such that [36]:

$$\frac{\partial u}{\partial z} = -\frac{\Lambda_x z}{\eta}, \qquad \Lambda_x \equiv -\frac{\partial P}{\partial x} \tag{29}$$

$$\frac{\partial v}{\partial z} = -\frac{\Lambda_y z}{\eta}, \qquad \Lambda_y \equiv -\frac{\partial P}{\partial y} \tag{30}$$

By substituting Eqs (29) and (30) into Eq (13), the shear rate $\dot{\gamma}$ is:

$$\dot{\gamma} = \frac{\Lambda z}{\eta}, \qquad \Lambda \equiv \sqrt{\Lambda^2{}_x + \Lambda^2{}_y} \tag{31}$$

The velocity components u and v are determined by integration of Eqs (29) and (30), respectively, using the boundary conditions $u(b) = v(b) = 0$:

$$u = \frac{\Lambda_x}{\Lambda}\int_z^b \dot{\gamma}\, d\tilde{z} \tag{32}$$

$$v = \frac{\Lambda_y}{\Lambda}\int_z^b \dot{\gamma}\, d\tilde{z} \tag{33}$$

By substituting Eq (32) into Eq (3), the average gapwise velocity $\bar{u}$ is:

$$\bar{u} = \frac{\Lambda_x S}{b} \tag{34}$$

where
$$S = \int_o^b \frac{z^2}{\eta} dz = \frac{1}{\Lambda} \int_o^b \dot{\gamma} z dz \tag{35}$$

Similarly,
$$\bar{v} = \frac{\Lambda_y S}{b} \tag{36}$$

Substituting $\bar{u}$ and $\bar{v}$ into the continuity Eq (2), results in [36]:

$$\frac{\partial}{\partial x}(S \frac{\partial P}{\partial x}) + \frac{\partial}{\partial y}(S \frac{\partial P}{\partial y}) = 0 \tag{37}$$

Similarly, for one-dimensional flow we have:

$$\frac{\partial u}{\partial r} = -\frac{\Lambda_\ell r}{2\eta} \qquad \Lambda_\ell \equiv -\frac{\partial P}{\partial \ell} \tag{38}$$

$$\dot{\gamma} = \frac{\Lambda r}{2\eta}, \qquad \Lambda = \left| \Lambda_\ell \right| = \sqrt{\Lambda^2{}_\ell} \tag{39}$$

$$u = \frac{\Lambda_\ell}{\Lambda} \int_r^R \dot{\gamma}\, dr \tag{40}$$

$$\bar{u} = \frac{\Lambda_\ell S}{R^2} \tag{41}$$

where
$$S = \frac{1}{2} \int_o^R \frac{r^3}{\eta} dr = \int_o^R \dot{\gamma} r^2 dr \tag{42}$$

Substituting $\bar{u}$ in the continuity Eq (8), results in:

$$\frac{\partial}{\partial \ell}\left[S \frac{\partial P}{\partial \ell} \right] = 0 \tag{43}$$

The term Λ_ℓ / Λ in Eq (40) determines the direction of the velocity vector u.

Solution of Eq (37) or Eq (43) will determine the magnitude of the pressure. The components of the velocity vectors within the melt domain can be determined from Eqs (34) and (36) or (41). A finite element or a finite difference method could be used for solution of these equations. However, finite difference methods are not convenient for complex geometries.

The Galerkin finite element formulation for simulation of the injection molding of plastic material has been formulated by Hieber and Shen [36]. This formulation, with some modification, was extended for cavity filling simulation of rubber compounds which is applicable for any viscosity model [50]. In this procedure, the mesh is first generated throughout the geometry of the cavity. Then the cavity filling simulation is done without creation of any new meshes.

The finite element form of Eq (37) can be derived as follows [58]:

$$\int_A \left[\frac{\partial}{\partial x}\left(S\frac{\partial P}{\partial x}\right) + \frac{\partial}{\partial y}\left(S\frac{\partial P}{\partial y}\right) \right] W_n \, dA = 0 \tag{44}$$

where W_n is a set of independent functions, so called weighting functions. Subscript n in W_n represents the nodal point number, and A is the area of the melt domain.

Using Gauss's theorem [58], Eq (44) takes the following form:

$$\int_C S W_n \frac{\partial P}{\partial n} dC - \int_A S\left(\frac{\partial P}{\partial x}\frac{\partial W_n}{\partial x} + \frac{\partial P}{\partial y}\frac{\partial W_n}{\partial y}\right) dA = 0 \tag{45}$$

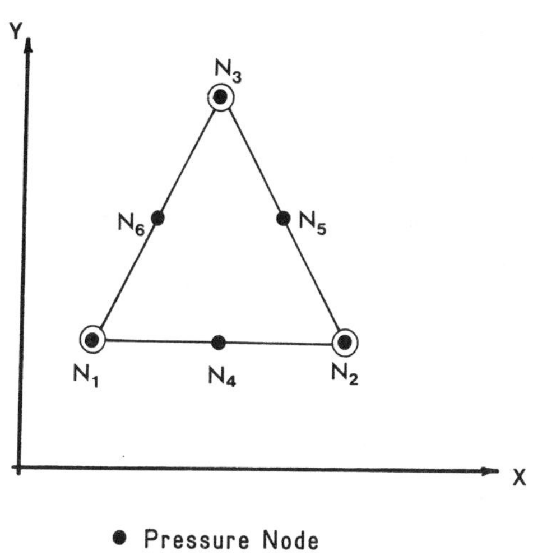

Figure 7.2 Triangular element used for the Galerkin finite element analysis.

where C is the boundary contour of the melt region in which nodal pressure is not specified, that is, the impermeable region. In the previous section, it was mentioned that the normal pressure gradient, $(\partial P/\partial n)$, at the impermeable boundary region is equal to zero. Therefore, the first integral vanishes in Eq (45) for this region. Pressure in the melt front region is assumed equal to atmospheric pressure $P = 0$, and we use this as a reference pressure. The pressure at the gate, P_g, depends on the delivery system. This pressure is specified or it could be related to the flow rate in the gate (see 7.3.5 Solution Algorithm). Therefore, field equations are not needed on these boundaries.

In particular, the flow domain is divided into a series of six nodes of triangular elements in the x-y plane (Fig. 7.2). Accordingly, for each element ℓ, the pressure field is assumed:

$$P(x,y,t) = \sum_{m=1}^{6} Q_m^{\ell}(x,y)P_m^{\ell}(t) \tag{46}$$

where P^{ℓ}_m is the nodal pressure, and Q^{ℓ}_m is the quadratic shape function [58] which is listed in Appendix.

In the context of the Galerkin finite element method, $W_n = Q_n$. By substituting Eq (46) into Eq (44) the continuity equation for any node N is:

$$\sum \sum_{j=1}^{6} S^{\ell} B^{\ell}_{i,j} P_{N'} = 0 \tag{47}$$

with

$$B^{\ell}_{i,j} = \int_{A^{\ell}} \left[\frac{\partial Q_i}{\partial x}\frac{\partial Q_j}{\partial x} + \frac{\partial Q_i}{\partial y}\frac{\partial Q_j}{\partial y} \right] dA^{\ell} \tag{48}$$

where A^{ℓ} is the area of an element ℓ, N = NELNOD(ℓ,i) and N' = NELNOD(ℓ,j), and the summation in Eq (47) is over all elements containing node N. The term NELNOD(ℓ,i) indicates the ith node of element ℓ, where $i = 1, 2, 3, \ldots 6$.

Alternatively, the governing equation for determination of the pressure profile could be derived by integrating the continuity equation. Let's assume the cavity domain is decomposed into a series of subdomains. In the literature, this subdomain is called a control volume, Figure 7.3. The volumetric flow rate for each region could be determined from the continuity equation:

$$2\int_A \left[\frac{\partial}{\partial x}\left[S\frac{\partial P}{\partial x} \right] + \frac{\partial}{\partial y}\left[S\frac{\partial P}{\partial y} \right] \right] dA = Q \tag{49}$$

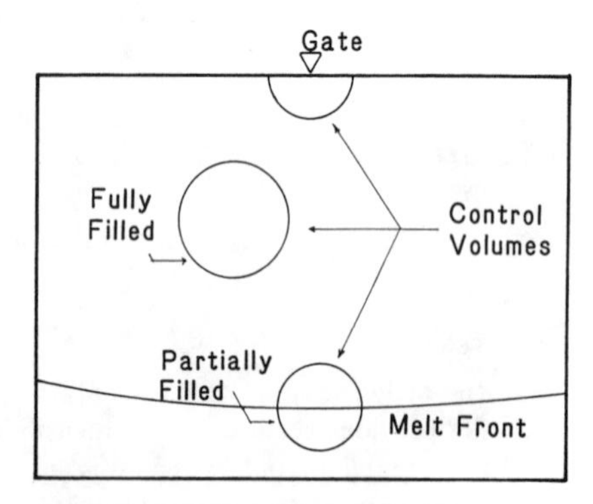

$q = q_0$ For Gate C. V.

$q = 0$ For Fully Filled C. V.

$q \neq 0$ For Partially Filled and $P = 0$

Figure 7.3 Schematic representation of the control-volume finite element method.

where A and Q are the area and volumetric flow rate entering across the boundary of each region.

Assuming that the thickness of the cavity remains constant with each control volume, and utilizing Green's theorem [58], Eq (49) will take the following form:

$$2\oint_C S\frac{\partial P}{\partial x}\cdot n_x \cdot dc + 2\oint_C S\frac{\partial P}{\partial y}\cdot n_y \cdot dc = Q \tag{50}$$

where $\oint$ means integration along contours which enclose the control volume. The term n is the unit vector normal to the line contour with n_x and n_y components in the x and y directions, respectively.

Equation (50) is the basis for determination of pressure and related quantities in the control volume finite element method. The flow domain is divided into a series of three-node triangular elements in the x-y plane, Figure 7.4. Accordingly, for each element, the pressure field may be assumed:

$$P(x,y,t) = \sum_{m=1}^{3} L_m^{\ell}(x,y)\, P_m^{\ell}(t) \tag{51}$$

with P^{ℓ}_m as defined in Eq (46) and:

$$L_m = \frac{1}{2A}(a_m + b_m x + c_m y) \tag{52}$$

is a set of linear shape functions where, A is the area of the triangular element and a_m, b_m, and c_m are listed in Appendix.

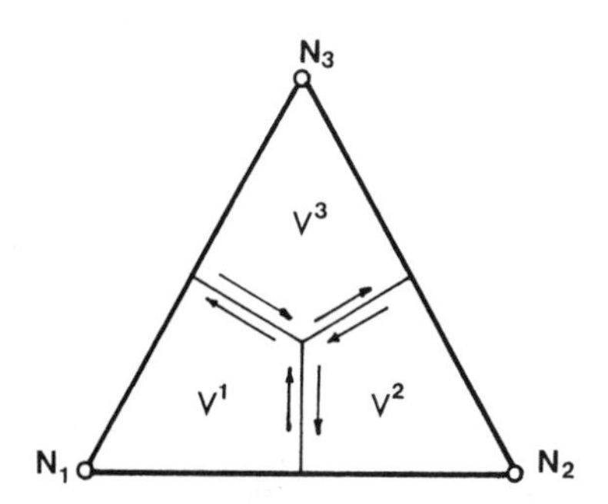

Figure 7.4 Subcontrol volume associated with each node of a triangular element in the control-volume finite element method.

For each node of a triangular element, an arbitrary volume is specified, for example, by connecting the centroid of each element to its midpoints [37]. The region which is enclosed by a contour in the counter-clockwise direction around each vertex node of a triangular element is called the subcontrol volume for this node. Subvolumes associated with nodes i=1,2,3 are indicated by v_i in Figure 7.4.

The control volume for each node is defined as the summation of subcontrol volumes which contains node N, and is enclosed by a contour C in the counter-clockwise direction (Fig. 7.5). Thus, the volumetric flow rate for each node is equal to the summation of all flow rates across the boundaries of the subcontrol volumes which contain node N.

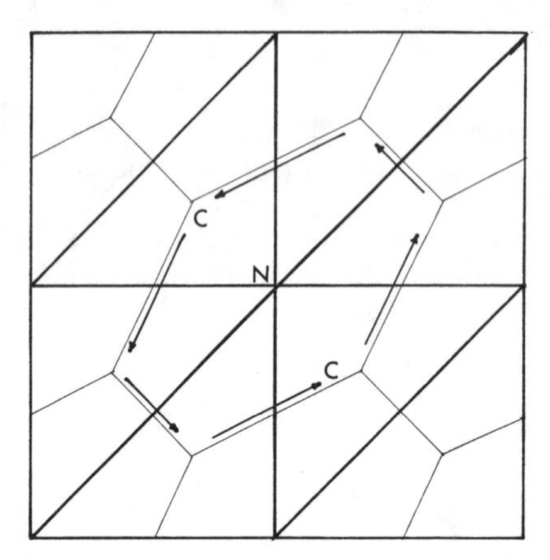

Figure 7.5 Schematic representation of the control-volume and the integration contour associated with a given node in the melt domain.

Let us assume q^{ℓ}_i represents the volumetric flow rate crossing the outer boundary of subcontrol volume associated with element ℓ at nodes i=1,2,3 (Fig. 7.4). By integration of Eq (50) along the paths which are shown in Figure 7.4, and using Eq (51), the volumetric flow rate in node i can be expressed as:

$$\begin{Bmatrix} q^{\ell}_1 \\ q^{\ell}_2 \\ q^{\ell}_3 \end{Bmatrix} = -\frac{S^{\ell}}{2A^{\ell}} \begin{bmatrix} b_1^2+c_1^2 & b_1\, b_2 + c_1 c_2 & b_1 b_3 + c_1 c_3 \\ & b_2^2 + c_2^2 & b_2 b_3 + c_2 c_3 \\ \text{symmetric} & & b_3^2 + c_3^2 \end{bmatrix}^{\ell} \begin{Bmatrix} P^{\ell}_1 \\ P^{\ell}_2 \\ P^{\ell}_3 \end{Bmatrix} \tag{53}$$

In general, the volumetric flow rate in a subcontrol volume which contains node N can then be expressed as:

$$q^{\ell}_N = \sum_{j=1}^{3} S^{\ell} \tilde{B}^{\ell}_{i,j} P_{N'} \tag{54}$$

where $\tilde{B}^{\ell}_{i,j}$ is defined via Eq (53) with N = NELNOD(ℓ,i) and N' = NELNOD(ℓ,j).

Similarly, the flow domain in one-dimensional flow is divided into a series of two-node tubular elements in the lengthwise direction (Fig. 7.6). The subcontrol volume for each node is

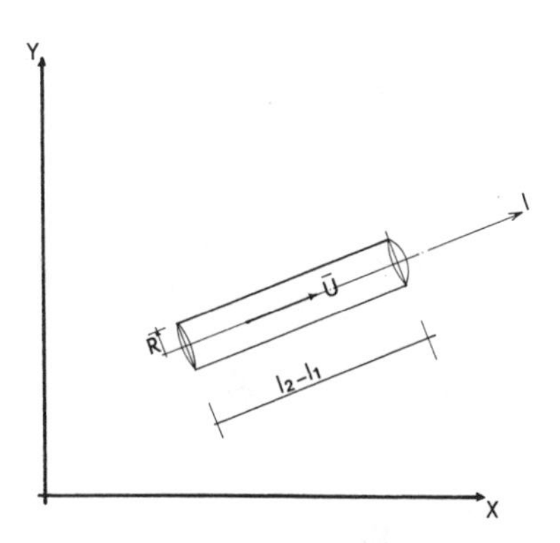

Figure 7.6 Tube element representation of the control-volume finite element method.

specified as half of the total volume of this element. The volumetric flow rate of nodes 1 and 2 can be derived in terms of average velocity, Eq (41), as follows:

$$q_1^{\ell} = -\bar{u}^{\ell} A^{\ell} = \pi S^{\ell} \frac{\partial P}{\partial \ell} \tag{55}$$

$$q_2^{\ell} = \bar{u}^{\ell} A^{\ell} = -\pi S^{\ell} \frac{\partial P}{\partial \ell} \tag{56}$$

where $\bar{u}^{\ell}$ and A^{ℓ} are the average axial velocity and the area of the cross section of element ℓ.

Similar to Eq (51), the pressure field may be represented as:

$$P(\ell,t) = \sum_{m=1}^{2} L_m(\ell) P_m^{\ell}(t) \tag{57}$$

where

$$L_m(\ell) = \frac{1}{\Delta \ell}(a_m + b_m \ell) \tag{58}$$

The parameters a_m and b_m are listed in Appendix, and $\Delta \ell$ is the length of element ℓ which can be expressed in terms of the nodal coordinate system as:

$$\Delta \ell = \sqrt{(x_2 - x_1)^2 + (y_2 - y_1)^2} \tag{59}$$

By substituting Eq (57) into Eqs (55) and (56), the volumetric flow rate in nodes 1 and 2 for tubular element ℓ is:

$$\begin{Bmatrix} q_1^{\ell} \\ q_2^{\ell} \end{Bmatrix} = -\frac{\pi S^{\ell}}{\Delta \ell} \begin{bmatrix} 1 & -1 \\ -1 & 1 \end{bmatrix} \begin{Bmatrix} P_1^{\ell} \\ P_2^{\ell} \end{Bmatrix} \tag{60}$$

Similar to Eq (54), for one-dimensional flow we have:

$$q_N^{\ell} = \sum_{j=1}^{2} S^{\ell} \tilde{B}_{i,j}^{\ell} P_{N'} \tag{61}$$

where N = NELNOD(ℓ,i), and N' = NELNOD(ℓ,j).

The control volume associated with each node, at any instant, may be filled, empty, or partially filled. For an incompressible flow, the volumetric flow rate for any completely filled node is equal to zero:

$$Q_N = \sum_{1}^{ku} q_N^{\ell} = \sum_{1}^{ku} \sum_{j=1}^{2\,or\,3} S^{\ell} \tilde{B}_{i,j}^{\ell} P_{N'} = 0 \tag{62}$$

where ku is the total number of elements containing node N, N' = NELNOD(ℓ,j), and i is such that N = NELNOD(ℓ,i), with j=1,2,3 for triangular elements and j=1,2 for tubular elements. By satisfying Eq (62) for all completely filled nodes, except entrance nodes, the magnitude of the nodal pressures can be determined. For entrance nodes, through which the fluid enters the system, the pressure is specified. This pressure is related to a specified volumetric flow rate through Eq (62) by setting Q_N equal to Q, where Q is the flow rate at the entrance. Partially-filled nodes are taken as melt-front nodes, with the pressure at these nodes assumed to be equal to atmospheric pressure, P = 0. The volumetric flow rate for a partially-filled node can be determined from Eq (62), where Q_N is different than zero and it corresponds to how fast its volume is getting filled up. The total volumetric flow rate into all the melt-front nodes should be equal to the specified flow rate at the entrance nodes.

Owing to the non-linearity of Eq (62), an iterative method must be adopted for solution of these equations. The pressure is assumed to converge when the residual of the continuity and the absolute difference between the calculated flow rate in the melt-front nodes and the specified flow rate is less than certain specified values.

Due to numerical stability, an under-relaxation method is used for calculation of the pressure:

$$P = (1 - \dot{\gamma}_p)\, P_{old} + \dot{\gamma}_p\, P_{new} \tag{63}$$

where $\dot{\gamma}_p$ is under-relaxation parameter (lying between 0 and 1), and P_{old} and P_{new} denote the pressure at the previous and present iteration.

The parameter S is a pressure dependent quantity (due to the shear-thinning behavior of the viscosity as well as a possible explicit pressure dependence); therefore, it should be updated at the end of every i^{th} iteration. Similar to Eq (63), an under-relaxation method is used for updating the parameter S. The explicit form of S is in terms of Λ and $\dot{\gamma}$. The value of Λ is constant within each element in the control volume method. In the Galerkin method, pressure gradients are linear, and the value of Λ at the centroid of each triangular element is used for computation of the shear rates and the parameter S.

The value of shear rate could be determined by substituting Eq (15) into Eq (39) or (31) for axisymmetric or two-dimensional analysis, respectively:

$$\dot{\gamma} = \frac{\Lambda r}{2\eta_0}\left[1 + \left[\frac{\eta_0}{\tau}\dot{\gamma}\right]^{1-n}\right] \quad \text{axisymmetric} \tag{64}$$

$$\dot{\gamma} = \frac{\Lambda z}{\eta_0}\left[1 + \left[\frac{\eta_0}{\tau}\dot{\gamma}\right]^{1-n}\right] \quad \text{two-dimensional} \tag{65}$$

It can be seen that an iterative method is required for solution of Eqs (64) and (65). The value of $\dot{\gamma}$ from the previous time or iteration can be taken as an initial guess.

So far, the governing equation for determination of pressure has been derived for a two-dimensional geometry. It is possible to expand the control volume formulation to a three-dimensional geometry using an appropriate coordinate transformation.

Let us assume an element ℓ is located in the coordinate system x, y, and z (Fig. 7.7). The volumetric flow rate, q_i, and nodal pressure, P_i, are scalar quantities and they are irrelevant to the position of a coordinate system. That is, for any arbitrary coordinate system, the volumetric flow rate at subcontrol volumes 1, 2, and 3, and the nodal pressure are the same. Therefore, a local coordinate system (x', and y') in the plane of the element is assumed. The coordinates of the vertex nodes of a triangular element are:

$$x_1' = 0, \quad x_2' = \ell_3, \quad x_3' = \ell_2 \cos\phi, \qquad y_1' = 0, \quad y_2' = 0, \quad y_3' = \sqrt{\ell_2^2 - x_3'^2} \tag{66}$$

where

$$\cos\phi = \frac{\ell_2^2 + \ell_3^2 - \ell_1^2}{2\,\ell_2\,\ell_3} \tag{67}$$

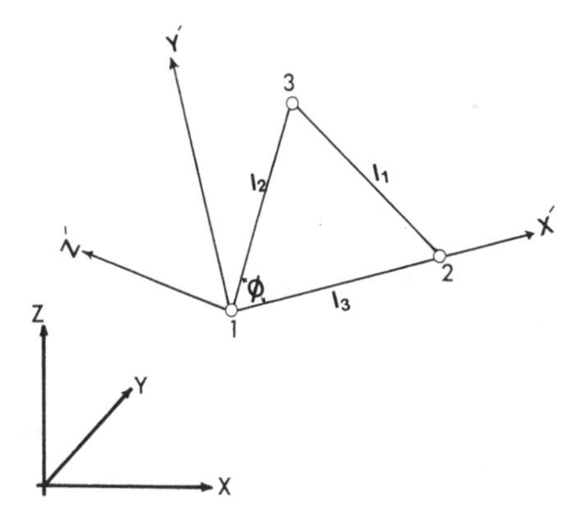

Figure 7.7 Three-dimensional representation of the control-volume finite element method.

and

$$\ell_1 = \sqrt{(x_2 - x_3)^2 + (y_2 - y_3)^2 + (z_2 - z_3)^2}$$

$$\ell_2 = \sqrt{(x_3 - x_1)^2 + (y_3 - y_1)^2 + (z_3 - z_1)^2} \tag{68}$$

$$\ell_3 = \sqrt{(x_1 - x_2)^2 + (y_1 - y_2)^2 + (z_1 - z_3)^2}$$

Accordingly, the volumetric flow rate crossing the boundary of each subcontrol volume of a triangular element is given by:

$$q_N^{\ell} = \sum_{j=1}^{2or3} S^{\ell} \tilde{B}'^{\ell}_{i,j} P_{N'} \tag{69}$$

where j=1,2 for tubular element and j=1,2,3 for triangular element and $\tilde{B}'_{i,j}$ is the $\tilde{B}_{i,j}$ evaluated based upon the local (x', y') coordinate system.

The governing equation for determination of pressure is then given by:

$$Q_N = \sum_{1}^{ku} q_N^{\ell} = \sum_{1}^{ku} \sum_{j=1}^{2or3} S^{\ell} \tilde{B}'^{\ell}_{i,j} P_{N'} = 0 \tag{70}$$

7.3.2. Temperature Profile

In order to solve the energy equation for determination of the temperature profile, a finite-difference method is employed [36]. In the finite-difference representation of the energy equation, an implicit form is used for the gapwise conduction term with the convective, viscous heating, and heat-of-reaction term being evaluated at the earlier time. Therefore, by knowing the velocity component, viscous heating, and heat-of-reaction terms at time t_k, the temperature profile can be determined at time $t_{k+1} = t_k + \Delta t$. The magnitude of Δt is selected such that only one partially-filled node gets filled during each time step [37].

Owing to a discontinuity in the pressure gradient across element boundaries, the shear rates are not continuous at a given node. Therefore, shear rates are evaluated at the centroid of each element, and an averaging process is used for determination of nodal shear rate which leads to calculation of nodal viscous heating.

Similarly, thermal convection terms $u(\partial T/\partial x)$ and $v(\partial T/\partial y)$ are evaluated at the centroid of each element. Then the corresponding nodal values are determined. However, due to numerical stability, only contributions from the adjacent upstream elements are considered. This can be effected by evaluating the dot product of the velocity and displacement vectors connecting the centroid of each element to a given node [36] (Fig. 7.8).

During the solution of energy equation, it was assumed that the temperature at the entrance node is equal to the injection melt temperature and the temperature at the contact between the melt and the wall remains constant and equal to the specified wall temperature. The temperature at the melt-front node was determined according to the temperature at the previous melt fronts. This will be elaborated on in Section 7.3.4.

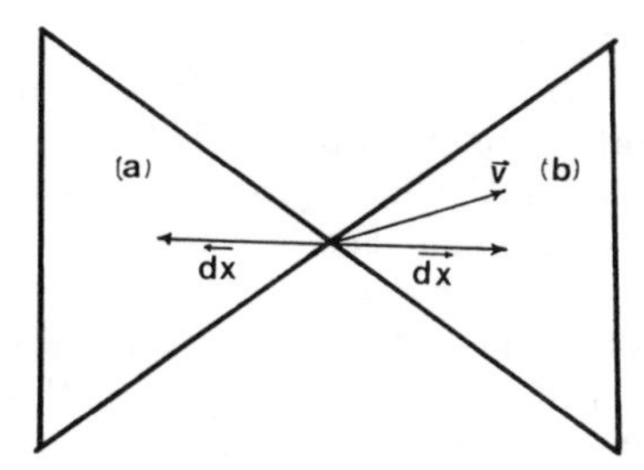

Figure 7.8 Determination of upstream and downstream elements. (a) upstream; (b) downstream.

7.3.3 Advancement of the Melt Front

In the control-volume finite-element method, the volumetric flow rate for a partially filled node can be determined from Eq (62) where $Q_N \neq 0$. The time increment is selected such that only one of the partially-filled nodes gets filled during each time step. Therefore, the melt-front location is defined in terms of the partially-filled nodes. The position of the fully-filled nodes can be determined without any ambiguity, whereas those of the partially-filled nodes are not well defined. In order to represent the melt front for partially-filled nodes, the vector $\Delta\vec{x} = \Sigma\vec{v}\Delta t$ is calculated at the centroid of the triangular elements containing these nodes (Fig. 7.9), where $\vec{v}$ is the average gapwise velocity vector. Subsequently, it is assumed that the vector $\vec{x}$ represents the advancement of the melt front for partially filled nodes. In addition, the melt front location at the mold boundary is defined orthogonal to the cavity boundary. Similar procedures were adopted for advancement of the melt front based upon the Galerkin finite-element method.

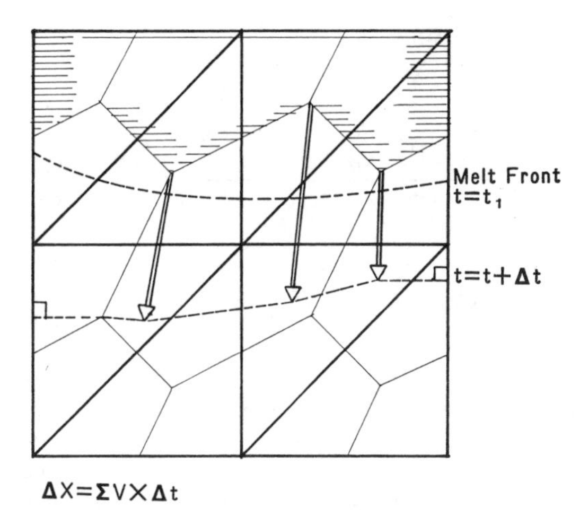

Figure 7.9 Advancement of the melt front.

It seems that this approximation of the melt front depends on the number of elements and their shape. However, studies have shown that for a reasonable number of elements, the overall shape and movement of the melt front are independent of the element number and shape [37]. The significant drawback in this approximation is oscillation of the melt front. This could be minimized by increasing the number of elements.

The calculated total flow rate at the melt front nodes may not be the same as the specified flow rate. This is due to the existence of an allowable error in the calculation of the nodal pressures. In order to satisfy the continuity equation, the volumetric flow rate along the melt front is scaled such that the conservation of mass is satisfied.

7.3.4 Treatment of the Melt Front

In the present simulation, temperature, induction time, and degree of cure at the melt-front nodes are assumed non-uniform in the gapwise direction. They are determined based on the respective values at the previous position of the melt front.

For example, let us assume the partially-filled node 1 in Figure 7.10 is filled by elements (1), (2), and (3). The temperature at this node is calculated according to a bulk temperature of the previous melt front:

$$T_{1,j,k+1} = \left\{ \frac{T'_{3,j}\,\bar{V}^1 + T'_{3,j}\,\bar{V}^2 + T'_{4,j}\,\bar{V}^2 + T'_{4,j}\,\bar{V}^3}{\bar{V}^1 + 2\bar{V}^2 + \bar{V}^3} \right\} \tag{71}$$

where

$$T'_{i,j} = \frac{T_{i,j,k+1} + T_{i,j,k}}{2} \tag{72}$$

and $\bar{V}^{\ell}$ is the magnitude of the velocity vector at the centroid of element ℓ.

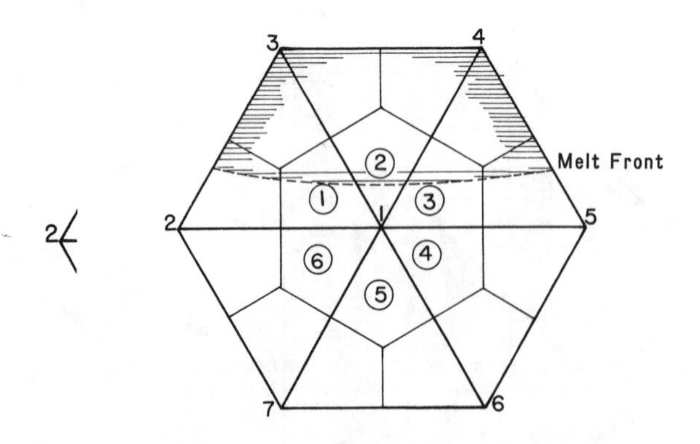

Figure 7.10 Determination of the values of variables in the melt front.

Similar treatments are utilized for determination of the induction time, state of cure, etc. for a newly filled node. However, since these values are not known for time t_{k+1}, these quantities are evaluated according to their magnitude at time t_k.

7.3.5 Solution Algorithm

It is assumed that all control volumes, associated with the entrance node, are filled isothermally at $t = t_1$. Thus, the temperature profile is known for time $t = t_1$. Furthermore, arbitrary nodal pressures are assigned for all completely-filled nodes. Based on these assumed nodal pressures, and given temperature profile, the parameter S is evaluated. Then with solving Eq (62) for every completely-filled node (except entrance nodes), the updated nodal pressures are obtained. If these updated nodal pressures do not satisfy the convergence criterion, the equation is resolved with these updated nodal pressures. Since the parameter S is a pressure dependent quantity, it is updated every i^{th} iteration during solution of Eq 62. In the Galerkin method, Eq (47) is used for determination of pressure.

In the control-volume formulation, the entrance pressure is calculated from Eq (62) by setting $Q_N = Q$, where Q is the specified entrance flow rate. In the Galerkin method, gate pressures are updated as:

$$\left[P_g\right]_{new} = \left[\frac{Q}{Q_c}\right]^n \left[P_g\right]_{old} \tag{73}$$

where P_g is the entrance pressure, subscripts new and old indicate present and previous iteration, Q_c is the calculated flow rate in the control volume associated with the entrance nodes, and n is a constant in the viscosity model, Eq (15). In the Galerkin method, Q_c is calculated according to the average gapwise velocity at the centroid of an element. This velocity profile is linear within each element, whereas it is constant in the control-volume technique.

In order to accelerate convergence of the pressure field, the nodal pressure is scaled after updating the nodal gate pressure:

$$(P)_{new} = \frac{(P_g)_{new}}{(P_g)_{old}}(P)_{old} \tag{74}$$

Following satisfaction of the pressure-convergence criterion, the velocity components are determined. The increment of Δt is selected such that only one partially-filled node gets filled during each time step. At the end, the temperature profile for time $t_2 = t_1 + \Delta t$ is calculated based on the velocity, viscous heating, and heat-of-reaction at time t_1. This procedure is continued until the cavity is filled.

7.3.6 Comparison of the Techniques

Both, the Galerkin method based upon linear shape function, and the control-volume approach, lead to the same nodal equations. The latter can be easily verified. Namely, the coefficients in Eq (53) are identical to those obtained using the Galerkin method with linear shape function. Accordingly, as far as the field equation for the pressure distribution is concerned, the control-volume method is equivalent to the Galerkin method, merely differing as to whether the shape function is linear, or quadratic. Hence, for a given melt domain, temperature field and viscosity model, both approaches must give the same pressure distribution upon mesh refinement. Specifically, for the same number of elements, the Galerkin method based upon the quadratic shape function must be more accurate (provided the problem is well posed) than the Galerkin method based upon the linear shape function.

For the purpose of a numerical evaluation and comparison of these techniques, an equilateral triangular and a square cavities (1 cm along each side) was considered for an isothermal simulation of cavity filling. The finite-element meshes for different mesh refinements, together with location of the gate and melt front, are given in Figures 7.11a-d. Calculations are for the volumetric flow rate 1.0 cm^3/sec and melt and mold temperature 393 K. The cavity thickness is $2b = 0.5$ cm. The material properties are given in Section 7.4.

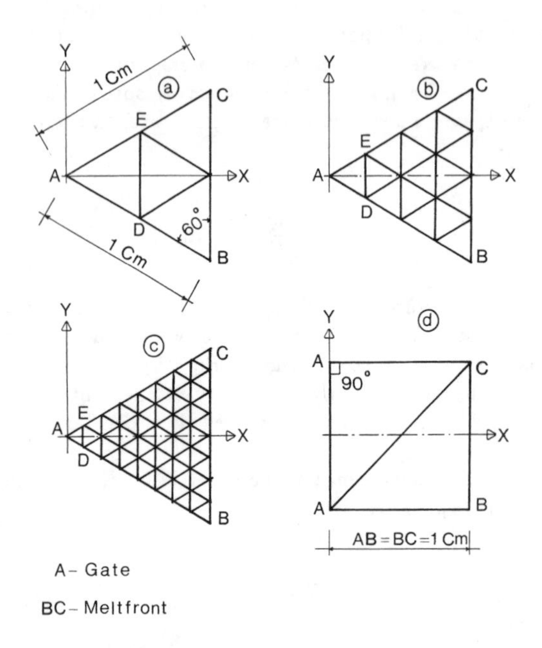

Figure 7.11 Triangular (a,b,c) and square (d) cavity with corresponding different number of meshes (a=4, b=16, c=64, and d=2) used for the comparison of the Galerkin and the control-volume finite element method.

The predicted pressure profile along x-axis for the triangular cavity for different mesh sizes depicted in Figures 7.11a-c are shown in Figures 7.12a-c. In general, the predicted pressures converge to a unique result with the mesh refinement except at the vicinity of the gate. The predicted gate pressures from the Galerkin method (curve 1) are higher than those from the control-volume method (curve 2). The mesh refinement leads to a reduction and an increase of pressure in the Galerkin method and the control-volume method, respectively. Relative change in the gate pressure is more noticeable in the Galerkin method. For example, when the number of elements is increased from 4 to 64 the relative change is respectively -13% and 4% for the Galerkin and the control-volume formulation. If the Galerkin method is more accurate than the control-volume method (as stated earlier), then the Galerkin formulation should converge faster with mesh refinement than the control-volume approach. The numerical results indicate contrary. The question, why does the divergence occur at the neighborhood of the gate, will be addressed in the following paragraph.

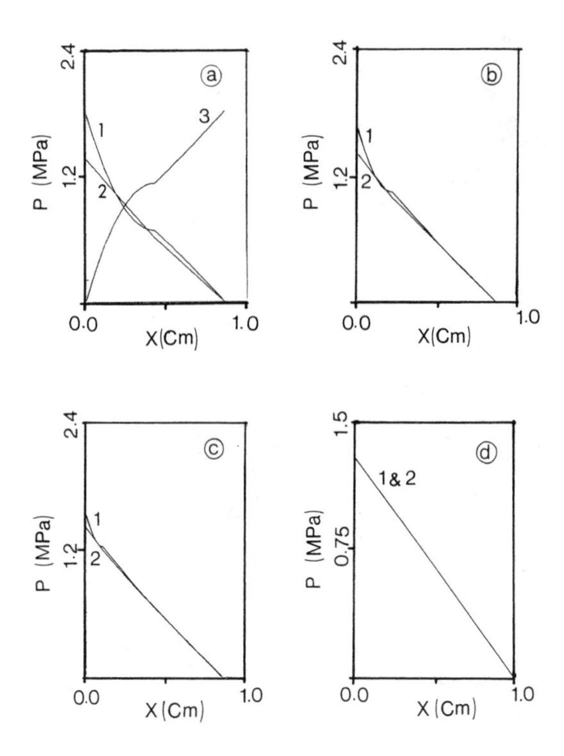

Figure 7.12 Comparison of the pressure profiles along x-axis obtained from the Galerkin (curves 1 and 3) and control-volume (curve 2) method, results correspond to a=4, b=16, c=64 elements for triangular geometry and d=2 elements for square geometry, shown in Figure 7.11.

The predicted pressure gradient ($\partial P/\partial x$) from the Galerkin formulation reduces linearly along x-axis from point A, Figures 7.12a-c, to zero along DE. Location of Point A and side DE are shown in Figures 7.11. As a consequence, the predicted velocity from the Galerkin method reduces linearly from the point A to zero along DE. This implies that the flow rate across DE is equal to zero. Similar results are obtained by reversing the flow direction (BC-gate, and A-melt front). The predicted pressure from the Galerkin method for an inverse flow with the mesh configuration depicted on Figure 7.11a, are shown in Figure 7.12a (curve 3). It is noted that the pressure gradient becomes zero along DE. The latter implies that the flow rate across DE is also equal to zero. This element, where ($\partial P/\partial x$) = 0 along DE, is defined as a singular element with singular source point (point-A).

This spurious pressure behavior in the first (left-most) element which contains node A is inherent to the Galerkin method with quadratic shape function. The results may be improved by imposing ($\partial P/\partial x$) $\neq 0$ along DE, such that the net flow rate will be non zero. The singular element can be avoided by using a line gate instead of a point gate. This gate should cover at least one side of an element. The same is true for the filling of the last node of a cavity.

In the absence of a singularity, both solutions predict the same results as can be seen from Figure 7.12d, where the pressure profile along x-axis for an isothermal filling of a square cavity, Figure 7.11d, is shown. Due to the simplicity of the cavity geometry, two elements were sufficient for modeling the melt domain. Of course, in the case of more complex geometry, a higher number of elements are required for getting identical results.

For further comparison, simulation of cavity filling for a quarter disk cavity using the Galerkin and the control-volume method was considered. The geometry of the cavity with corresponding dimensions and finite-element mesh for different mesh sizes with the location of the gate is given in Figure 7.13. The material properties, the boundary, and the processing conditions are given in Section 7.4.

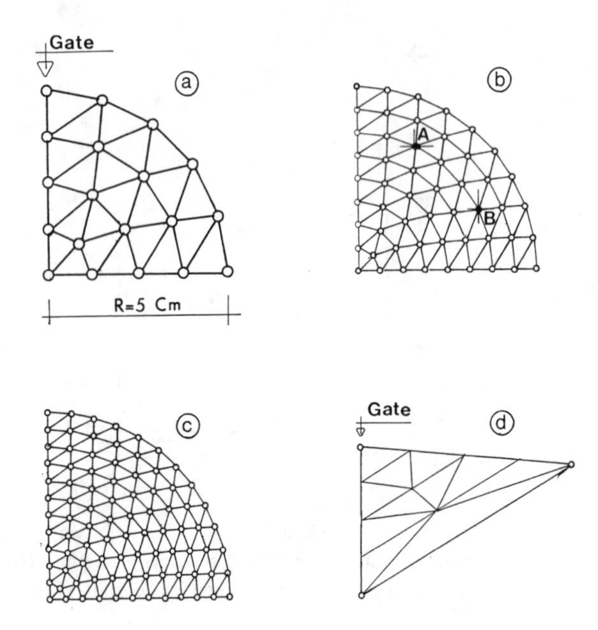

Figure 7.13 Same as Figure 7.11, but for a quarter of a disk cavity, a=23, b=79, c=142 elements, d=refined mesh in the triangular element which contains the gate node (total number of element is 152 in this case).

The predicted gate pressure during filling stage from these two methods for different mesh sizes are shown in Figure 7.14. In particular, for 23 elements, the predicted pressure from the Galerkin method is much higher than those from the control-volume method. By increasing the number of elements from 23 to 79, the predicted gate pressure decreases for both solutions. However, these decreases are more noticeable in the Galerkin solution. For the first time step, predicted gate pressure for 79 elements using the Galerkin method is 1247 kPa, whereas, this is 793 kPa for the control-volume method. The difference between gate pressures remains almost constant during filling stage, and it becomes almost double upon the completion of the filling stage, $P_g = 5357$ kPa for the Galerkin method, and $P_g = 4400$ kPa for the control-volume solution.

The predicted pressures during filling stage at two internal nodes A and B (locations of these points are given in Fig. 7.13b), are also shown in Figure 7.14. Contrary to the gate pressure, the predicted results are almost identical during the filling stage, but diverge from each other upon completion of the filling stage. That is, both solutions at these points, agree during filling except near the end of the filling.

In order to investigate the convergence of these solutions for 79 elements, the number of elements were increased to 142 elements. Few changes were observed in the predicted gate pressures, Figure 7.14. That is, both solutions have converged to different results, and further refining the mesh size did not alter the results significantly. The higher pressure obtained in the Galerkin formulation is due to the presence of a singularity point at two ends of the cavity. The first singular point is located at the gate. It is similar to the example shown in Figures 7.11a-c with the gate A and meltfront BC. The second singular point is located at the end of the cavity. The latter is similar to an inverse flow situation with the gate BC and meltfront A. Due to the presence of singularity at the end of the cavity the predicted pressure at the internal nodes (for example, nodes A and B in Fig. 7.13b) based on Galerkin method are also increased at the fill time. In order to reduce the effect of the singularity, the triangular element which contains the gate node is subdivided into 11 elements. The predicted gate pressure up to 8 seconds with this new mesh configuration (152 elements) is shown in Figure 7.14. The predicted gate pressure from the Galerkin method using 152 elements is shifted significantly toward the results obtained from the control-volume method, whereas the changes in the gate pressure are very small in the control-volume solution.

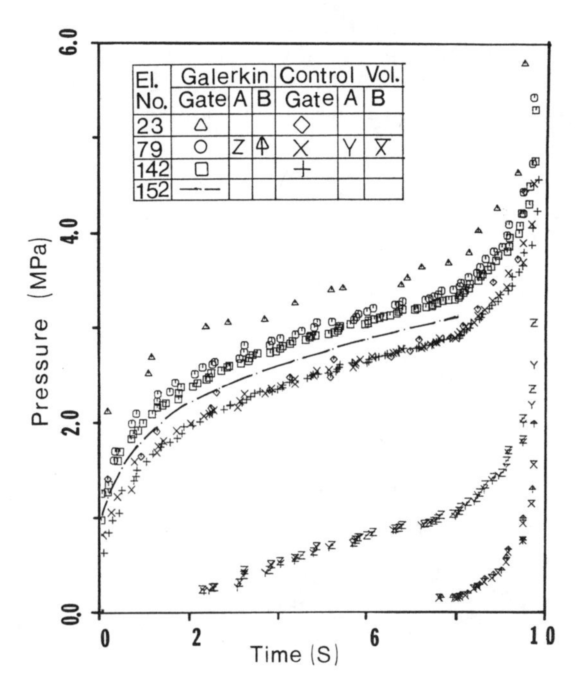

Figure 7.14 Comparison of the pressure development obtained from the Galerkin and control-volume method using different mesh sizes (shown in Fig. 7.13) for filling of a quarter disk cavity.

Finally, these examples lead to the conclusions that: (1) the Galerkin and the control-volume method leads to the same results upon mesh refinement; (2) the predicted pressure from the Galerkin method is higher than that from control-volume approach in the presence of a singular point; (3) the effects of singularity can be reduced by refining the mesh in the case of the Galerkin formulation; (4) the control-volume method is not sensitive to the presence of a singularity.

7.4 NUMERICAL EXAMPLE

7.4.1 Processing Conditions and Cavity Dimensions

In this section, an example of a numerical simulation of cavity filling of a rubber compound using the control-volume finite-element procedure will be presented. Results for a single-gated cavity, using the Galerkin finite-element method, have been presented earlier [50]. The cavity is a quarter of a circular disk with radius, $R = 50$ mm, and half thickness, $b = 2.5$ mm with two edge gates. The melt is transferred through a system of circular tubes, segment $\overline{ABCD}$ in Figure 7.15, to the gates of the cavity. The radius of the runners is equal to 5.0 mm, being reduced linearly to 2.5 mm from C to point D. The geometry of the cavity and the runners has been represented in terms of 142 three-node triangular elements and 15 two-node one-dimensional tubular elements, respectively. The gapwise direction (b) is divided into 16 equal segments. The volumetric flow rate is assumed equal to 1.0 cm^3/sec with the inlet melt temperature 393 K and mold wall temperature 453 K, respectively.

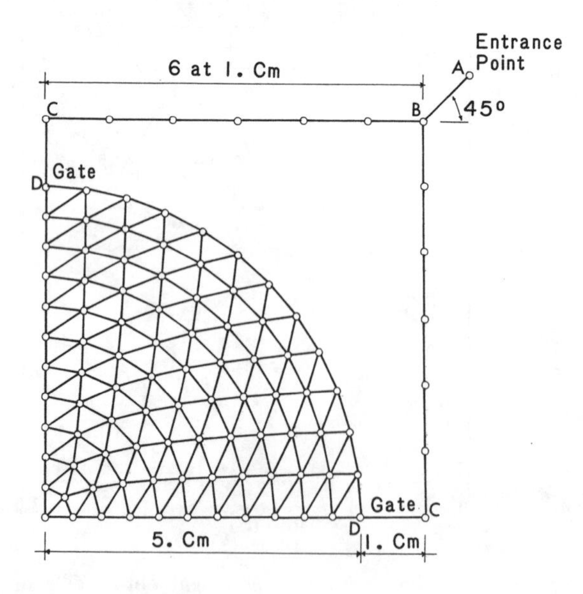

Figure 7.15 The geometry of the cavity and runner system with corresponding finite element meshes.

7.4.2 Physical Properties of Rubber

The rubber compound 19E was considered for this simulation. The physical properties of this rubber are: density ρ = 950 kg/m^3, specific heat C_p = 1.574 x 10^3 J/kg K, and thermal conductivity k_{th} = 0.23 w/m. K. The recipe for this rubber compound is listed in Table 7.1.

Table 7.1 Recipes of rubber compound, 19E.

SBR 1204	100.0
CAB-0-Sil	20.0
Diethylene Glycol	1.5
Active Zinc Oxide	0.8
Stearic Acid	1.0
Wingstay L	0.5
Butyl Zimate	0.3
Tetrabutylthiuram Disulfide	1.0
Sulfur	2.0
Si 69 Coupling Agent	1.5
	128.6

7.4.3 Rheological Constants

A Monsanto Processibility Tester (MPT) has been used to determine the rheological properties of the rubber compound 19E. Three temperatures, namely 71, 90, and 110° C, and shear rates from 10^0 to 10^3 sec^{-1} have been employed. By curve fitting the experimental data (see Fig. 7.16), one can obtain the viscosity constants for the modified Cross model, Eq (15). This has been done by a computer program specifically developed for this purpose. The viscosity constants are: n = 0.0814, τ = 328.5 kPa, T_b = 3965 K, and A = 2.047 Pas.

7.4.4 Curing Kinetic Constants

Differential Scanning Calorimetry (DSC) has been employed to determine the curing kinetic constants. The DSC instrument used in the present investigation was a DuPont 9900 thermal analyzer with a 910 DSC module. The DSC method is based on the differential enthalpic analysis. The reactive sample is tested with an inert reference at the same temperature level. The curing constants for an isothermal kinetic model can be determined from a set of isothermal tests by curve fitting Eq (17) to the experimental data; $d\alpha/dt$ versus time (Fig. 7.17). In addition, from these experimental data the induction time, t_i, can be measured for different temperatures. Then, the constants in Eq (20) are determined by fitting a straight line to Ln t_i versus 1/T, (Fig. 7.18). The constants related to the induction time are t_0 = 8.336 x 10^{-13} sec, and T_0 = 14060 K. The curing kinetic constants for the empirical model, Eq (17), has been determined to be k_0 = 1.16 x 10^{13} sec$^{-\bar{n}}$, E = 1.66 x 10^5 J/gmol, and $\bar{n}$ = 3.178. The total heat of reaction measured by the DSC method is Q_∞ = 20 J/cm^3.

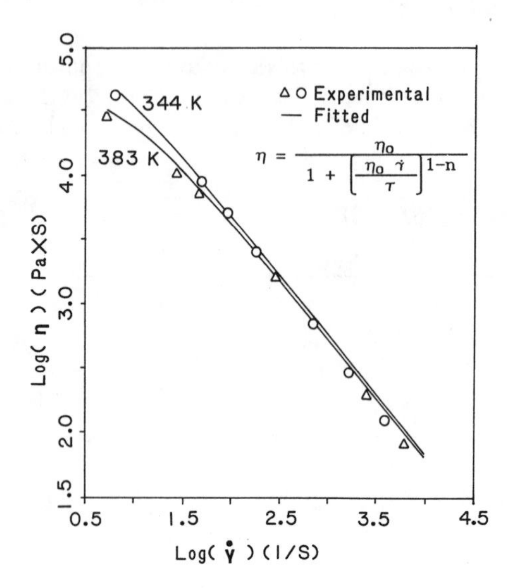

Figure 7.16 Apparent viscosity versus shear rate for sample 19E [11].

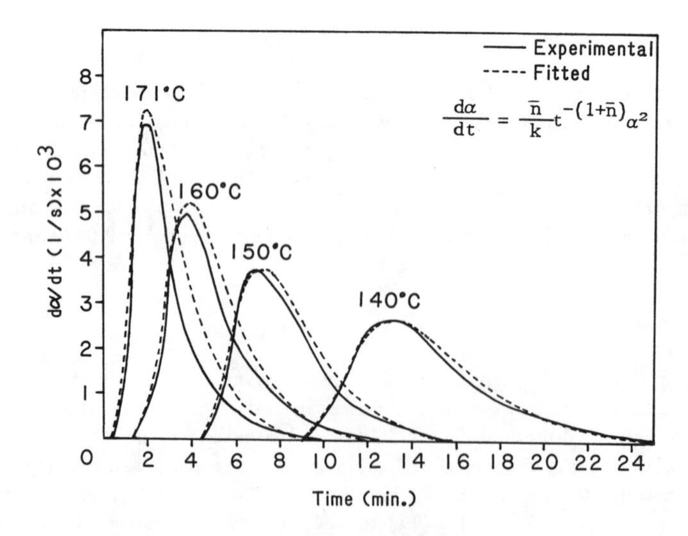

Figure 7.17 Rate of isothermal vulcanization reaction versus time [11].

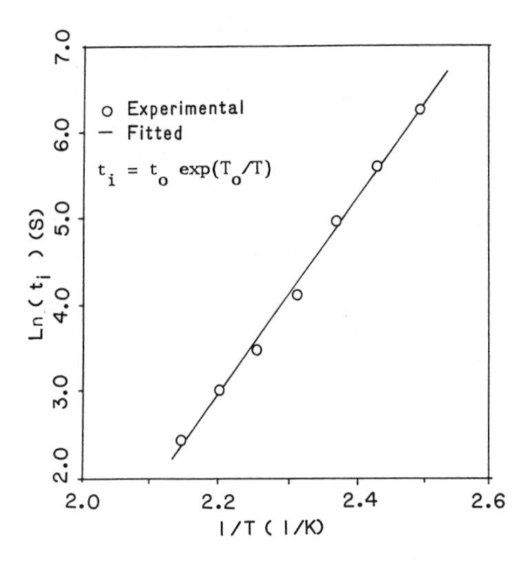

Figure 7.18 Natural logarithm of induction time versus reciprocal temperature [11].

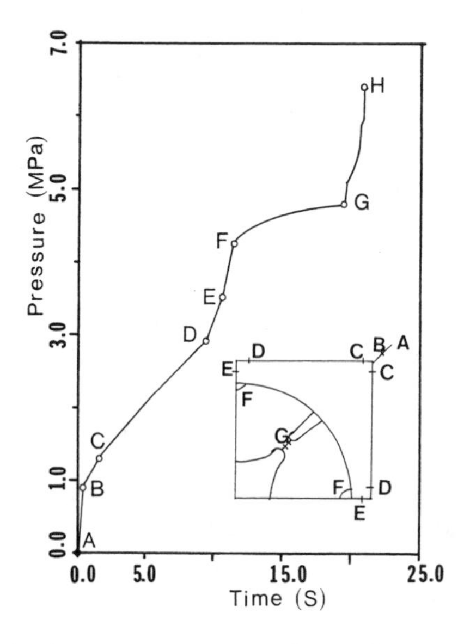

Figure 7.19 Predicted entrance pressure versus time and corresponding melt front locations.

7.4.5 Predictions For Filling Stage

Pressure. Predicted entrance pressure versus time during cavity filling is shown in Figure 7.19. Three distinct segments may be observed in this Figure. The location of the melt front corresponding to this pressure is also shown in Figure 7.19. The segments AB, BC, CD, DE, and EF correspond to the movement of the melt front in the runners. Point F is the first location of the melt front in the cavity, and point G is the first location of the weld-line. During the movement of the melt front from point F to G, the entrance pressure increases gradually. Whereas, beyond point G, the entrance pressure increases very rapidly.

Predicted contours of constant pressure at the instant of fill, $t_f = 20.9$ sec, are shown in Figure 7.20. Since the cavity is filled from both sides with the same flow rate, the contours of pressure are symmetric with respect to the line passing through the center and making a 45° angle with respect to the x-axis. This line will be referred to as the symmetry line from now on.

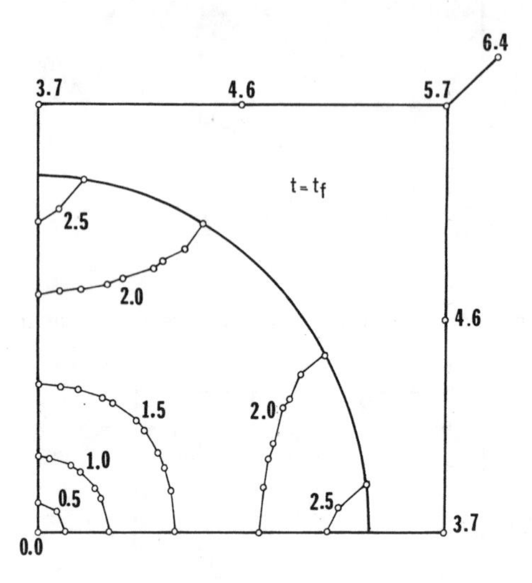

Figure 7.20 Predicted contours of constant pressure (MPa) at the time of fill, $t_f = 20.9$ sec.

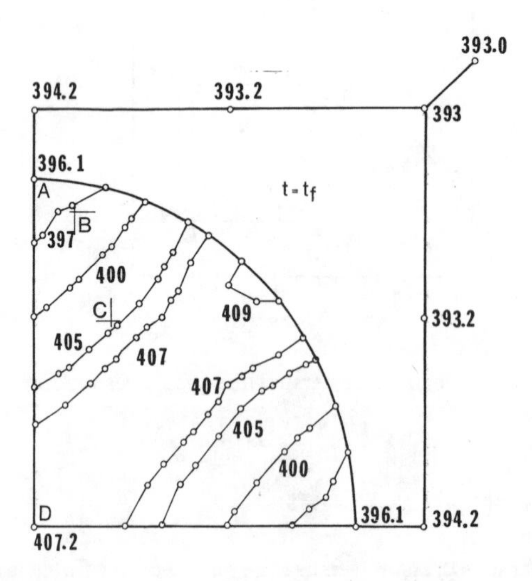

Figure 7.21 Predicted contours of constant centerline temperature (K) at the time of fill $t_f = 20.9$ sec.

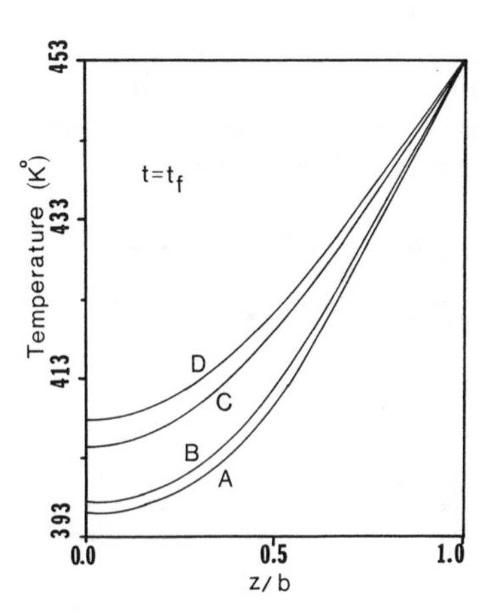

Figure 7.22 Temperature distribution in gapwise direction at the time of fill at different locations in the cavity. Locations A, B, C, and D are indicated in Figure 7.21.

Temperature. The predicted contours of constant temperature, at the instant of fill along the center line, are shown in Figure 7.21. Similar to the pressure, the temperature contours are symmetric with respect to the symmetry line. The temperature distribution at the time of fill in the gapwise direction for points A, B, C, and D are shown in Figure 7.22. The location of these points are given in Figure 7.21.

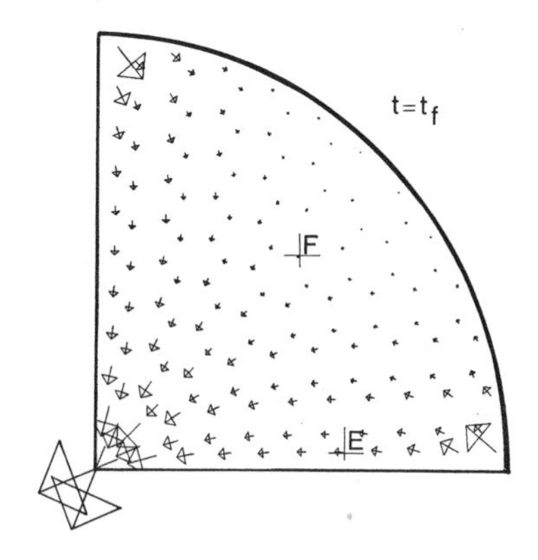

Figure 7.23 Gapwise average velocity vector at the time of fill.

Velocity. In order to visualize the flow movement inside the cavity, the gapwise average velocity vectors at the centroid of each element, at the time of fill, are shown in Figure 7.23. As expected, due to reduced flow areas at the cavity end, higher velocity vectors are observed in these two regions.

Velocity distributions in the gapwise direction for points E and F are shown in Figure 7.24. The locations of points E and F are given in Figure 7.23. For point E, located in the proximity of the cavity edge, the velocity component u is higher than v. For point F, located in the vicinity of the cavity midpoint, velocity components u and v are almost equal.

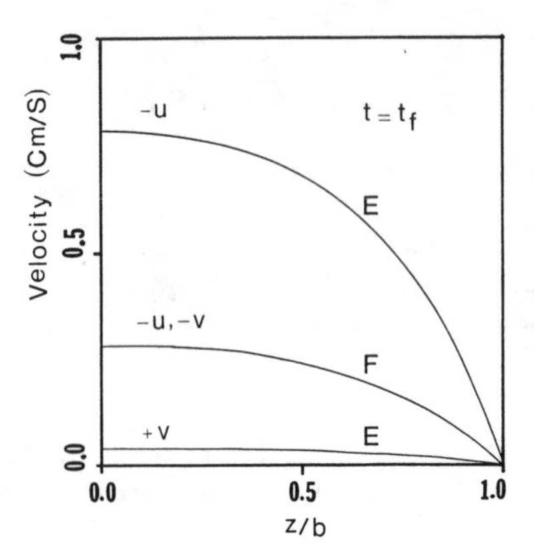

Figure 7.24 Velocity distribution in gapwise direction at the time of fill at two locations in the cavity. Locations E and F are indicated in Figure 7.23.

Melt Front. Predicted melt front, as explained in Section 7.3.3, is shown in Figure 7.25. The predicted melt front slightly oscillates. By increasing the number of elements the degree of oscillation reduces, and the melt front can be predicted more accurately. As expected, the location of the weld-line is at the center line of the cavity. The weld-line is defined when two completely-filled nodes join together or a node is filled from two different melt fronts.

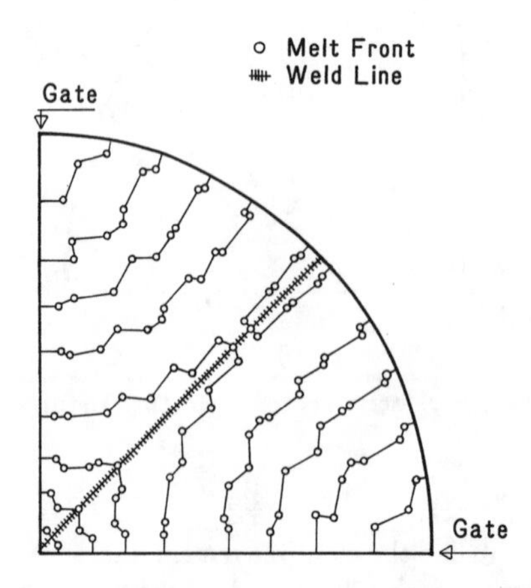

Figure 7.25 Predicted melt front location at various times during cavity filling until fill time, $t_f = 20.9$ sec.

7.4.6 Predictions for Post-Filling Stage

Center line temperature trace for point C during the curing process is shown in Figure 7.26. The locations of this point is indicated in Figure 7.21. Due to the heat of reaction, the temperature trace shows an overshoot. The maximum value of the temperature overshoot is about 3 K. Temperature distributions for point C in the gapwise direction at a time of 150 seconds are shown in Figure 7.27. This indicates that the temperature overshoot corresponds to the center line.

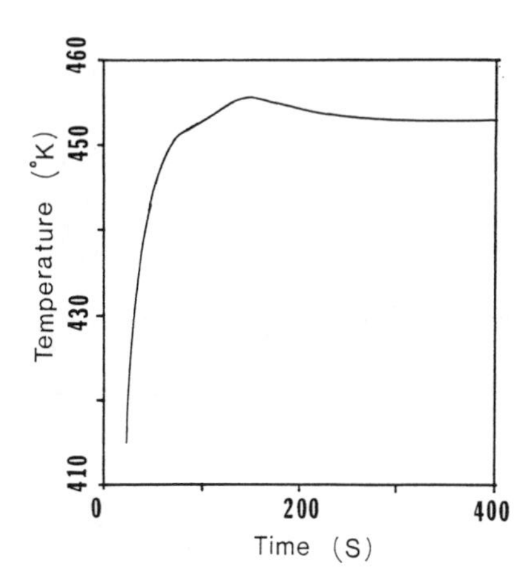

Figure 7.26 Center line temperature traces in post-filling stage at point C in the cavity. Location is given in Figure 7.21.

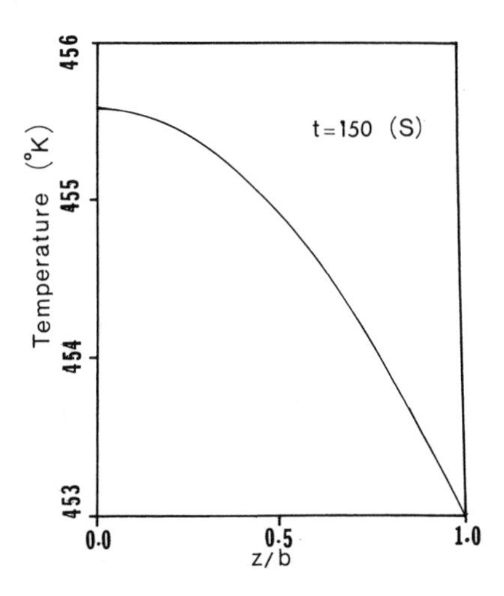

Figure 7.27 Gapwise temperature distribution at point C during post-filling stage at time t = 150.0 sec. Location of this point is shown in Figure 7.21.

The predicted state of cure development for point C during the post-filling process at $z = 0$ cm is shown in Figure 7.28. Induction time for this point is about 80 seconds, which is greater than the fill time. With adjusting the processing conditions, one can optimize injection molding process such that curing starts right after the filling is completed.

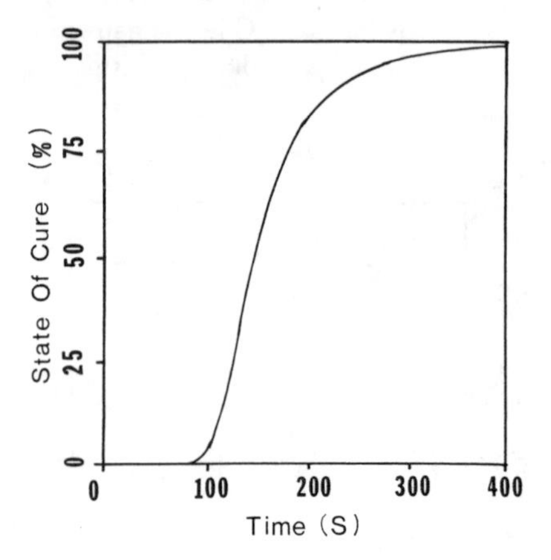

Figure 7.28 Temporal development of state of cure (%) in the center line at point C during post-filling process. Location of this point is indicated in Figure 7.21.

Contours of the state of cure at a time of 150.0 seconds for $z = 0$ are shown in Figure 7.29. The predicted state of cure for points A, B, C, and D in the gapwise direction at time $t = 150.0$ seconds are shown in Figure 7.30. The state of cure is highly non-uniform in the gapwise direction and increases as one approaches the solid wall.

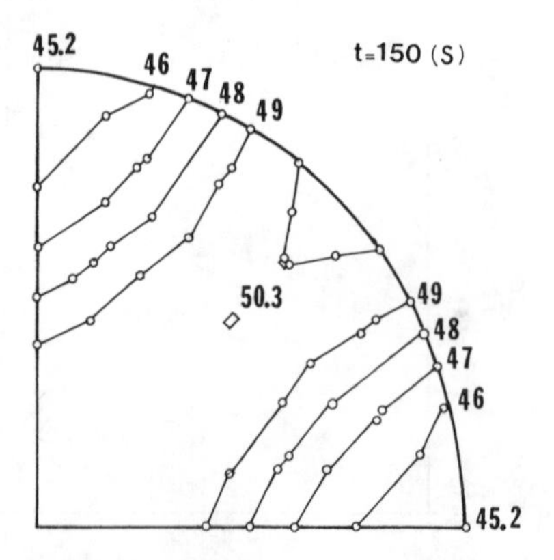

Figure 7.29 Contour of state of cure (%) at the center line for $t = 150.0$ sec.

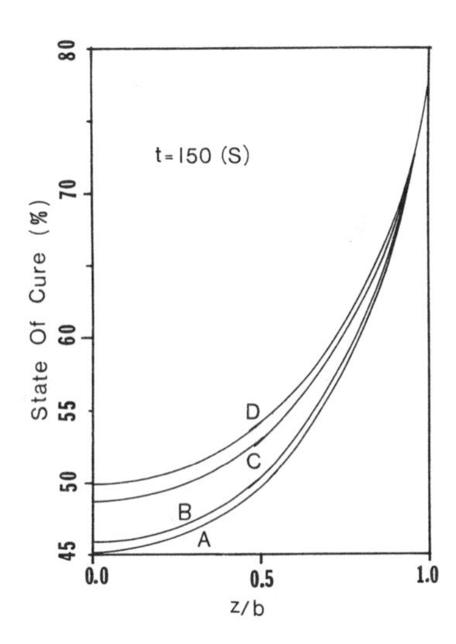

Figure 7.30 Predicted state of cure distribution in the gapwise direction at t = 150.0 sec. Locations are indicated in Figure 7.21.

7.5 CONCLUDING REMARKS

A general finite-element and finite-difference computer program has been developed for the filling of a thin cavity with an arbitrary three-dimensional geometry. This program employs a control-volume finite-element method for solution of the continuity and momentum equations, and a finite-difference method for solution of the energy equation in the gapwise direction; it is able to accommodate any type of rheological and vulcanization model. A detailed example of application of this program is presented for the filling of a quarter disk cavity with two gates.

This program consists of the filling and post-filling stages. In the filling stage, the geometry of the cavity is divided into a series of three-node triangular elements in the planar coordinates. The runners are modeled with a series of one-dimensional two-node tubular elements in the lengthwise direction. The gap or radial direction is divided into a number of equal segments. The quantities such as pressure, velocity, temperature, induction time, melt front, and weld-line are determined during the filling stage. In this simulation, the induction-time calculation starts from the time, $t = 0$, when the rubber enters the mold. Thus, at the end of filling time, the induction time is non-uniform and specified at each point in the mold.

In the post-filling stage, the evolution of cure distribution in the rubber molding is predicted from solution of the energy equation with heat generation due to reaction. The initial conditions of temperature and induction time distribution are determined from the filling stage at the time of fill.

With the aid of this program, one can avoid scorch problems by optimizing the processing design such that the filling time is equal to the non-isothermal induction time.

The Galerkin and control-volume methods are equivalent, merely differing as to whether the pressure shape function is quadratic (Galerkin) or linear (control-volume). The predicted results from these two techniques are compared. It is found that the Galerkin method and the control-volume technique converge to the same results upon refining the mesh sizes. However, in the presence of a singularity, the predicted pressure from the Galerkin method is higher than that from the control-volume approach. The effect of the singularity can be reduced by refining the mesh in the Galerkin method. The control-volume method is not sensitive to the presence of a singularity.

ACKNOWLEDGEMENT

This work has been supported by a grant MSM 8514952 from the National Science Foundation, Division of Engineering. The authors wish to thank GenCorp for providing the rubber compound samples. Valuable comments on the manuscript made by Dr. C.A. Hieber are highly appreciated.

NOMENCLATURE

A: Area of melt domain or preexponent in an Arrhenius type temperature dependent viscosity
b: Half gap thickness of the cavity
C: Boundary contour
C_p: Specific heat
E: Activation energy of reaction of vulcanization
i: Node number
j: Node number
k: Rate of vulcanization
k_{th}: Thermal conductivity
k_0: Preexponent in an Arrhenius type temperature dependent rate of vulcanization
l: Coordinate axis in length direction of runner
L_m: Linear shape functions
n: Power-law index or normal unit vector
$\bar{n}$: Kinetic constant
N: Node number
P: Pressure
P_g: Pressure at the entrance
P_m: Nodal pressure
P_{old}: Pressure at the previous iteration
P_{new}: Pressure at the present iteration
q^1_i: Volumetric flow rate crossing the outer boundary of sub-control volume associated with finite element 1 at node i
$\dot{Q}$: Rate of heat generated by reaction of vulcanization
Q: Heat released up to time t
Q_c: Calculated flow rate in the control volume
Q_∞: Total heat of vulcanization
Q_m: Quadratic shape functions
r: Radial coordinate
R: Radius of tube
t_0: Preexponent in an Arrhenius type temperature dependent induction time

t_i: Induction time
$\bar{t}$: Dimensionless time
T: Temperature
T_b: Temperature sensitivity of viscosity
T_o: Temperature sensitivity of induction time
T_w: Mold wall temperature
u: Velocity component in x direction
$\bar{u}$: Gapwise-averaged velocity component in x direction
v: Velocity component in y direction
$\bar{v}$: Gapwise-averaged velocity component in y direction
$\vec{v}$: Gapwise-averaged velocity vector
w: Velocity component in z direction
W_n: Weighting functions
x: Coordinate axis in length direction
y: Coordinate axis in width direction
z: Coordinate axis in gapwise direction
α: Degree or state of cure
α_i: Cumulative state of cure
η: Apparent viscosity
η_o: Zero-shear rate viscosity
Φ: Viscous heating term
$\dot{\gamma}$: Shear rate
τ: Constant in modified-Cross model
Λ: Pressure gradient
ρ: Density

REFERENCES

1. Isayev, A.I., Ed., "Injection and Compression Molding Fundamentals," Marcel Dekker, New York (1987).
2. Kamal, M.R., Ryan, M.E., in "Injection and Compression Molding Fundamentals," Isayev, A.I., Ed., Marcel Dekker, New York (1987).
3. Piloyan, G.O., Ryahchikkov, I.D., Novikova, O.S., *Nature*, *212*, 1229 (1966).
4. Kamal, M.R., Sourour, S., *Polym. Eng. Sci.*, *13*, 1 (1973).
5. Hsich, H.S.Y., *J. Appl. Polym. Sci.*, *27*, 3265 (1982).
6. MacCallum, J.R., Tanner, J., *Nature*, *225*, 1127 (1970).
7. Dutta, A., Ryan, M.E., *Thermochim. Acta*, *33*, 87 (1979).
8. Stevenson, J.F., *Polym. Process. Eng.*, *1*, 203 (1983-84).
9. Lee, L.J., *Polym. Eng. Sci.*, *21*, 483 (1981).
10. Claxton, W.E., Liska, J.W., *Rubber Age*, 237, May 1964.
11. Isayev, A.I., Deng, J.S., *Rubber Chem. Technol.*, *61*, 340 (1988).
12. Manzione, L.T., Ed., "Applications of Computer Aided Engineering in Injection Molding," Hanser Publishers, Munich (1987).
13. Bernhardt, E.C., Ed., "Computer Aided Engineering for Injection Molding," Hanser Publishers, Munich (1983).
14. Toor, H.L., Ballman, R.L., Cooper, L., *Mod. Plast.*, *38*, (Dec.) 117 (1960).
15. Harry, D.H., Parrott, R.G., *Polym. Eng. Sci.*, *10*, 209 (1970).

16. Lord, H.A., Williams, S.G., *Polym. Eng. Sci.*, *15*, 569 (1975).
17. Pearson, J.R.A., "Mechanical Principles of Polymer Melt Processing," Pergamon Press, Oxford (1976).
18. Kamal, M.R., Kenig, S., *Polym. Eng. Sci.*, *12*, 294 (1972).
19. Kamal, M.R., Kenig, S., *Polym. Eng. Sci.*, *12*, 302 (1972).
20. Berger, J.L., Gogos, C.G., *Polym. Eng. Sci.*, *13*, 102 (1973).
21. Wu, P.C., Huang, C.F., Gogos, C.G., *Polym. Eng. Sci.*, *14*, 223 (1975).
22. Stevenson, J.F., *SPE Tech. Papers*, *22*, 282 (1976).
23. Stevenson, J.F., Hauptfleish, R.A., Hieber, C.A., *Plast. Eng.*, *32*(12), 34 (1976).
24. Stevenson, J.F., *Polym. Eng. Sci.*, *18*, 577 (1978).
25. Stevenson, J.F., *Polym. Eng. Sci.*, *19*, 849 (1979).
26. Nunn, R.E., Fenner, R.T., *Polym. Eng. Sci.*, *17*, 811 (1977).
27. Williams, G., Lord, H.A., *Polym. Eng. Sci.*, *15*, 553 (1975).
28. Hieber, C.A., Upadhyay, R.K., Isayev, A.I., *SPE Tech. Papers*, *29*, 698 (1983).
29. Richardson, S.M., Pearson, H.J., Pearson, J.R.A., *Plast. Rubber Process.*, *5*, 55 (1980).
30. Kuo, Y., *SPE Tech. Papers*, *24*, 135 (1978).
31. Austin, C., in "Computer Aided Engineering for Injection Molding," Bernhardt, E.C., Ed., Hanser Publishers, Munich (1983).
32. Hieber, C.A., *SPE Tech. Papers*, *28*, 356 (1982).
33. Richardson, S., *J. Fluid Mech.*, *56*, 609 (1972).
34. Kamal, M.R., Kuo, Y., Doan, P.H., *Polym. Eng. Sci.*, *15*, 863 (1975).
35. Kuo, Y., Kamal, M.R., *AIChE J.*, *22*, 661 (1976).
36. Hieber, C.A., Shen, S.F., *J. Non–Newtonian Fluid Mech.*, *7*, 1 (1980).
37. Wang, V.W., Hieber, C.A., Wang, K.K., *J. Polym. Eng. Sci.*, *7*, 21 (1986).
38. Broyer, E., Tadmor, Z., Gutfinger, C., *Israel J. Technol.*, *11*, 189 (1973).
39. Tadmor, Z., Broyer, E., Gutfinger, C., *Polym. Eng. Sci.*, *14*, 660 (1974).
40. Broyer, E., Gutfinger, C. Tadmor, Z., *Trans. Soc. Rheol.*, *19*, 423 (1975).
41. Krueger, W.L., Tadmor, Z., *Polym. Eng. Sci.*, *20*, 426 (1980).
42. White, J.L., *Polym. Eng. Sci.*, *15*, 44 (1975).
43. Schlichting, H., "Boundary-Layer Theory," Sixth Ed., McGraw-Hill, New York (1969).
44. Castro, J.M., Macosko, C.W., *AIChE J.*, *28*, 250 (1982).
45. Castro, J.M., Gonzales, V.M. Macosko, C.W., *SPE Tech. Papers*, *27*, 363 (1981).
46. Manzione, L.T., *SPE Tech. Papers*, *27*, 338 (1981).
47. Domine, J.D., Gogos, C.G., *Polym. Eng. Sci.*, *20*, 847 (1980).
48. Bowers, S., *Rubber Plast. News*, *70*, May 18, 1987.
49. Hsich, H.S.Y., Ambrose, R.J., *Rubber World*, *29*, July (1987).
50. Isayev, A.I., Sobhanie, M., Deng, J.S., *Rubber Chem. Technol.*, *61*, 906 (1988).
51. Cross, M.M., *Rheol. Acta*, *18*, 609 (1979).
52. Gogos, C.G., Huang, C.F., Schmidt, L.R., *Polym. Eng. Sci.*, *26*, 1457 (1986).
53. Mavridis, H., Hrymak, A.N., Vlachopoulos, J., *Polym. Eng. Sci.*, *26*, 449 (1986).
54. Behrens, R.A., Crochet, M.J., Denson, C.D., Metzner, A.B., *AIChE J.*, *33*, 1178 (1987).
55. Coyle, D.J., Blake, J.W., Macosko, C.W., *AIChE J.*, *33*, 1168 (1987).
56. Tadmor, Z., *J. Appl. Polym. Sci.*, *18*, 1753 (1974).
57. Hieber, C.A., in "Injection and Compression Molding Fundamentals," Isayev, A.I., Ed., Marcel Dekker, New York (1987).

58. Sobhanie, M., Deng, J.S., Isayev, A.I., *J. Appl. Polym. Sci.*, Appl. Polym. Symp., *44*, 115 (1989).
59. Zienkiewicz, O.C., "Finite Element Method," 3rd Ed., McGraw Hill, New York (1977).

APPENDIX

Mathematical Formulation of Shape Functions

According to [59] for a triangular element, the quadratic shape function, Q_i, and linear shape function, L_i, are:

$$\left.\begin{aligned} Q_1 &= (2L_1 - 1)L_1 & Q_4 &= 4L_1L_2 \\ Q_2 &= (2L_2 - 1)L_2 & Q_5 &= 4L_2L_3 \\ Q_3 &= (2L_3 - 1)L_3 & Q_6 &= 4L_3L_1 \end{aligned}\right\} \tag{A1}$$

and

$$L_i = \frac{a_i + b_i x + c_i y}{2A}; \quad i = 1, 2, 3 \tag{A2}$$

$$a_1 = x_2 y_3 - x_3 y_2 \tag{A3}$$

$$b_1 = y_2 - y_3 \tag{A4}$$

$$c_1 = x_3 - x_2 \tag{A5}$$

with cycle permutation for a_2, etc.

In this equation, A is the area of a triangle, and x_i, y_i are coordinates of nodal point i.
For one-dimensional tube element, we have:

$$L_1 = \frac{\ell_2 - \ell}{\Delta\ell} \tag{A6}$$

$$L_2 = \frac{\ell - \ell_1}{\Delta\ell} \tag{A7}$$

CHAPTER 8

VISCOELASTIC MODELING OF INJECTION MOLDING OF A STRIP AND A CENTER GATED DISK CAVITY

by N. Famili and A.I. Isayev

Institute of Polymer Engineering
University of Akron
Akron, Ohio 44325
U.S.A.

ABSTRACT
8.1 INTRODUCTION
8.2 THEORETICAL ANALYSIS
8.2.1 Governing Equations
8.2.2 Problem Setup
8.3 NUMERICAL SOLUTION
8.4 NUMERICAL RESULTS
8.5 CONCLUSIONS
NOMENCLATURE
REFERENCES

ABSTRACT

The chapter describes a simulation method for non-isothermal flow of viscoelastic polymeric materials during filling a center-gated disk or strip cavity, as well as the stress relaxation during the cooling stage. The method permits evaluation of the time and spatial development of, e.g., velocity, shear and normal stresses, birefringence and thickness of the boundary layer. The simulation uses of Leonov rheological constitutive equation. The data, computed for three-dimensional case, indicate that the normal stress effects are important. The results under-predict the observed levels of birefringence.

8.1 INTRODUCTION

Injection molding is a process widely used in the plastics industry. The process consists of heating the polymer to above its melting point, injecting the melt into a cold mold and packing extra melt to ensure the geometrical integrity of the part. Hence, there are the three fundamental stages of the process: filling, packing, and cooling. Although the first patent in injection molding was granted to Hyatt in 1875 [1], the first cited scientific investigation is the work of Gilmore and Spencer in the early 1950s [2,3]. Since then numerous investigations have been performed relating various aspects of the process. During the last twenty years significant progress has been made in development of analytical models describing injection molding. A finite difference numerical simulation introduced by Pearson [4] was first implemented by Kamal and Kenig [5] and was later considered by various investigators (for references see [6-8]). Use of computers has resulted in numerous commercialized packages in CAE/CAD/CAM dealing with all aspects of the process, from mold and cooling system design to digital control of injection [6-8]. Up to the present time, however most analyses available for cavity filling have been based on non-isothermal flow of a generalized Newtonian fluid.

Although these analyses could provide information on velocity and pressure development and melt front advancement during one-dimensional cavity filling, they may give erroneous results when applied to with a cavity having a sudden change in geometry. However, even in one-dimensional cavity filling, these inelastic analyses cannot provide information about frozen-in stresses and orientations. Thus, researchers have started to pay more attention to viscoelastic modeling of injection molding (for references see [6]). In the early 1980s, the first attempt to incorporate the effects of viscoelasticity into the process was made by Isayev and Hieber [9]. Their work was based on one-dimensional non-isothermal flow of a viscoelastic melt described by the Leonov model [10,11]. Flow in a strip cavity was considered with heat transfer based on one-dimensional heat conduction. Later, Isayev and Famili [12] incorporated the effects of convective heat transfer and viscous heating upon flow of the viscoelastic fluid into a slit channel, with special attention on development of the frozen layer, stresses, and orientation. Recently, Kamal and Lafleur [13] described simulation of strip cavity filling using a White-Metzner viscoelastic equation. However, the concept of frozen layer development during the cavity filling omitted in their investigation. Further, Mavridis et al. [14] have investigated the viscoelastic effects as described by the Leonov model with special attention to the flow front region during filling of a strip cavity.

The importance of viscoelasticity and induced orientation on the final part has been realized for some time [2,3], but in recent years, due to the development of injection molding technology for compact disks, these effects have received more attention. There have been numerous investigations on different aspects of optical disks [15-19]. In spite of the strict requirements on optical retardation in optical disks (< 100 nm by IEC 60-96 of July 1983), there have been no modeling efforts to predict frozen-in birefringence governing optical retardation. The present work is a first step in that direction. Thus, the main goal of the present investigation is to simulate injection molding of a viscoelastic melt into a strip cavity and a center gated disk, including filling and cooling stages. This is a continuation of our previous efforts [9,12], where non-isothermal injection and extrusion of a viscoelastic melt through a slit die was analyzed. In the present simulation, transverse velocity and the moving melt front have been incorporated. Complete kinematics, dynamics of the process and frozen layer development have been obtained. The simulation is carried out using the Leonov nonlinear viscoelastic constitutive equation. Some results on viscoelastic simulation of injection molding (based on finite element method) were reported in [20,21].

8.2 THEORETICAL ANALYSIS

8.2.1 Governing Equations

The governing equations for an incompressible, non-isothermal fluid flow with no body forces are:

$$\rho\left(\frac{\partial \tilde{U}}{\partial t}+\tilde{U}\cdot\nabla\tilde{U}\right)=-\nabla p-\nabla\cdot\tilde{\tilde{\tau}} \tag{1}$$

$$\nabla\cdot\tilde{U}=0 \tag{2}$$

$$\rho C_p\left(\frac{\partial T}{\partial t}+\tilde{U}\cdot\nabla T\right)=k\nabla^2 T+\phi \tag{3}$$

The stress field and mechanical dissipation are related to the flow field according to the Leonov model:

$$\tilde{\tilde{\tau}}=\eta_0 s(\nabla\tilde{U}+\nabla\tilde{U}^T)+\sum_{k=1}^{N}\frac{\eta_k}{\theta_k}\tilde{\tilde{C}}_k \tag{4}$$

$$\frac{\partial \tilde{\tilde{C}}_k}{\partial t}+\tilde{U}\cdot\nabla\tilde{\tilde{C}}_k-\nabla\tilde{U}^T\cdot\tilde{\tilde{C}}_k-\tilde{\tilde{C}}_k\cdot\nabla\tilde{U}+\frac{1}{2\theta_k}(\tilde{\tilde{C}}_k\cdot\tilde{\tilde{C}}_k-\tilde{\tilde{\delta}})=0 \tag{5}$$

$$\phi=2\eta_0 s\,\mathrm{tr}(\nabla\tilde{U}.\nabla\tilde{U})+\sum_{k1}^{N}\frac{\mu_k}{2\theta_k}\mathrm{tr}(\tilde{\tilde{C}}_k.\tilde{\tilde{C}}_k) \tag{6}$$

Where $\tilde{U}$ is the velocity vector, $\tilde{\tilde{\tau}}$ is the stress tensor, p is pressure, $\tilde{\tilde{\delta}}$ is the unit tensor, $\tilde{\tilde{C}}_k$ is the elastic strain tensor, ϕ is the dissipation function [11], $\theta_k(T)$ and $\eta_k(T)$ are the relaxation time and shear viscosity of the kth mode, respectively and η_0 is the zero-shear-rate viscosity defined as:

$$\eta_0 = \sum_{k=1}^{N} \eta_k/(1-s),$$

with s being a rheological parameter between 0 and 1. It should be noted that in the case of viscoelastic fluid, the energy equation includes the dissipation function term, ϕ, which is different from viscous heating term usually employed in the case of flow generalized Newtonian fluids.

8.2.2 Problem Setup

Strip Cavity

An idealized problem is considered in which a polymer melt with a uniform temperature, T_0, is injected into a strip cavity having a wall temperature $T_w \lesssim T_g$. The cavity has a thickness 2b and length L (Figs. 8.1a,c). The melt enters the cavity under a fully developed velocity with an

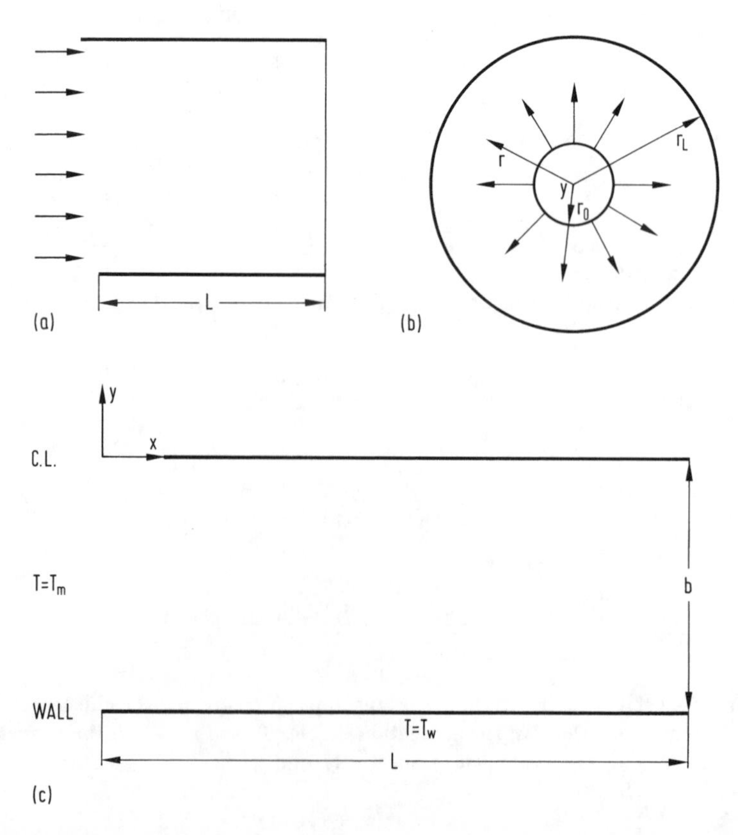

Figure 8.1 Top view of the strip cavity (a), the center gated disk (b) and the gapwise view of both cavities (c).

average velocity U. The problem is then to determine the effects of the developing temperature field described in terms of conduction, convection, and viscous heating, upon the viscoelastic temperature dependent medium as it fills the cavity and subsequently relaxes following the cessation of the flow.

For this flow problem, the components of $\tilde{\tilde{C}}_k$, $\tilde{\tilde{\tau}}$ and $\tilde{U}$ in Eqs (1-6) are:

$$\tilde{\tilde{C}}_k = \begin{bmatrix} C_{11,k} & C_{12,k} & 0 \\ C_{12,k} & C_{22,k} & 0 \\ 0 & 0 & 1 \end{bmatrix}; \quad \tilde{\tilde{\tau}} = \tau_{ij} = \begin{bmatrix} \tau_{11} & \tau_{12} & 0 \\ \tau_{12} & \tau_{22} & 0 \\ 0 & 0 & \tau_{33} \end{bmatrix}; \quad \tilde{U} = (u_1, u_2, 0) \qquad (7)$$

where subscripts 1, 2 and 3 denote flow, gapwise, and width directions, respectively.

To non-dimensionalize Eqs (1-6), the following variables are introduced:

$$\epsilon = b/L, \; u = u_1/U, \quad v = u_2/\epsilon U, \quad x = x_1/L, \quad y = x_2/b$$

$$t' = t/\theta_1^0, \quad T' = \frac{T - T_w}{T_m - T_g}, \quad \xi = \frac{b}{\epsilon U \theta_1^0}$$

$$\nu_k = \frac{\eta_k}{\eta_1^0} = \nu_k^0 a_T, \quad \lambda_k = \frac{\theta_k}{\theta_1^0} = \lambda_k^0 a_T$$

$$n = \frac{2s}{1-s} \frac{\epsilon U}{b} \eta_1 \sum_k \nu_k^0$$

$$N_k = \frac{\lambda_k^0 \theta_k^0}{\nu_k^0 \eta_k^0} n, \quad M_k = \frac{b a_T}{\epsilon U \theta_k^0} N_k, \quad N_k' = \frac{N_k}{2\epsilon}, \quad n' = \frac{n}{2\epsilon} \qquad (8)$$

$$d = \frac{s}{1-s} \left(\frac{\epsilon U}{b}\right)^2 \eta_1 \sum_k \nu_k^0, \quad D_k = \frac{(1-s)\left(\frac{\epsilon U}{b}\right)^{-2} \nu_k^0 N_k^2}{s\,(\theta_1^0 \lambda_k^0)^2}$$

$$\tau_{11} = n\tau_{11}', \quad \tau_{22} = n\tau_{22}', \quad \tau_{12} = n'\tau_{12}', \quad \pi = np, \quad C' = \frac{nC}{\epsilon}$$

$$C_{11,k} = N_k C_{11,k}', \quad C_{22,k} = N_k C_{22,k}', \quad C_{12,k} = N_k' C_{12,k}'$$

Superscript "0" denotes the values at the temperature where the model parameters are given.

Employing variables defined in Eq (8) and dropping the primes from the non-dimensionalized terms, Eqs (1-6) are reduced to:

$$\frac{\rho\epsilon^2 U^2}{n}\frac{Du}{Dt} = \frac{\partial\pi}{\partial x} + \frac{\partial\tau_{11}}{\partial x} + \frac{1}{2\epsilon^2}\frac{\partial\tau_{12}}{\partial y} \tag{9}$$

$$\frac{\rho\epsilon^2 U^2}{n}\frac{Dv}{Dt} = \frac{\partial\pi}{\partial y} + \frac{1}{2}\frac{\partial\tau_{12}}{\partial x} + \frac{\partial\tau_{22}}{\partial y} \tag{10}$$

$$\frac{\partial u}{\partial x} + \frac{\partial v}{\partial y} = 0 \tag{11}$$

$$\frac{DT}{Dt} = \frac{\alpha}{\epsilon U b}\left(\epsilon^2\frac{\partial^2 T}{\partial x^2} + \frac{\partial^2 T}{\partial y^2}\right) + \phi \tag{12}$$

$$\phi = \frac{d\left(\frac{\epsilon U}{b}\right)}{\rho C_p(T_m - T_g)}\left\{a_T\left[2\left(\frac{\partial u}{\partial x}\right)^2 + \frac{1}{\epsilon^2}\left(\frac{\partial u}{\partial y} + \frac{\partial v}{\partial x}\right) + \left(\frac{\partial v}{\partial y}\right)^2\right] + \right.$$

$$\left.\sum_{k=1}^{N}\frac{D_k}{a_T^2}\left(C_{11,k}^2 - \frac{1}{\epsilon^2}C_{12,k}^2 + C_{22,k}^2 + \frac{1}{N_k^2}\right)\right\} \tag{13}$$

$$\tau_{ij} = \sum_{k=1}^{N}\begin{bmatrix} C_{11,k} & C_{12,k} & 0 \\ C_{12,k} & C_{22,k} & 0 \\ 0 & 0 & 1 \end{bmatrix} +$$

$$a_T\begin{bmatrix} \frac{\partial u}{\partial x} & \left(\epsilon^2\frac{\partial v}{\partial x} + \frac{\partial u}{\partial y}\right) & 0 \\ \left(\frac{\partial u}{\partial y} + \epsilon^2\frac{\partial v}{\partial x}\right) & \frac{\partial v}{\partial y} & 0 \\ 0 & 0 & 0 \end{bmatrix} \tag{14}$$

$$\frac{DC_{11,k}}{Dt} = \left(2\frac{\partial u}{\partial x}C_{11,k} + \frac{1}{\epsilon^2}\frac{\partial u}{\partial y}C_{12,k}\right) - \frac{M}{2\,a_T}\left(C_{11,k}^2 - C_{11,k}C_{22,k} + \frac{1}{2\epsilon^2}C_{12,k}^2\right)$$

(15)

$$\frac{DC_{12,k}}{Dt} = \left(2\epsilon^2\frac{\partial v}{\partial y}C_{11,k} + 2\frac{\partial u}{\partial y}C_{22,k}\right) - \frac{M}{2\,a_T}C_{12,k}\left(C_{11,k} + C_{22,k}\right) \tag{16}$$

$$C_{22,k} = \frac{1 + N'^2 C_{12,k}}{N^2\, C_{11,k}} \tag{17}$$

where D( )/Dt is the total derivative defined as:

$$\frac{D(\ )}{Dt} = \xi\frac{\partial(\ )}{\partial t} + u\frac{\partial(\ )}{\partial x} + v\frac{\partial(\ )}{\partial y} \tag{18}$$

Center Gated disk

We have assumed that the melt enters a center gated disk of thickness 2b and radius of r_L (Figs. 8.1b,c), under fully developed flow conditions of average velocity U at uniform temperature T_0 through a gate having the same thickness as the disk and radius r_0(where $r_L - r_0 = L$, $x_0 = r_0/L$). Disk wall temperature is T_w, which is below T_g.

Employing variables in Eqs (7) and (8), the governing Eqs (1-6) are non-dimensionalized. Dropping the primes from the non-dimensionalized terms and using total derivatives D( )/Dt as defined in Eq (18), the obtained equations are:

$$\frac{\rho\epsilon^2U^2}{n}\frac{Du}{Dt} = \frac{\partial\pi}{\partial x} + \frac{\tau_{11}}{x+x_0} + \frac{\partial\tau_{11}}{\partial x} + \frac{1}{2\epsilon^2}\frac{\partial\tau_{12}}{\partial y} \tag{19}$$

$$\frac{\rho\epsilon^2U^2}{n}\frac{Dv}{Dt} = \frac{\partial\pi}{\partial y} + \frac{1}{2}\left(\frac{\tau_{12}}{x+x_0} + \frac{\partial\tau_{12}}{\partial x}\right) + \frac{\partial\tau_{22}}{\partial y} \tag{20}$$

$$\frac{u}{x+x_0} + \frac{\partial u}{\partial x} + \frac{\partial v}{\partial y} = 0 \tag{21}$$

$$\frac{DT}{Dt} = \frac{\alpha}{\epsilon Ub}\left(\epsilon^2\left(\frac{1}{x+x_0}\frac{\partial T}{\partial x} + \frac{\partial^2 T}{\partial x^2}\right) + \frac{\partial^2 T}{\partial y^2}\right) + \phi \tag{22}$$

where ϕ is defined in Eq (13). For $r/b \gg 1$, it could be assumed that hoop strain $C_{\theta\theta} \cong 1$. Thus, the equations for the stress tensor and governing equations for the elastic strain tensor in the case of the center gated disk are the same as the Eqs (14-17) derived for the strip cavity.

8.3 NUMERICAL SOLUTION

In solving Eqs (9-17) for the strip cavity and Eqs (19-22) together with Eqs (13-17) for the center gated disk cavity the following assumptions are made:

a. The melt front is flat with a uniform temperature corresponding to that of the inlet, and a fully developed longitudinal velocity profile, based on volumetric flow rate.
b. Shear stress is linear in the gapwise direction.
c. Transverse velocity is assumed to be of a general third order polynomial of the form $v = ay^3 + by^2 + cy$, to satisfy the boundary conditions on v, such that v vanishes at the center line and the wall.
d. The boundary and initial conditions on the temperature (T(x,y,t)) are:

$$T(0,y,t) = \frac{T_0 - T_W}{T_m - T_g}, \quad T(x>0,1,t) = 0, \quad \frac{\partial T}{\partial y}(x \geq 0,0,t) = 0$$

e. The boundary and initial conditions on the flow velocities ($u(x,y,t)$ and $v(x,y,t)$) are:

$$u(x,1,t) = v(x,1,t) = 0, \quad \frac{\partial u}{\partial y}(x,0,t) = 0, \quad v(x,0,t) = 0.$$

f. Associated with Eqs (15-17) are initial conditions given by:

$$C_{ij,k}(0,y,t) = \overset{0}{C}_{ij,k}(y)$$

where $C^0{}_{ij,k}$ is the appropriate solution of the steady state form of these equations corresponding to fully developed Poiseuille type flow at $t = 0$, namely:

$$\left.\begin{aligned} C_{11,k}(y) &= \frac{\sqrt{2}\; z(y)}{N_k[1+z(y)]^{0.5}} \\ C_{12,k}(y) &= \frac{4\,\dfrac{\partial u}{\partial y}\,a_T}{M[1+z(x_2)]} \\ C_{22,k}(y) &= \frac{\sqrt{2}}{N[1+z(y)]^{0.5}} \end{aligned}\right\} \tag{24}$$

where

$$z(y) = \{1 + \frac{4N_k^2}{\epsilon^2 M_k^2}[\dot{\gamma}^0(y)\,a_T]^2\}^{0.5} \tag{25}$$

where $\dot{\gamma}^0(y) = (\partial u/\partial y)^0$. Thus, at the gate $C_{ij,k}$ are known in terms of $\dot{\gamma}^0(y)$ according to Eq (24). Hence, $\dot{\gamma}^0(y)$, $u^0(y)$ and $C_{ij,k}$ can be determined by numerical iteration, outlined in [9].

The Eqs (9-17) for the strip cavity, Eqs (19-22) and Eqs (13-17) for the center gated disk cavity, with the stated boundary and initial conditions were solved employing a modified explicit Eulerian finite difference scheme with variable mesh sizes (Fig. 8.2). Since an explicit Eulerian scheme is in general globally unstable, the following modifications were made. In order to advance in time, the procedure was modified employing a relaxation scheme. For

$$f_s^n = f(t_n, x_i, y_j),$$

the scheme is represented as:

$$f_s^n = f_s^{n-1} + [\omega(\partial f_s^n/\partial t) + (1-\omega)(\partial f_s^{n-1}/\partial t)].\Delta t \tag{26}$$

where subscript s denotes space, superscript n denotes time and ω is the relaxation parameter. Mesh sizes were modified using von Neumann stability criteria [22], where it is assumed that the independent solutions of the difference equations are all of the form:

$$f_s^n = \xi^n(\cos(ks) + \iota \sin(ks)) \tag{27}$$

where k is an arbitrary real spatial number and $\xi^n(k)$ is a complex number depending on k. To have an unconditional stability at each time step n, a value of Δx in space s is found such that in Eq (27) $|\xi^n| \leq 1$.

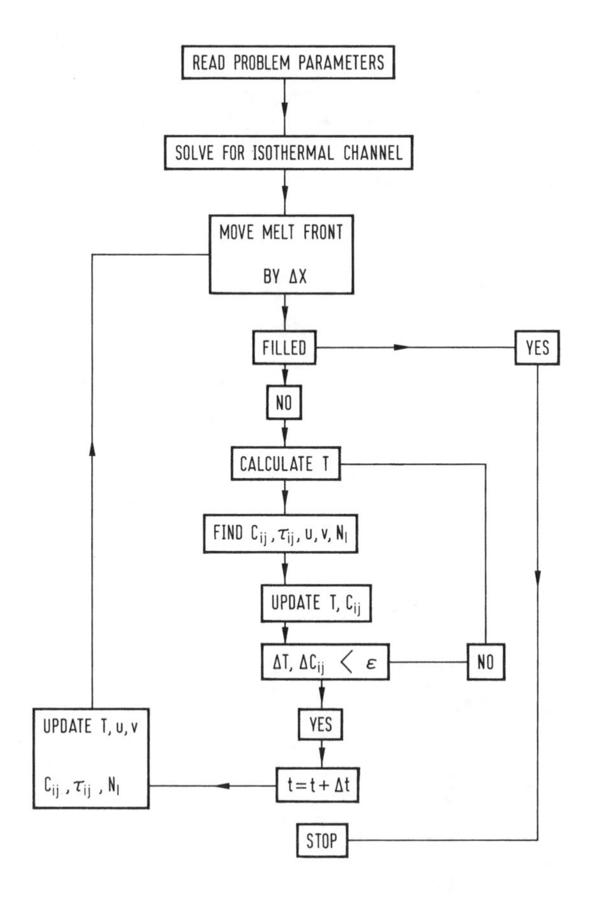

Figure 8.2 Computational flow chart for a general explicit finite difference method.

Temperature dependence of $\theta_k(T)$ and $\eta_k(T)$ was based on a WLF temperature shift factor, namely $\theta_k(T) = \theta_k(T_o).a_T/a_{To}$. For $T \leq T_g$, $a_T = a_{Tg}$ and for $T > T_g$,

$\ln a_T = -C_1(T - T_{ref})/(C_2 + T - T_{ref})$ in which $C_1 = 20.4$, $C_2 = 101.6$ K and T_{ref} is a reference temperature which depends on the polymeric material.

The calculation was continued until $t = t_{fill}$ at which the flow stopped and the polymer proceed to relax under the continuous cooling. That is, for $t \geq t_{fill}$, Eqs (9-17) with $\dot{\gamma} \equiv 0$ were solved until the maximum polymeric temperature dropped below T_g, resulting in frozen-in stresses.

8.4 NUMERICAL RESULTS

For a numerical solution of the problem, model parameters in the constitutive equation must be evaluated first. In the present investigation, a two-mode formulation (N = 2) was used. For polystyrene melt at 463 K, the following parametric values were given in [9] (corresponding to superscript " o " in Eq (8)): $\eta_1 = 5.44$ kPa.s, $\eta_2 = 1.5$ kPa.s, $\theta_1 = 0.8$ s, $\theta_2 = 0.027$ s, $s = 9 \times 10^{-3}$, $\rho = 940$ kg/m^3, $p = 2050$ J/kg K, $K = 0.122$ J/sm K, and $C = 4.8 \times 10^{-9}$ Pa^{-1}.

In the case of the strip cavity, the results are given for the following conditions: $U = 0.12$ m/s, $b = 0.001$ m, $L = 0.3$ m, $T_o = 503$ K, and $T_w = 323$ K, which represent the injection molding experiments by Wales et al. [23], with T_g and T_{ref} taken as: $T_g = 373$ K and $T_{ref} = 407$ K.

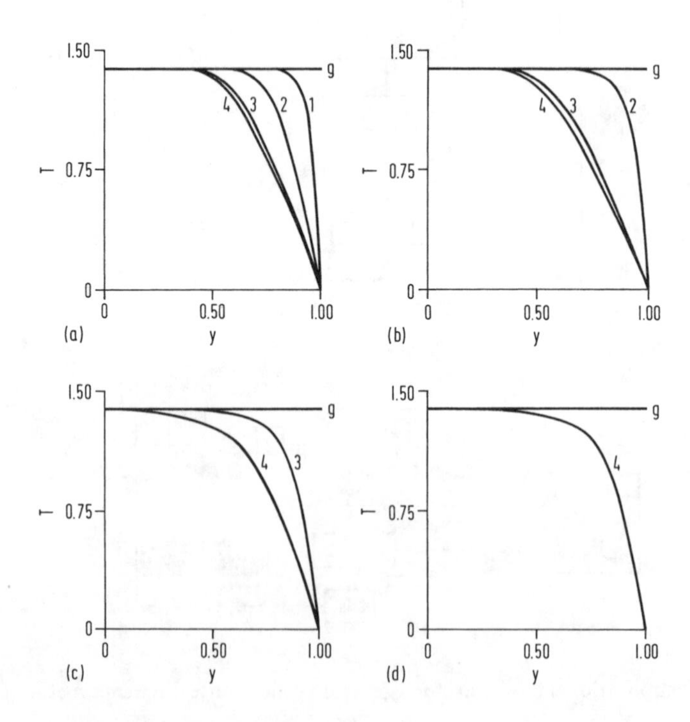

Figure 8.3 Gapwise distributions of the temperature in the strip cavity at x = 0.23 (a), 0.48 (b), 0.73 (c) and 0.97 (d) for the flow times t = 0.8, 1.6, 2.3, 3.1 (curves 1-4 respectively); g denotes the gapwise distribution at the entrance.

The center gated disk has a thickness 2b of 2 mm and total injection length L of 52.5 mm where outer radius $r_L = 60$ mm and inner radius $r_0 = 7.5$ mm (these dimensions are typical of an optical disk). Processing conditions on U, T_0 and T_W are the same as in the case of the strip cavity (hence, flow rate $Q = 4b\pi r_0 U = 1.13 \times 10^{-5} m^3/s$).

The present results are based on 20 uniform mesh increments in the gap direction, variable meshes in the flow direction and a constant $\Delta t = 2.5 \times 10^{-4}$. The results presented below are for temperature (T), axial velocity (u), transverse velocity (v), shear rate ($\dot{\gamma}$), shear stress (τ_{12}), first normal stress difference (N_1) and birefringence Δn. The birefringence is determined according to

$$\Delta n = C\sqrt{\epsilon^2 N_1^2 + \tau_{12}^2} \tag{28}$$

where C is the non-dimensional stress optical coefficient.

Figure 8.3 shows the gapwise distribution of temperature (T) in the strip cavity during filling at various cross sections from the gate x = 0.23(a), 0.48(b), 0.73(c) and 0.97(d) for different flow times 0.78, 1.56, 2.34 and 3.12 (note that position of melt front x^* in the channel is calculated from $x^* = U.t$). At a cross section near the gate and the melt front (curve 1, Fig. 8.3a), most of the melt is at the inlet temperature. Temperature decay is restricted to the wall region. As the flow time increases and the melt progresses inside the cavity (curves 2-4), the temperature decay propagates further away from the wall at this cross section. Calculations of temperature profiles are also performed by using the viscous heating term in the energy equation instead of the dissipation function. The results obtained indicated that the difference in temperature between these two sets of calculations was negligible.

The effects of melt front on the development of temperature in the cavity can be seen by comparing the curves of the same flow time at different cross sections (e.g. curve 2 of Figs. 8.3a and 8.3b or curve 3 of Figs. 8.3a, 8.3b, and 8.3c). These comparisons show that in the regions behind the melt front, the gapwise development of temperature is retarded. This is due to the movement of hot melt from the core toward the wall in these regions (fountain flow effect).

Figure 8.4 presents the gapwise distributions of temperature in the center gated disk at four radial distances from the gate x = 0.23(a), 0.48(b), 0.73(c) and 0.97(d) for different flow times 0.26, 0.75, 1.49, 2.46. The gapwise temperature developments show the same tendency as those of Figures 8.3a-d. Note that in the center gated disk cavity the position of the melt front x^* is calculated from:

$$x^* = \sqrt{2r_0Ut + r_0^2} - r_0$$

A direct comparison between the disk and the strip cavity is not possible. However, comparison of Figures 8.3 and 8.4 shows that melt front progression in the disk cavity affects the temperature development behind melt front more than in the strip cavity. This could be due to the fact that the melt front progression in the disk cavity is slower than that of the strip cavity.

Figure 8.5 shows the growth of the frozen layer in the strip cavity (a) and center gated disk cavity (b) based on glass transition temperature for four different flow times. Close to the gate the growth of this layer is retarded due to the input of the hot melt. Thickness of this layer

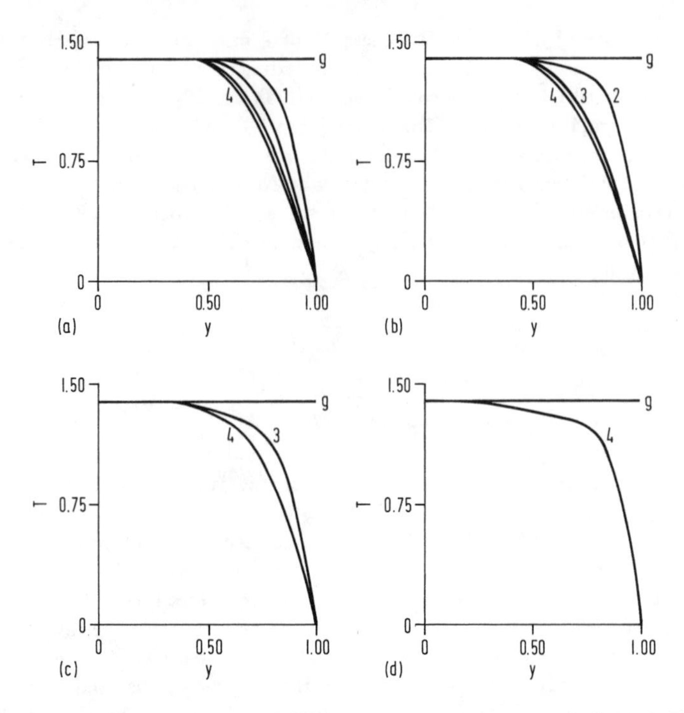

Figure 8.4 Gapwise distributions of temperature in the center gated disk cavity at $x = 0.23$ (a), 0.48 (b), 0.73 (c) and 0.97 (d) for the flow times $t = 0.26$, 0.75, 1.49, 2.46 (curves 1-4 respectively); g denotes the gapwise distribution at the entrance.

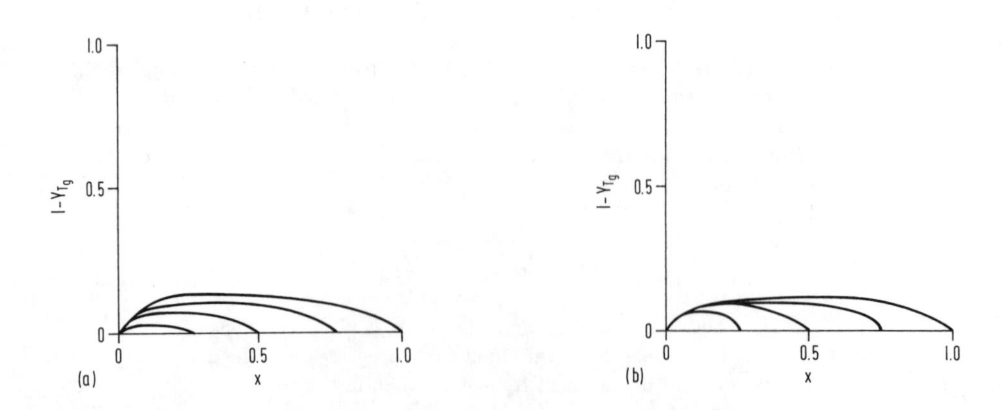

Figure 8.5 Thickness of the solidified layer in the strip cavity (a) and in the center gated disk cavity (b) based on the glass transition temperature vs. distance from the gate for four longitudinal positions of the melt front.

reaches a maximum at a position between the gate and the melt front, closer to the gate. It reduces in thickness toward the melt front due to movement of the hot melt from the core region toward the wall. In the case of the center gated disk, the layer grows faster at the beginning, but the growth slows down drastically as the melt progresses further inside the cavity. A comparison of the growth of the frozen layers in the strip cavity and the center gated disk at the same flow times (different distances from the gate), shows that in the case of equal entrance velocity the frozen layer grows faster in the disk than in the strip cavity due to decay of velocity in the center gated disk.

Figures 8.6a,b are the three-dimensional representation of the temperature distribution in the strip cavity at $t = 1.56$ and 3.12, respectively. Figures 8.6c,d are three-dimensional distributions of temperature at $t = 0.75$ and 2.46 in the center gated disk cavity. Development of the temperature during the flow is clearly illustrated by these figures.

Kinematics of the flow in the strip cavity and the center gated disk are depicted in Figures 8.7 - 8.16, in terms of longitudinal velocity (u), transverse velocity (v), and shear rate ($\dot{\gamma}$). Figure 8.7 shows the gapwise distributions of longitudinal velocity (u) in the strip cavity at various cross sections from the gate $x = 0.23$(a), 0.48(b), 0.73(c), 0.97(d) for different flow times 0.78, 1.56, 2.34 and 3.12. Figure 8.8 presents the gapwise distribution of longitudinal velocity in the center gated disk at various radial positions $x = 0.23$(a), 0.48(b), 0.73(c), 0.97(d) for different flow times 0.26, 0.75, 1.49, 2.46. Figure 8.9 illustrates three-dimensional profiles of longitudinal velocity in the strip cavity for $t = 1.56$(a) and 3.12(b), and in the center gated disk for $t = 0.75$(c) and 2.46(d). For any cross section, an increase in the flow time leads to a significant increase of the longitudinal velocity (u) in the core region, with its maximum value at the center line. This is accompanied by a decrease in the longitudinal velocity near the wall, caused by cooling during cavity filling, and development of a frozen layer near the wall. In the center gated disk, the value of the longitudinal velocity at the center line increases abruptly, reaching a maximum at the first node after the gate and then decreasing gradually toward the melt front. In the strip cavity the longitudinal velocity increases gradually to a maximum away from the gate and then reduces up to the melt front.

Gapwise distributions of the transverse velocity (v) at various cross sections from the gate $x = 0.23$(a), 0.48(b), 0.73(c), 0.97(d) for different flow times 0.78, 1.56, 2.34 and 3.12 are given in Figure 8.10. Similar distributions in the center gated disk at various radial positions $x = 0.23$(a), 0.48(b), 0.73(c), 0.97(d) for different flow times 0.26, 0.75, 1.49, 2.46 are given in Figure 8.11. Three dimensional representations of the transverse velocity distributions for $t = 1.56$(a) and 3.12(b) in the strip cavity and for $t = 0.75$(c) and 2.46(d) in the center gated disk are given in Figure 8.12.

The melt enters the gate with a fully developed velocity profile, hence no transverse velocity exists ($v = 0$). As soon as the melt enters the cold cavity, v increases abruptly to a maximum positive value (directed toward the core region); after that, it reduces continuously up to a region where again there is no transverse velocity ($v = 0$). Further downstream a region starts where v becomes negative (directed toward the cavity wall) with its absolute value increasing toward the melt front, where it reaches its maximum value. Throughout these regions the gapwise distributions of the transverse velocity is zero at the wall and the center line and has a maximum in an intermediate position, closer to the center line. The largest contribution of the transverse velocity during the cavity filling has been found to occur in the vicinity of the cavity entrance, where the ratio of the maximum longitudinal over maximum transverse velocities has been found to be approximately 15 and 2 for the strip and center gated disk cavities, respectively. However, this ratio increases quickly as distance from the gate increases, more so in the case of the center gated disk than the strip cavity.

Gapwise distributions of shear rate in the strip cavity at various cross sections from the

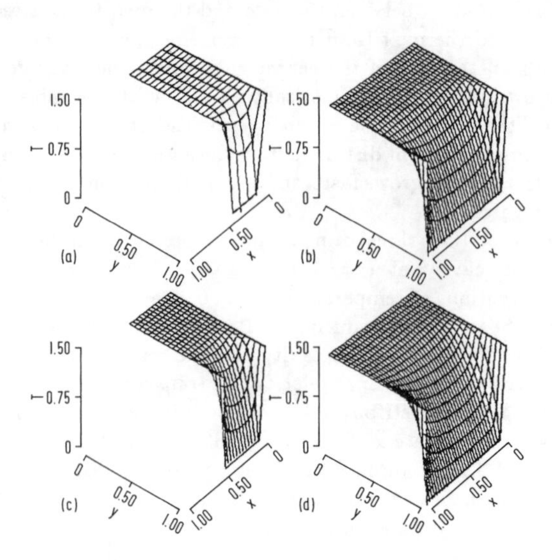

Figure 8.6 Three dimensional profiles of the temperature distribution in the strip cavity for the flow times t = 1.6 (a), 3.1 (b) and in the center gated disk cavity for the flow times t = 0.75 (c), 2.46 (d).

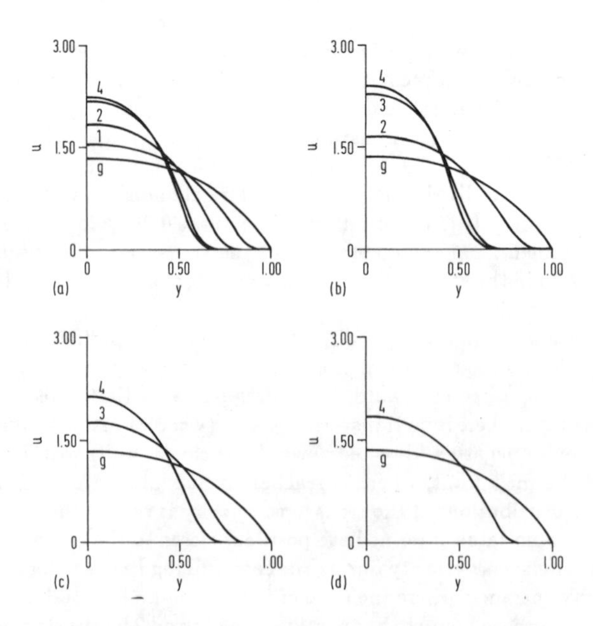

Figure 8.7 Gapwise distributions of the longitudinal velocity in the strip cavity at x = 0.23 (a), 0.48 (b), 0.73 (c) and 0.97 (d) for the flow times t = 0.8, 1.6, 2.3, 3.1 (curves 1-4 respectively); g denotes the gapwise distribution at the entrance.

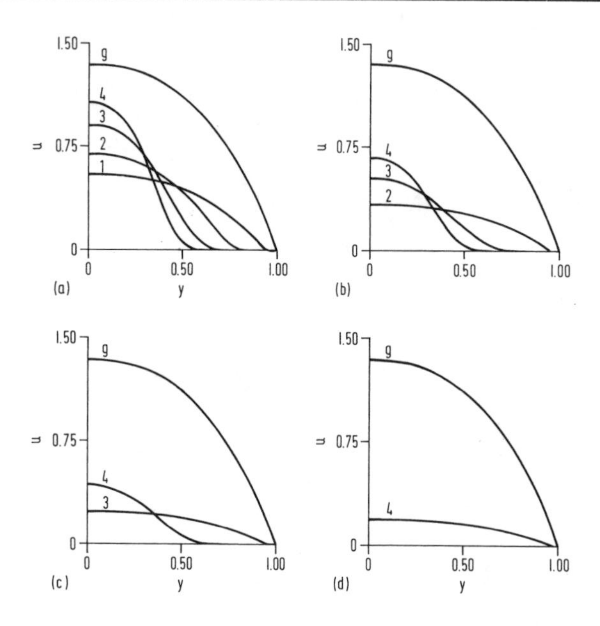

Figure 8.8 Gapwise distributions of longitudinal velocity in the center gated disk cavity at x = 0.23 (a), 0.48 (b), 0.73 (c) and 0.97 (d) for the flow times t = 0.26, 0.75, 1.49, 2.46 (curves 1-4 respectively); g denotes the gapwise distribution at the entrance.

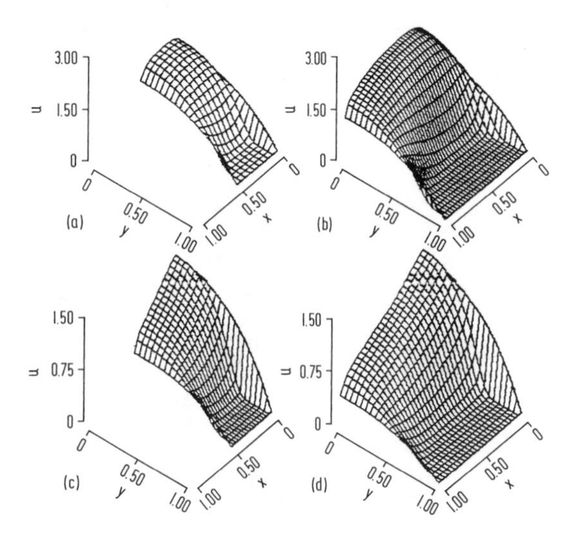

Figure 8.9 Three dimensional profiles of the longitudinal velocity distribution in the strip cavity for the flow times t = 1.6 (a), 3.1 (b) and in the center gated disk cavity for the flow times t = 0.75 (c), 2.46 (d).

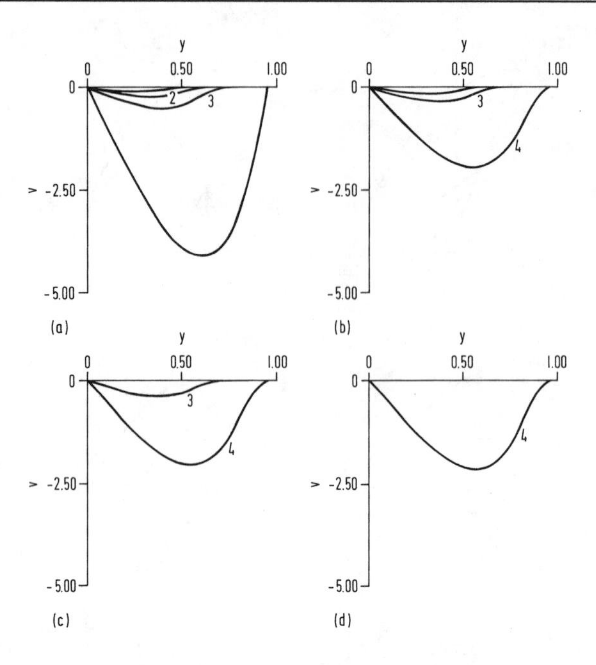

Figure 8.10 Gapwise distributions of the transverse velocity in the strip cavity at x = 0.23 (a), 0.48 (b), 0.73 (c) and 0.97 (d) for the flow times t = 0.8, 1.6, 2.3, 3.1 (curves 1-4 respectively).

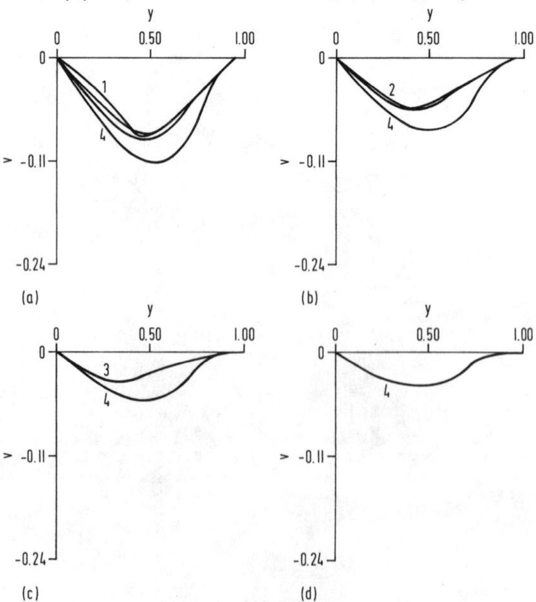

Figure 8.11 Gapwise distributions of transverse velocity in the center gated disk cavity at x = 0.23 (a), 0.48 (b), 0.73 (c) and 0.97 (d) for the flow times t = 0.26, 0.75, 1.49, 2.46 (curves 1-4 respectively).

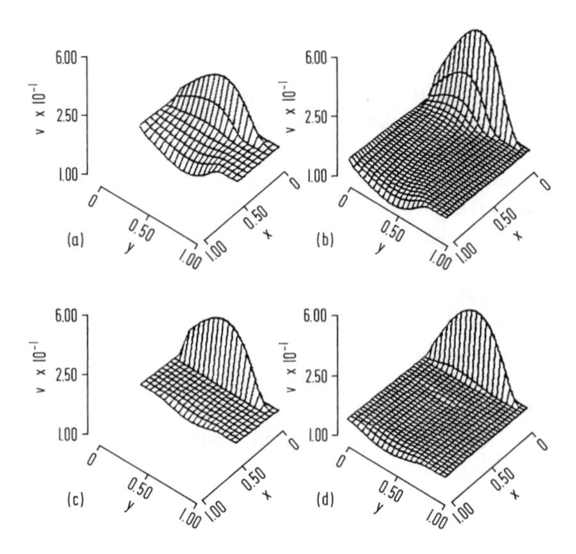

Figure 8.12 Three dimensional profiles of the transverse velocity distribution in the strip cavity for the flow times $t = 1.6$ (a), 3.1 (b) and in the center gated disk cavity for the flow times $t = 0.75$ (c), 2.46 (d).

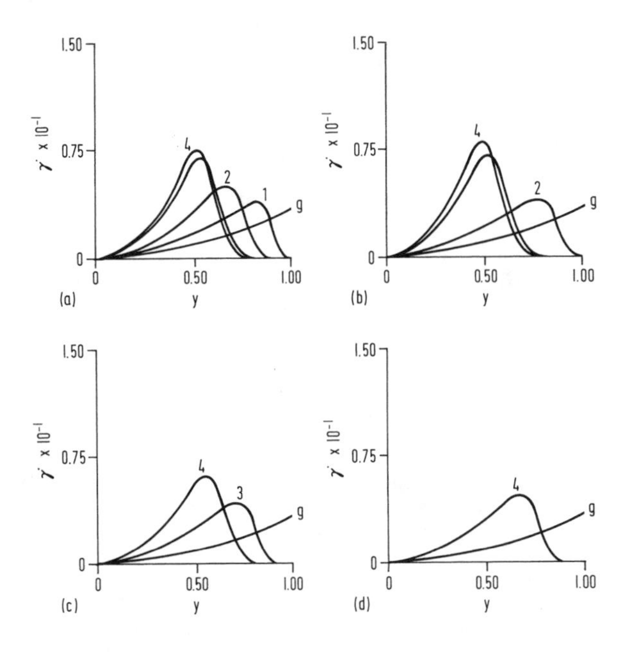

Figure 8.13 Gapwise distributions of the shear rate in the strip cavity at $x = 0.23$ (a), 0.48 (b), 0.73 (c) and 0.97 (d) for the flow times $t = 0.8, 1.6, 2.3, 3.1$ (curves 1-4 respectively); g denotes the gapwise distribution at the entrance.

gate x = 0.23(a), 0.48(b), 0.73(c), 0.97(d) for different flow times 0.78, 1.56, 2.34 and 3.12 are given in Figure 8.13. Similar distributions in the center gated disk at various radial positions x = 0.23(a), 0.48(b), 0.73(c), 0.97(d) for different flow times 0.26, 0.75, 1.49, 2.46 are given in Figure 8.14. Shear rate reduces to zero near the wall and has a maximum in between the wall and the center line. This is due to the cooling effect during cavity filling. The value of the maximum at each cross section not only depends on the flow time but also on the proximity to the melt front. In particular, the maximum increases as the flow time increases but it reduces as the position is closer to the melt front. The gapwise positions of these maxima $(1-y_{\dot{\gamma}max})$ vs. longitudinal distance from the gate are given in Figure 8.15 for the strip cavity (a) and the center gated disk cavity (b). In the strip cavity the value of $(1-y_{\dot{\gamma}max})$ increases monotonically until it reaches a maximum close to the gate, then decreases gradually toward the vicinity of the melt front and abruptly at the melt front. At the melt front, $(1-y_{\dot{\gamma}max})$ is zero, since the flow is assumed to be isothermal and has a fully developed velocity profile. In the center gated disk cavity, the value of $(1-y_{\dot{\gamma}max})$ increases monotonically until it reaches a steady plateau and it decreases abruptly to zero behind the melt front. The value and the length of this plateau increases with the flow times. Figure 8.16 illustrates the three dimensional profile of shear rate in the strip cavity for t = 1.56(a) and 3.12(b), and in the center gated disk cavity for t = 0.75(c) and 2.46(d).

Figure 8.17 represents the gapwise distributions of shear stress in the strip cavity at various cross sections from the gate x = 0.23(a), 0.48(b), 0.73(c), 0.97(d) for different flow times 0.78, 1.56, 2.34 and 3.12. Similar distributions in the center gated disk at various radial positions x = 0.23(a), 0.48(b), 0.73(c), 0.97(d) for different flow times 0.26, 0.75, 1.49, 2.46 are given in Figure 8.18. Figure 8.19 illustrates the three-dimensional profile of shear stress in the strip cavity for t = 1.56(a) and 3.12(b), and in the center gated disk cavity for t = 0.75(c) and 2.46(d). Shear stress has a linear gapwise distribution with a slope depending on the flow times and proximity to the melt front. It increases as the flow time increases and at the same flow, time it decreases closer to the melt front. Figures 8.19a,b shows that in the strip cavity the slope increases gradually to a maximum and decreases monotonically toward the melt front, where it has the minimum slope. Figures 8.19c,d show that in the center gated disk the slope increases abruptly to its maximum value, adjacent to the gate, and decreases gradually toward the melt front, where it has the minimum slope.

Figure 8.20 shows the gapwise distribution of the first normal stress difference N_1 ($N_1 = \tau_{11} - \tau_{22}$) in the strip cavity at various cross sections from the gate x = 0.23(a), 0.48(b), 0.73(c), 0.97(d) for different flow times 0.78, 1.56, 2.34 and 3.12. Similar distributions in the center gated disk at various radial positions x = 0.23(a), 0.48(b), 0.73(c), 0.97(d) for different flow times 0.26, 0.75, 1.49, 2.46 are given in Figure 8.21. The results indicate that the value of N_1 becomes frozen in the vicinity of the wall and shows a pronounced maximum in the intermediate region with subsequent decay in the core region. The magnitude of the maximum depends on the flow times and proximity to the melt front. The maximum increases in magnitude as the flow time increases, but for the same flow time, it decreases for the cross sections closer to the melt front. The gapwise positions of these maxima $(1-y_{N1,max})$ vs. axial distance from the gate are given in Figure 8.22 for the strip cavity (a) and the center gated disk (b). The value of $(1-y_{N1,max})$ increases monotonically until it reaches a steady plateau and decreases to zero behind the melt front. As the flow time increases the value and the length of this plateau increases too. Figure 8.23 illustrates the three-dimensional profile of N_1 in the strip cavity for t = 1.56(a) and 3.12(b) and in the center gated disk for t = 0.75(c) and 2.46(d).

Based on the numerical results for τ_{12} and N_1, the gapwise distributions of birefringence Δn is obtained using Eq (28). Figure 8.24 represents the gapwise distributions of Δn in the strip cavity at various cross sections from the gate x = 0.23(a), 0.48(b), 0.73(c),

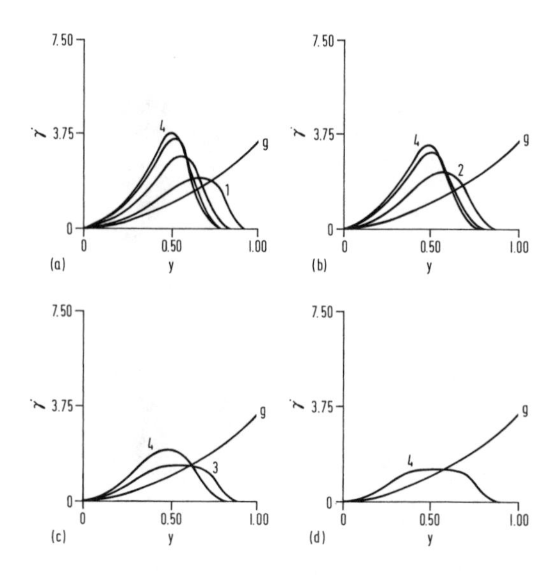

Figure 8.14 Gapwise distributions of the shear rate in the center gated disk cavity at $x = 0.23$ (a), 0.48 (b), 0.73 (c) and 0.97 (d) for the flow times $t = 0.26$, 0.75, 1.49, 2.46 (curves 1-4 respectively); g denotes the gapwise distribution at the entrance.

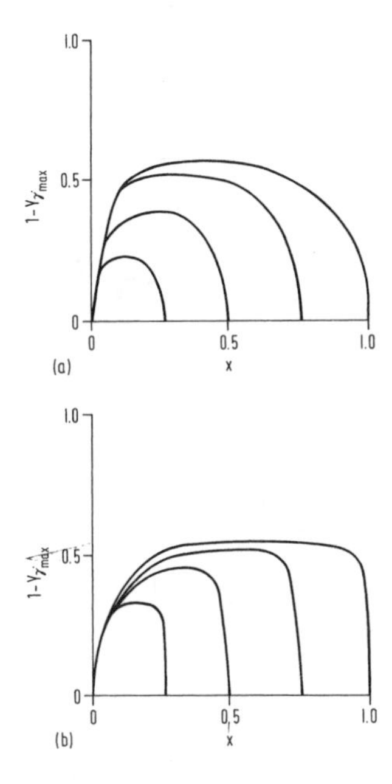

Figure 8.15 Gapwise position of the maximum shear rate ($1\text{-}y_{\dot{\gamma}max}$) in the strip cavity (a) and in the center gated disk cavity (b) vs. distance from the entry for four longitudinal positions of the melt front.

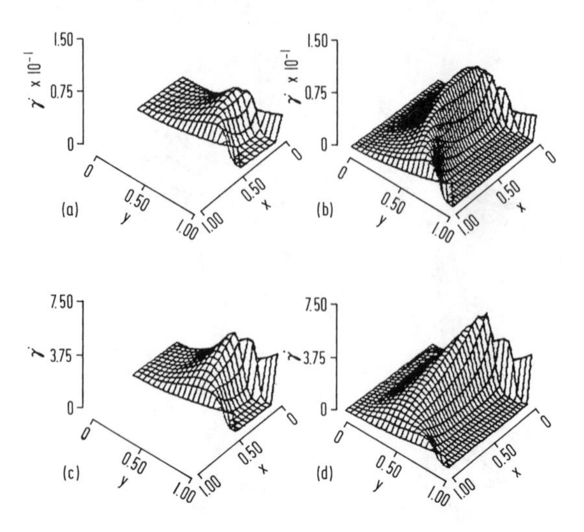

Figure 8.16 Three dimensional profiles of the shear rate in the strip cavity for the flow times $t = 1.6$ (a), 3.1 (b) and in the center gated disk cavity for the flow times $t = 0.75$ (c), 2.46 (d).

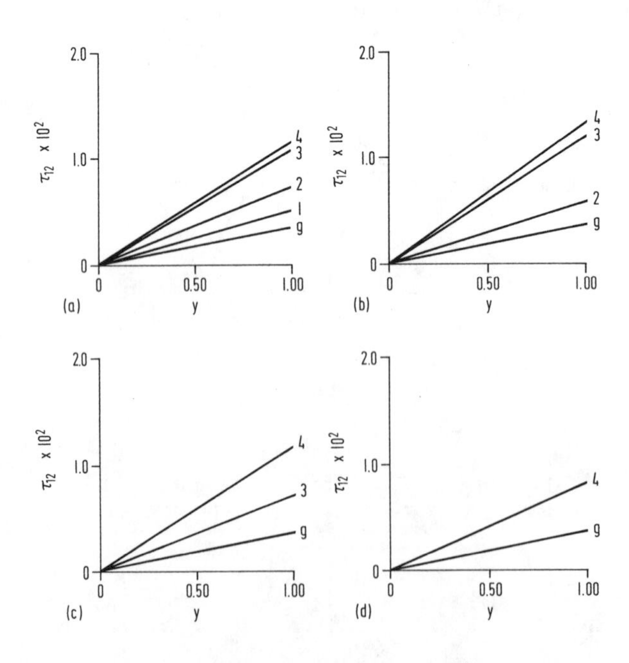

Figure 8.17 Gapwise distributions of the shear stress in the strip cavity at $x = 0.23$ (a), 0.48 (b), 0.73 (c) and 0.97 (d) for the flow times $t = 0.8, 1.6, 2.3, 3.1$ (curves 1-4 respectively); g denotes the gapwise distribution at the entrance.

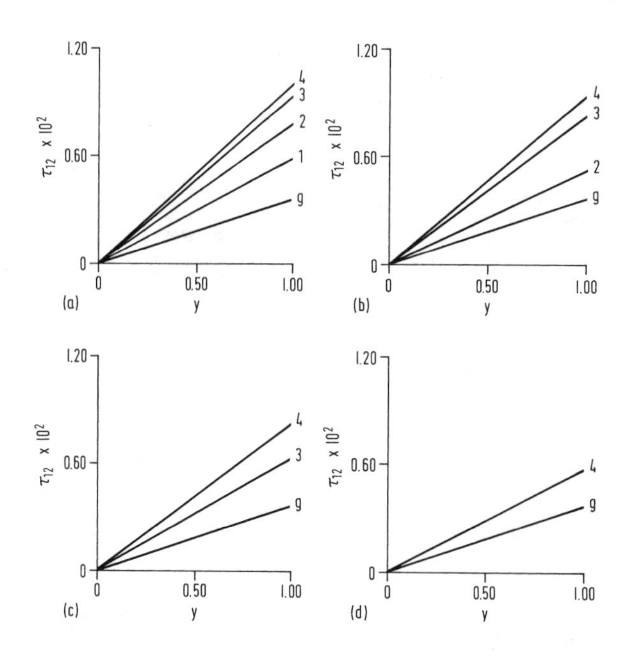

Figure 8.18 Gapwise distributions of the shear stress in the center gated disk cavity at x = 0.23 (a), 0.48 (b), 0.73 (c) and 0.97 (d) for the flow times t = 0.26, 0.75, 1.49, 2.46 (curves 1-4 respectively); g denotes the gapwise distribution at the entrance.

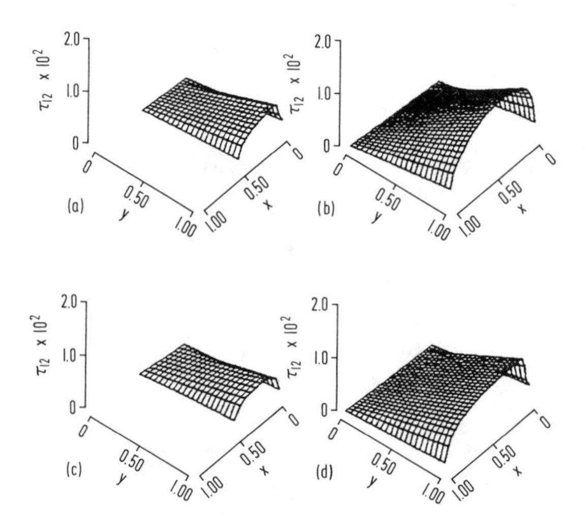

Figure 8.19 Three-dimensional profiles of the shear stress in the strip cavity for the flow times t = 1.6 (a), 3.1 (b) and in the center gated disk cavity for the flow times t = 0.75 (c), 2.46 (d).

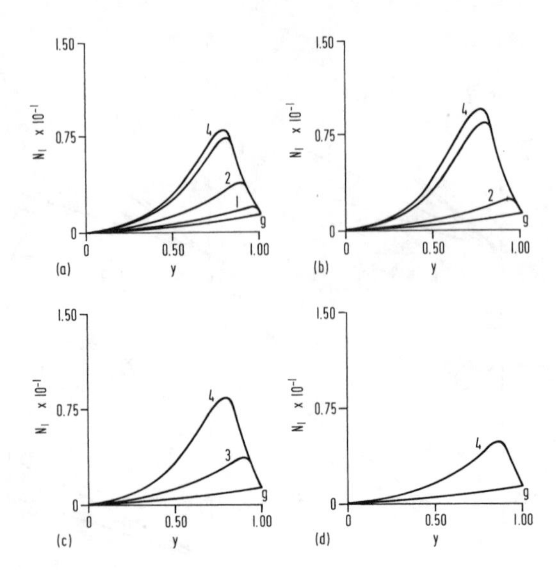

Figure 8.20 Gapwise distributions of the first normal stress difference in the strip cavity at x = 0.23 (a), 0.48 (b), 0.73 (c) and 0.97 (d) for the flow times t = 0.8, 1.6, 2.3, 3.1 (curves 1-4 respectively); g denotes the gapwise distribution at the entrance.

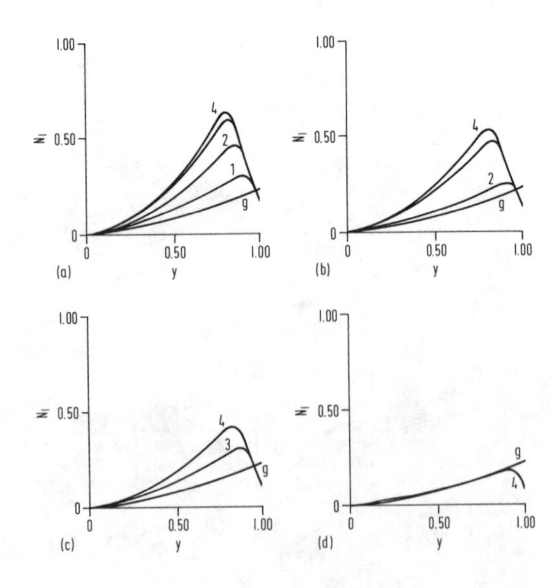

Figure 8.21 Gapwise distributions of the first normal stress difference in the center gated disk cavity at x = 0.23 (a), 0.48 (b), 0.73 (c) and 0.97 (d) for the flow times t = 0.26, 0.75, 1.49, 2.46 (curves 1-4 respectively); g denotes the gapwise distribution at the entrance.

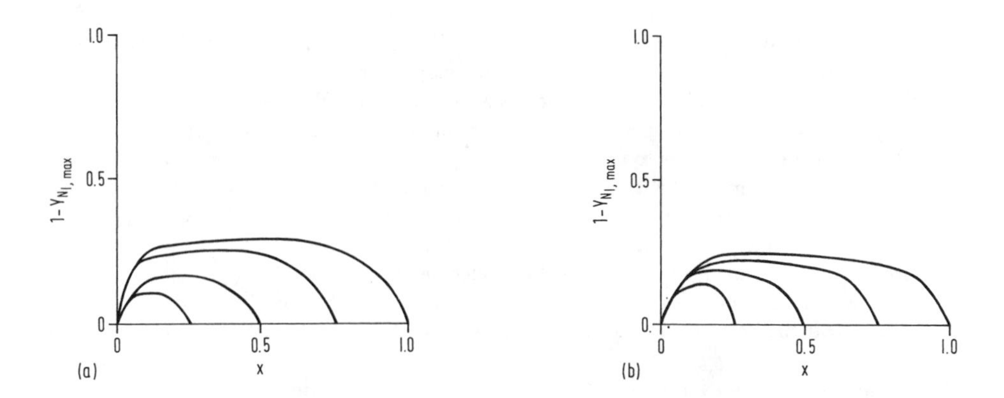

Figure 8.22 Gapwise position of the maximum first normal stress difference ($1-y_{N1,max}$) in the strip cavity (a) and in the center gated disk cavity (b) vs. distance from the entry for four longitudinal positions of the melt front.

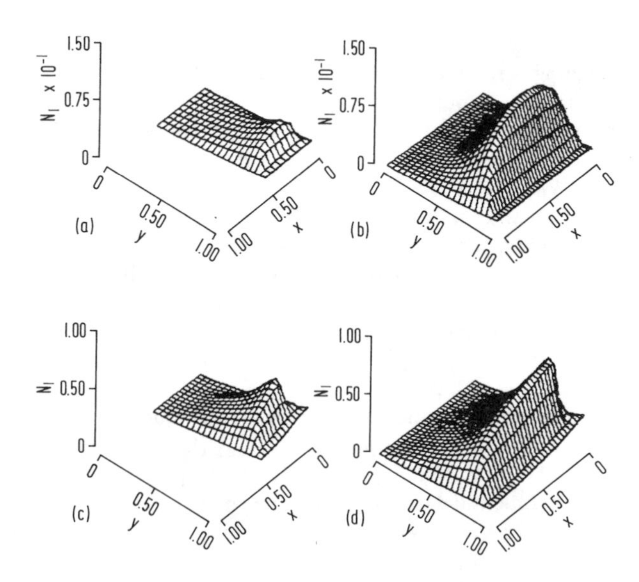

Figure 8.23 Three-dimensional profiles of the first normal stress difference in the strip cavity for the flow times $t = 1.6$ (a), 3.1 (b) and in the center gated disk cavity for the flow times $t = 0.75$ (c), 2.46 (d).

0.97(d) for different flow times 0.78, 1.56, 2.34 and 3.12. Similar distributions in the center gated disk at various radial positions x = 0.23(a), 0.48(b), 0.73(c), 0.97(d) for different flow times 0.26, 0.75, 1.49, 2.46 are given in Figure 8.25. The gapwise distribution of birefringence exhibit a maximum away from the wall. This maximum is mainly dominated by the contribution by the normal stresses to the total birefringence. However, an increase in the birefringence near the wall (especially in the case of the strip cavity) is dominated by the contribution of the shear stress to the total birefringence. This is due to a continuous increase in the shear stresses combined with a freezing of the normal stresses at the wall. The magnitude of the birefringence maximum increases with increasing flow times at fixed cross sections. On the other hand, at fixed flow time, the magnitude of this maximum decreases as the melt front is approached. The gapwise positions of the birefringence maxima $(1-y_{\Delta n,max})$ vs. distance from the gate are presented in Figure 8.26, for the strip cavity (a) and the center gated disk cavity (b). Near the entrance to the cavity the value of $(1-y_{\Delta n,max})$ monotonically increases until it reaches a steady plateau. Further downstream behind the melt front, the value of $(1-y_{\Delta n,max})$ decreases until it becomes zero at the melt front. The thickness of this layer could be assumed to represent the thermal boundary layer for non-isothermal flow of the viscoelastic melt in the cavity. It is seen that the thickness of this layer is substantially larger than that determined using the glass transition temperature (Fig. 8.5). Figure 8.27 shows the three-dimensional profiles of birefringence distributions in the strip cavity for t = 1.56(a) and 3.12(b) and in the center gated disk for t = 0.75(c) and 2.46(d).

As the cavity is filled, the flow stops and the cooling stage begins, leading to the stress relaxation under highly non-isothermal cooling conditions. Due to an exponential increase of relaxation times of the melt during cooling, stresses accumulated during flow do not completely relax, leading to residual orientation and stresses.

Figure 8.28 shows a comparison of the theoretical gapwise distribution of birefringence in the strip cavity at the flow time t = 3.12 and at the end of relaxation (frozen in birefringence) at various cross sections from the gate x = 0.23(a), 0.48(b), 0.73(c) with experimental data of Wales et al. [21] for frozen-in birefringence. The theoretical frozen-in birefringence results show that in the core region, where the temperature remained well above T_g at the end of the filling, the birefringence relaxes to zero. Close to the wall, stresses are frozen in and the birefringence retains most of its value; however during the relaxation during the cooling stage, the gapwise position of the maximum shifts toward the wall. The discrepancy between the experimental and the theoretical results could be partly explained by omission of residual thermal birefringence and residual birefringence generated by packing pressures in the theoretical calculations.

Figure 8.29 shows the theoretical and experimental values of the maximum frozen-in birefringence Δn_{max} (a) and its gapwise position $y_{\Delta n,max}$ (b) in the molded strip vs. distance from the gate. It can be seen from Figure 8.29 that the present predictions are in good agreement with the experimental data for the gapwise position of the maximum birefringence, but the computation under-predicts the values, especially at the cross sections close to the gate. The observed discrepancy is due to the effect of packing pressure on the frozen-in birefringence, which is neglected in the present calculations. In particular, packing pressure affects the magnitude of the gapwise maximum of the residual birefringence and does not affect the gapwise positions of these maxima to the same degree [6,24]. Furthermore, packing mainly affects the value of birefringence at the cross sections located close to the gate and to a lesser extent at the locations close to the end of the strip.

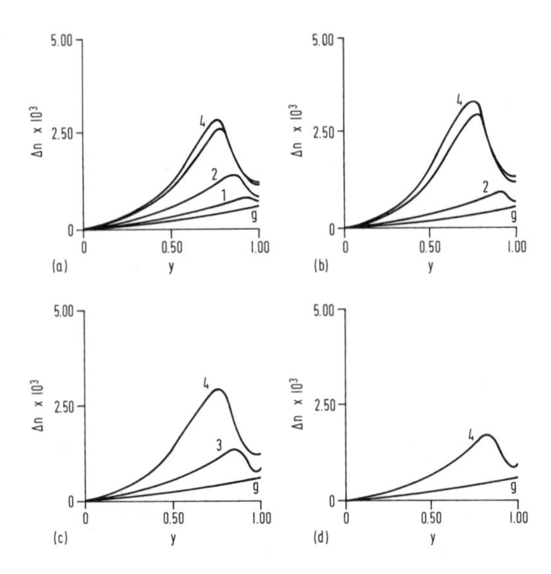

Figure 8.24 Gapwise distributions of the birefringence in the strip cavity at x = 0.23 (a), 0.48 (b), 0.73 (c) and 0.97 (d) for the flow times t = 0.8, 1.6, 2.3, 3.1 (curves 1-4 respectively); g denotes the gapwise distribution at the entrance.

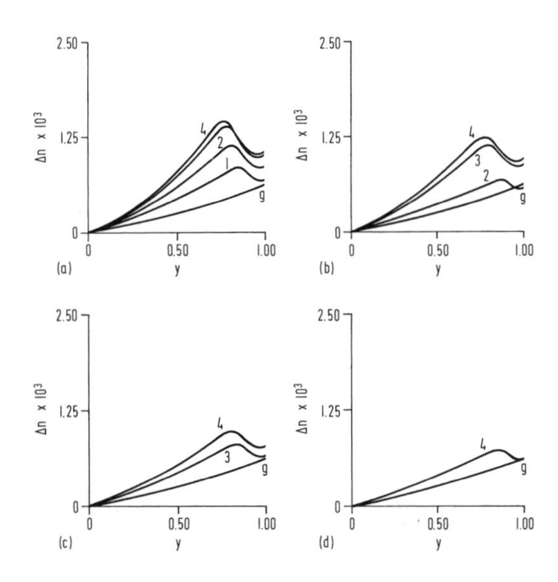

Figure 8.25 Gapwise distributions of the birefringence in the center gated disk cavity at x = 0.23 (a), 0.48 (b), 0.73 (c) and 0.97 (d) for the flow times t = 0.26, 0.75, 1.49, 2.46 (curves 1-4 respectively); g denotes the gapwise distribution at the entrance.

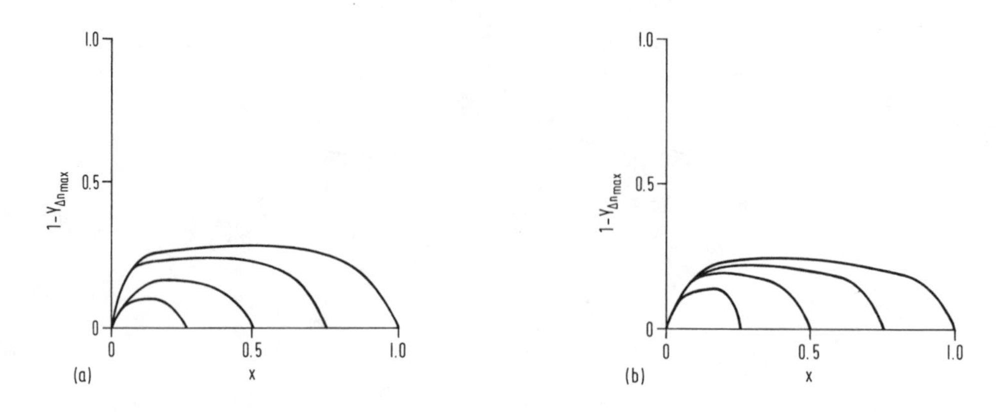

Figure 8.26 Gapwise position of the maximum birefringence (1-$y_{\Delta n,max}$) in the strip cavity (a) and in the center gated disk cavity (b) vs. distance from the entry for four longitudinal positions of the melt front.

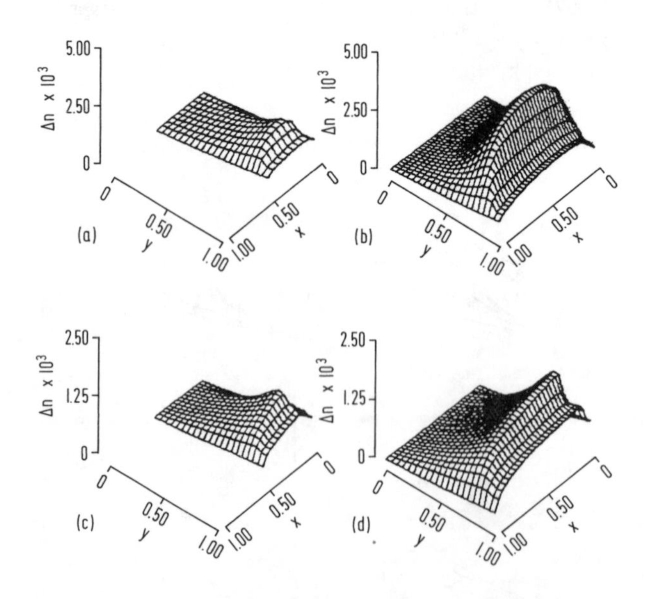

Figure 8.27 Three-dimensional profiles of the birefringence in the strip cavity for the flow times t = 1.6 (a), 3.1 (b) and in the center gated disk cavity for the flow times t = 0.75 (c), 2.46 (d).

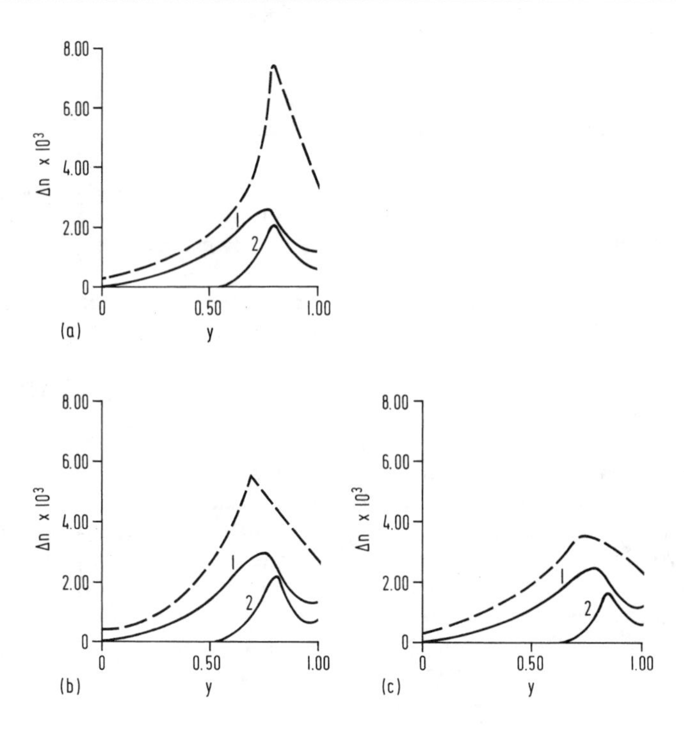

Figure 8.28 Gapwise distributions of the birefringence in the strip cavity at $x = 0.23$ (a), 0.47 (b) and 0.73 (c) for flow time $t = 3.1$ (curve 1) and at the end of relaxation (curve 2). Dashed lines are experimental results of Wales et al. [23] for frozen-in birefringence.

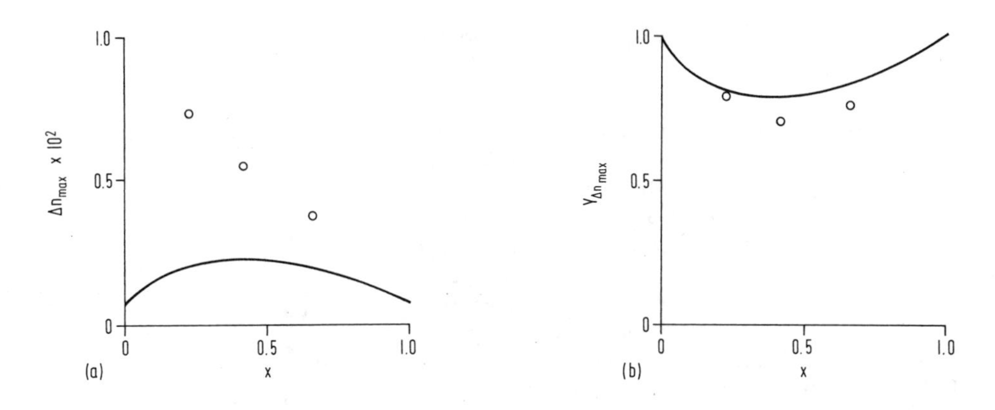

Figure 8.29 Value of the maximum frozen-in birefringence Δn_{max} (a) and its gapwise position (b) in the molded strip vs. distance from the gate. Curve: theoretical predictions; Symbol: experimental results of Wales et al. [23].

8.5 CONCLUSIONS

In the present study, a simulation of non-isothermal flow of a viscoelastic melt during filling of a center gated disk and a strip cavity has been performed. In addition, after the cavity has been filled, relaxation of the stresses during the cooling stage has been simulated. This type of flow gives rise to a solidified thermal boundary layer at the cold wall, which develops in time. These simulations have been done using the Leonov viscoelastic constitutive equation. Application of the viscoelastic model to the injection molding of polymer melts permits evaluation of the kinematic and dynamic fields during the process. In contrast to the previously published papers related to modeling of the frozen-in birefringence in the molded products, and based only on transient conductions, the present study includes the longitudinal and transverse convection, conduction, and viscous dissipation terms in the energy equation. Time and spatial development of velocity, shear rate, shear and normal stresses, birefringence and thickness of the thermal boundary layer have been determined. In particular, the thickness of this layer was determined based upon the gapwise position of the birefringence maximum and glass transition temperature. It was found, that the layer calculated based on the birefringence maximum is thicker than the one based on T_g. This simulation indicates that the normal stress effect is significant. The predicted results have been compared with experimental data on frozen-in birefringence in a molded strip indicating that the theory predicts the thickness of the birefringence layer near the wall quite well but under-predicts the observed maximum of the birefringence. The latter seems to be due to the neglect, by the present simulation, of the packing stage of injection molding, which is known to increase the frozen-in birefringence. Further work to experimentally verify the theoretical results is in progress.

Finally, a comparison of the present predictions of the frozen-in birefringence with those reported earlier [9], based on one-dimensional heat conduction alone, indicates that the latter predicts higher levels of birefringence and is accordingly closer to the available experimental data. Although the assumption of a continuous increase of the frozen-in layer thickness toward the gate is unrealistic, due to the neglect of the convective and viscous heating terms in the energy equations, this rather fortunate good agreement between experimental data and theoretical predictions was a result of the over-prediction of the cooling effect during cavity filling.

NOMENCLATURE

a_T: Shift factor of the WLF equation
2b: Cavity thickness
C_1, C_2: Constants of the WLF equation
$\underset{\sim}{C}$: Stress optical coefficient
$\underset{\approx}{C}_{,k}$: Elastic strain tensor of the kth relaxation mode
C_p: Heat capacity
d: Non-dimensional parameter in the dissipation function
D_k: Non-dimensional parameter in the dissipation function
k: Thermal conductivity
L: Effective flow length of the cavity
M_k: Relaxation time normalization parameter for the kth relaxation mode
N: Number of mode in the Leonov model
N_1: First normal stress difference ($N_1 = \tau_{11} - \tau_{22}$)
N_k, N'_k: Viscosity normalization parameters for the kth relaxation mode

n, n': Stress non-dimensional parameters
p: Isotropic pressure
$\tilde{r}$: Position vector in the cylindrical coordinate system
r_0: Gate radius of the center-gated disk
r_L: Outer radius of the center-gated disk ($L = r_L - r_0$)
s: A rheological parameter of the Leonov model ($0 < s < 1$)
t: Time
t_{fill}: Filling time
T: Temperature
T_g: Glass transition temperature
T_m: Inlet melt temperature
T_0: Temperature at which the parameters are evaluated
T_{ref}: Reference temperature in the WLF equation
T_w: Wall temperature
U: Average velocity at the cavity entrance
$\tilde{U}$: Velocity vector
u: Longitudinal component of the non-dimensional velocity vector
v: Transverse component of the non-dimensional velocity vector
$\tilde{x}$: Position vector in the Cartesian coordinate system
x: Longitudinal component of the non-dimensional position vector
x^*: Longitudinal position of the melt front
y: Transverse component of the non-dimensional position vector
α: Thermal diffusivity
$\tilde{\tilde{\delta}}$: Unit tensor
Δn: Birefringence
ϵ: Non-dimensional half gap thickness of the cavity
$\tilde{\tilde{\dot{\gamma}}}$: Strain rate tensor
ϕ: Dissipation function
η_k: Shear viscosity of the kth relaxation mode
η_0: Zero shear rate viscosity
λ_k: Non-dimensional relaxation time of the kth relaxation mode
μ_k: Shear modulus in the kth relaxation mode
ν_k: Non-dimensional shear viscosity of the kth relaxation mode
ω: Relaxation parameter
θ_k: Relaxation time of the kth relaxation mode
ρ: Density
$\tilde{\tilde{\sigma}}$: Total stress tensor
$\tilde{\tilde{\tau}}$: Deviatoric stress tensor
ξ: An arbitrary complex number in the numerical stability criteria
ζ: Time normalization factor

REFERENCES

1. Rosato, D.V., Rosato, D.V., "Injection Molding Handbook," Van Nostrand Reinhold, New York (1986).
2. Gilmore, G.D., Spencer, R.S., *Mod. Plast.*, *27*, 143 (1950).
3. Gilmore, G.D., Spencer, R.S., *J. Colloid Sci.*, *6*, 118 (1951).
4. Pearson, J.R.A., "Mechanical Principles of Polymer Processing," Pergamon Press,

Oxford (1966).

5. Kamal, M.R., Kenig, S., *Polym. Eng. Sci.*, *12*, 294 (1972).
6. Isayev, A.I., Ed., "Injection and Compression Molding Fundamentals," Marcel Dekker, New York (1987).
7. Manzione, L.T., Ed., "Application of Computer Aided Engineering in Injection Molding," Hanser Publishers, Munich (1987).
8. Bernhardt, E.C., Ed., "Computer Aided Engineering for Injection Molding," Hanser Publishers, Munich (1983).
9. Isayev, A.I., Hieber, C.A., *Rheol. Acta*, *19*, 168 (1980).
10. Leonov, A.I., Lipkina, E.H., Paskhin, E.D., Prokunin, A.N., *Rheol. Acta*, *15*, 411 (1976).
11. Leonov, A.I., *Rheol. Acta*, *15*, 85 (1976).
12. Isayev, A.I., Famili, N., *J. Plast. Film Sheeting*, *2*, 269 (1986).
13. Kamal, M.R., Lafleur, P.G., *Polym. Eng. Sci.*, *26*, 92, (1986).
14. Mavridis, H., Hrymak, A.N., Vlachopoulos, J., *J. Rheol.*, *32*, 639 (1988).
15. Takeshima, M., Funakoshi, N., *Kobun. Ronbun.*, *42*, 191 (1985).
16. Anders, S., *Kunststoffe*, *77*, 21 (1987).
17. Nazim, A., *Elec. Op. Pub. Rev.*, *7*, 78 (1987).
18. Schmidt, L., Maxam, J., *SPE Tech. Papers.*, *34*, 334 (1988).
19. Greener, J., *SPE Tech. Papers.*, *33*, 238 (1987).
20. Sobhanie, M., Isayev, A.I., *Rubber Chem. Technol.*, *62*, 939 (1989).
21. Sobhanie, M., Isayev, A.I., *SPE Tech. Papers.*, *35*, 286 (1989).
22. Press, W.H., "Numerical Recipes: The Art of Scientific Computing," Cambridge University Press, New York (1986).
23. Wales, J.L.S., Van Leeuwen, J.,Van der Vijgh, R., *Polym. Eng. Sci.*, *12*, 358 (1972).
24. Isayev, A.I., Hariharan, T., *Polym. Eng. Sci.*, *25*, 271 (1985).

CHAPTER 9

AN INTERACTIVE PC-BASED DATA ACQUISITION SYSTEM FOR INJECTION MOLDING

by G.D. Kiss

Bell Communications Research
445 South St.
Morristown NJ 07962-1910
U.S.A.

and

by R.C. Progelhof

University of South Carolina
Mechanical Engineering
Colombia SC 29208
U.S.A.

ABSTRACT
9.1 INTRODUCTION
9.1.1 History
9.1.2 Field Signals
9.1.3 Computer Interface: A/D Converter
9.1.4 Computer Characteristics
9.2 DATA ACQUISITION SYSTEM
9.2.1 System Preface
9.2.2 System Goals
9.2.3 Hardware
9.2.4 Software
9.2.5 Examples of Application
9.3 SUMMARY AND CONCLUSIONS
REFERENCES

ABSTRACT

The chapter focuses on methods of data acquisition for injection molding. First, the field signal, computer interface, and computer characteristics are discussed from a historical perspective. The main purpose of the text is presentation of new and inexpensive data acquisition system developed and tested by the author. The system monitors critical parameters of closed loop control, precision injection molding. The system is self-contained and portable, with software built upon ASYST^{TM} high-level computer language. The program has a provision for two-level usage: standard via a series of menus, and advanced via possibility of program modification. Examples oa application illustrate a wide range of information generated by the system.

9.1 INTRODUCTION

9.1.1 History

For many years the principal method of data acquisition was the visual reading and recording on paper of the output of an instrument; height of the column of mercury in a thermometer, the position of the needle of a bourdon tube pressure gage, the visual output of a measuring device used to display the signal from the sensor or thermocouple mv reading on a potentiometer. With the advent of the chart recorder, the output readings from the sensors in terms of signal voltages were marked on a paper chart. Typically, the method of marking the paper was an ink pen. To overcome the many problems associated with ink pens, many unique marking techniques were developed. However, the end result was still a line or a series of points on a chart.

With the advent of the digital computer, this process was significantly changed. Such sensor output signals (commonly referred to as field signals) as voltage, current, capacitance, electrical pulse, etc.) were conditioned or converted to digital signals to be read by the computer, and stored in memory. The computer could also be used to perform prescribed arithmetic operations and to reduce the recorded data to a prescribed format. During the 1960s, the Digital Equipment Corporation's PDP-8 family computer had the greatest influence on data acquisition. This unit was relatively large in size, required a controlled environment, had limited memory, and was relatively expensive. These factors significantly limited the broad application of this system into production facilities, and hence, most applications were limited to Research and Development studies [1-3]. The centralized computer philosophy is used extensively today to monitor and control an automated manufacturing facility [4,5].

During the 1970s, a revolution was taking place in the computer industry. The integrated circuit and memory chips were being produced at significantly reduced cost. A new line of small computers based upon this new chip technology evolved at the end of the 1970s and the beginning of the 1980s. These systems were called Personal Computers. The two most notable PCs at this time were the 8-bit IBM system based on the Intel 8088 and the Apple system using the Motorola 5802. Initially, both systems were limited in random access memory, RAM, of from 64 to 128K. Within a few years, the memory was increased to 256K and ultimately to 640K. Special math co-processor chips were developed to increase the rate of computation. Two operating systems, DOS and CPM, evolved as the de-facto industry standards.

Although the initial intent of the PC manufacturers was not to design these as data acquisition systems, third party vendors [6,7] developed series of add-on boards that enabled the PC to interact with other computers, send digital or analog signals, and most importantly to "read" digital or analog signals.

The initial PC-based data acquisition systems were limited in speed, number of channels, and storage capacity. Today's commercially available PCs are significantly faster, have more main memory, have hard disk drives to store large quantities of data, can perform multitasking operations, have been environmentally hardened, and are relatively inexpensive [8-10]. These are the attributes of a true data acquisition system.

Each of these hardware and software factors affects the choice of the system for a particular application. Before considering these individual features of the PC system, the type of field signals encountered and the effects of the workplace environment on these signals will be discussed.

9.1.2 Field Signals

The individual piece of instrumentation mounted on a system to measure its performance will give either a digital or analog signal. In most applications both types of signals will be encountered.

Digital signals are presented to the computer as either pulses or steady state values. The voltage magnitude of the digital signal has been standardized around the voltages originally encountered in transistor logic levels circuits. These are referred to as transistor-transistor logic (TTL) levels or five volt signals. The TTL standard for passing digital data is further classified into two groups:

LOW LEVEL 0 - 0.8V and HIGH LEVEL 2.4 - 5.5V

Typically, the digital signal is less prone to environmental interaction. Hence, this method of transmitting data from a sensor to the PC has several advantages. With the exception of a pulse type counter, the sensing unit requires a conditioning device to generate a signal. Through micro miniaturization of electrical circuits, these conditioning devices are being directly integrated into a sensor package, and are referred to as smart sensors. It is expected that this trend will continue and a greater variety of digital sensors will be made available commercially.

The three most common analog signals encountered in data acquisition are voltage, current, and resistance. Of these, voltage signals are the most prevalent. However, most signals are preconditioned, that is the primary sensor output and is directed to a preamplifier (or "signal conditioner") before presentation to the recording device. The most common output voltages are ± 1V, ± 5V, ± 10V or 0-10V. Traditionally, low voltage signals, 0-5 mV, are also encountered when using thermocouples. Past experience indicates that low voltage signals, i.e. thermocouple signals, should be amplified at or near the sensor and be presented to the acquisition system at significantly higher voltage. For most applications, an amplification of 100 is desired.

When setting up a data acquisition system to measure the response of a piece of machinery that is electrically driven or heated, such as the case of an injection molding press, electrical grounding is of paramount importance. Both the field signal and the input to the data acquisition system are referenced to the ground. For electrically driven or resistively heated systems there can be a slight difference in potential between the two ground signals. For a small voltage signal, i.e. thermocouple output, this small ground difference may result in significant temperature error.

With the problems associated with field interference, the current signal has become popular. Typically, the current signal is not a primary sensor output but the results of a conditioning circuit. A pair of twisted wires is run from the signal conditioning circuit to the input of the data acquisition system. This approach is used since current loops are very insensitive to electrical noise.

The last type of field signal is a resistance signal. Typically, these signals are encountered when using strain gages and RTD temperature devices. With the recent advances in high precision microelectronics, there is a trend to use a signal conditioner at the sensor location. In this manner, the resistance field signal is effectively converted to a voltage signal.

9.1.3 Computer Interface: A/D Converter

The field signals, whether voltage, current or resistance, must be converted to digital impulses for input to the PC. When data acquisition systems were first sold for PC systems, two approaches were used. Firstly, the A/D converter was mounted on a PC board. This board was inserted into one of the expansion slots in the PC, enabling the A/D system to have direct access to the PC bus. This approach enabled high data transfer speeds, eliminated the need for an external power supply and had lower cost. Conversely, this approach reduced the expansion capabilities of the PC and required that PC be located near the source of signal.

The second approach used was to mount the A/D system in an external enclosure with a communication capability through the RS-232C or IEE 488 bus of the PC. This approach required use of a separate processor, power supply and sometimes memory. Units of this type were considerably more expensive than the single board approach. However, the processor used in many of these systems had significantly higher clock speeds than the PCs central processor. Today, this is not necessarily true. "Smart Boards" with high speed processors can now be obtained commercially.

Irrespective of the approach taken, the major two factors that must be considered when choosing an A/D system is the acquisition rate and bit resolution. When considering the rate of data acquisition, the determination should be made on the rate per channel for the number of channels. The single channel frequency in MHz only characterizes the processor frequency, but of equal importance is the circuit's dwell time when switching between channels. The bit resolution relates to the systems overall accuracy. For example, an 8-bit A/D converter provides a resolution of one part in 256 or approximately 0.39% of full scale. Shown in Table 9.1 is a comparison of the accuracy of different bit converters.

Table 9.1 Comparison of A/D accuracy.

Bit Size	Steps	Accuracy
8	256	0.39
10	1024	0.0977
12	4096	0.0244
14	16384	0.0061

With respect to problems associated with ground loop circuits, most A/D systems can be operated as either single ended or as differential inputs. It is recommended that for low voltage signals the system be operated in the differential mode.

9.1.4 Computer Characteristics

The initial PCs developed had relatively slow internal clock speeds, about 4 to 5 MHz, small memory (less than 256 kbyte memory), and used floppy disks to store large quantities of data. Data acquisition software had developed to a relatively high degree of sophistication, that is programs were being written in high level languages to take advantage of faster processing times. However, these early systems had severe limitations in both the rate and quantity of data that could be taken. With relatively slow data stream rates, the data was passed into memory and then onto the floppy disks for storage. High speed data acquisition was limited to the speed of the A/D board and the available memory for storage. Today, these restrictions are not as severe since today's PCs operate at significantly higher clock speeds, up to 33 MHz, have considerably more memory and the storage media can be a hard disk drive or a memory board expansion that emulates storage media [11-13].

9.2 DATA ACQUISITION SYSTEM

9.2.1 System Preface

The remainder of this document is a fairly detailed explanation of the system we have constructed, with emphasis on the aspects which make it unusual, and an examination of several applications of the system which demonstrate its utility.

9.2.2 System Goals

The purpose of this data acquisition system is to monitor the critical process parameters of a closed-loop injection molding machine which is being used in the study of ultra-high precision injection molding. The process parameters incorporated so far are the screw and clamp position (or velocity), the mold temperature, and the melt temperature.

There is a natural tendency when using microprocessor controlled equipment to assume that the machine's operation accurately conforms to the instructions which one programs into the controller. After all, look at that digital display! However, it does not hurt to be sceptical and to actually monitor what the machine is doing. This is particularly important when trying to achieve extreme precision in the molded product, since slight deviations from an established process can have a significant impact [5].

The system was designed to be self-contained and portable for two reasons. Firstly, by being portable it is available for use at off-site molding trials. This is extremely helpful when trying to reproduce a successful molding protocol on different equipment, e.g. during evaluation of new equipment for purchase. Secondly, during a day of operation one accumulates a considerable amount of data. It is often desirable to physically transport the data on the computer's hard disk to another site for analysis, rather than trying to send it over the phone at 1200 baud. We also wanted to base the system on readily available and inexpensive hardware, hence the choice of IBM-PC compatible MS-DOS computers.

Another system goal was the provision for a two-level user interface. We wanted a system which could be used by relatively inexperienced and untrained operators, via a series of menus and with access to context-sensitive help. On the other hand, we wanted to avoid hampering an experienced user, and therefore provided a means of bypassing the menus. Also, since the system is built upon ASYST™, which is a high-level language with numerous primitive functions, the user is free to modify and extend the system on-the-fly, if he knows how to use ASYST and is aware of the internals of the acquisition system.

9.2.3 Hardware

As shown schematically in Figure 9.1, the equipment consists of a Compaq 286 transportable computer, a Data Translation DT-2801 A/D card, and Omega Omni-Amp battery operated type J thermocouple amplifiers. Peripheral devices can be attached to the computer for convenience but are not required for the operation of the system. The computer cost is by far the major component. Note that the signal voltages are obtained from the injection machine's controller; there was no need to fit the machine with external position transducers.

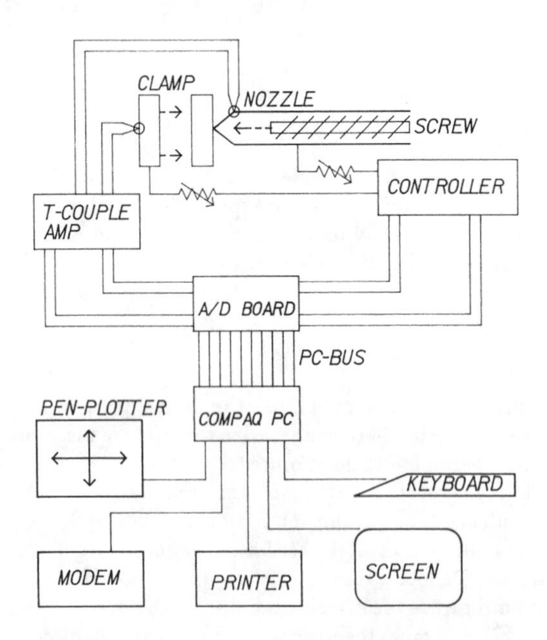

Figure 9.1 System hardware configuration.

The A/D board allows a number of options which are jumper-selectable. It has an amplifier which can monitor up to 16 channels in single-ended configuration or 8 channels in differential configuration. During single-ended operation, the return sides of all of the channels are connected together in a single return path and only the high side is amplified. In differential mode, the return sides of each channel are also amplified. The advantage to differential operation is that it is much less sensitive to noise. Since measurements are taken as differences between amplified signals, any noise common to both paths is in effect subtracted out. The 80 dB common mode rejection ratio of this board means that the inputs are 10,000 times less sensitive to voltages common to both paths than to voltage differences. Use of differential inputs is particularly important when working in a high-noise environment, such as that produced by electric motors, solenoid valves, etc. The board was configured for bipolar input, that is, it can accept both positive and negative voltages, from -10V to +10V. If one is sure that negative voltages will not be present, one can also configure for unipolar input, from 0 to +10V, and double the resolution within that range.

This board is also capable of applying software-controlled gain to the signals prior to digitizing. Gains of 1, 2, 4, and 8 are available, but gain is applied to all channels. A newer version of the board, the DT-2820 series, is capable of independent gain control for each channel. Unfortunately ASYST did not support this board until the release of version 2.0, late in 1987.

The sampling rate of the DT-2801 board is once every 2.5 microseconds. This is far faster than needed for our purposes and in fact our system uses the PC's clock to set the data acquisition rate. The sampling rate is the same for all channels.

The board has features which we have not as yet exploited, such as DMA (direct memory access) transfer of data. This allows the board to write directly into the memory of the PC, bypassing the CPU. This reduces the overhead substantially and is required for data acquisition at very high speeds. Other features which we did not exploit are digital I/O and digital-to-analog conversion.

9.2.4 Software

Implementation Language: the ASYST Programming Environment

The ASYST package is best described as an "environment." It is at the same time a programming language, a set of utilities for various tasks such as relaying instructions to software controllable devices (in particular the Data Translation DT-2801 A/D converter board used in this system), file I/O, file maintenance, designing screens, etc., and an incredibly versatile "calculator" to perform all of these operations interactively.

As a programming language, ASYST has a number of extremely unusual characteristics which make it powerful but at the same time confusing and difficult to use for persons familiar with traditional programming languages. However, these disadvantages are outweighed by the speed of execution, ability to use arrays as data primitives, extensibility, built-in hardware support, and analytical tools.

First, it is totally stack oriented and consequently operations utilize Reverse Polish notation. In other words, any operator will pull the appropriate operands off the stacks as required. It is up to the programmer to ensure that the desired operands are available. There are several independently maintained stacks. The one most often used is the number stack which is capable of storing single or double precision integer, real or complex numbers, and arrays of these numbers. Another stack is the symbol stack, which stores strings and logical values. Yet another stack is the template stack, which stores templates defined to control the operation of software programmable devices (A/D templates) and file I/O (FILE templates).

The stack orientation has its disconcerting aspects. One problem is to make sure that the proper operands are available when an operation is invoked. One often has to undertake some non-obvious shuffling of operands within the stacks. However, an "algebraic parser" is also available which allows one to define expressions in the usual precedence order of operations.

Another difficulty with the stack orientation is the issue of program control. All of the familiar structures are there (IF...THEN, IF...ELSE...ELSE...ENDIF, DO...LOOP, DO...WHILE, BEGIN...AGAIN, etc.), but the semantics are quite confusing. While on the subject of execution control structures, it should be pointed out that these structures can only be used within colon definitions (discussed below) and cannot be used interactively. This ordinarily would not present much of a problem, except that one often wants code to execute as it is being LOADed into the system ("pseudo-interactively"). Any control structures can only be implemented in this situation by defining a word to do the required operation, then executing the word immediately. These trivial examples show the price one pays for the power of the stack oriented structure. It should be clear that doing real programming in this environment takes a shift in point of view which takes a lot of getting used to.

A second feature of this language, which makes it extremely powerful, is the concept of "extensibility" via "colon definitions" (a colon indicates the beginning of a new word, a semi-colon indicates the end). As in FORTH, one can define new "words" or operations which can use primitive operations as they exist in the language, or previously defined words. This can be done interactively on-the-fly, or else in a pre-written program. An important point to note is that the words defined in this way are compiled immediately. Thus, execution is extremely rapid because the code making up the word does not have to be re-interpreted every time it is encountered. This feature is known as an "incremental compiler." The ability to define functions ad-hoc is extremely reminiscent of APL.

A side-effect of this, however, is that if a word used within a colon definition is redefined, the definition uses the old meaning, not the new meaning. That is, ASYST does not "check back" to see that any word which uses a redefined word is automatically updated. This can really get one into trouble until one develops a new set of programming standards and disciplines. The result is a characteristic "reverse-hierarchical" style of programming.

An ASYST program generally consists of a series of colon definitions which are then combined into further definitions until finally one gets to the main program which is just a list of words to execute in sequence.

```
OK : BLOCK.1 .......................... ;
OK : BLOCK.2 .......................... ;
OK : BLOCK.3 .......................... ;
OK : BLOCK.4 .......................... ;
OK : MAIN.PROG
  BLOCK.1
  BLOCK.2
  BLOCK.3
  BLOCK.4
;
```

Typing the word MAIN.PROG at the OK system prompts then executes the four blocks one after the other. It should be obvious how this enables one to write very clearly structured code and furthermore to use code fragments in a variety of programs with great versatility. A real example is given in the discussion of the user interface.

The third unusual and extremely powerful feature of the language is its ability to operate on arrays directly, another feature reminiscent of APL. If, for example, the top two entries in the number stack are arrays of equal dimensions, any operation will operate on both arrays on an element-by-element basis. On the other hand, if one is an array and the other is a scalar, the scalar will operate on each element of the array. In addition, a variety of methods are provided to extract sub-arrays. This third feature allows one to bypass the looping which is required to handle arrays in most conventional languages. It is particularly important in a data acquisition system which almost invariably acquires arrays of data and must often process these arrays at high speeds. These three features (stack orientation, extensibility, and use of arrays as data primitives) give the language its character and flavor. However, ASYST also includes a wide variety of pre-defined words which are utilities for use either interactively or within programs.

Of the many types of utilities available, four are of particular importance to this system. First, there are the utilities for software control of data acquisition parameters. ASYST can be informed of the identity of the particular A/D expansion board (in this case a Data Translation 2801) and then one can set up a library of board-independent A/D templates which store the data acquisition parameters (conversion delay, number of channels scanned, gain, repetition factor, buffer array, etc.). Once this library is set up, the board can be re-configured within a program as appropriate for the process being monitored. Second, there are file I/O utilities. These are of course important for editing, storing, and loading programs onto disks from within ASYST, but in the context of data acquisition they are important for buffering the input. Data acquisition can be directed to the number stack or to a pre-defined buffer array. When data is collected rapidly, it is advantageous to write it directly to an array in main memory, rather than using the limited number stack. This array can then be emptied periodically by writing its contents to a file on disc. Third, there is a large variety of analysis routines for statistics, data smoothing, curve fitting, etc. Error handling utilities might also be placed in this category. Fourth, there are graphics utilities for plotting data on linear or logarithmic scales, placing error bars, using different line styles, interactively scaling, etc. Furthermore, utilities are available for setting up text windows and viewports on the screen to customize the appearance, place prompts, and output in desired locations. An important member of this family of utilities is the set of pre-defined words which allows one to direct output to a pen-plotter and to produce an aesthetically pleasing end-product (e.g. control of color, line type, text slant, etc.). These come into play in this system as an entire set of modules devoted to giving the user easy control over the graphical display and output of his data.

User Interface

An unusual feature of this system was the approach taken to the user interface. It was desirable to have a system which is usable by relatively inexpert users, such as molding technicians, but on the other hand a system responsive to the expert user. The strategy employed was to design a "beginner mode" user interface which consists of a hierarchical series of nested function key menus. Figure 9.2 shows the major modules of the system in terms of the function key menus. Figure 9.3 shows details of one sub-menu and an associated sub-sub-menu. Certain keys in each function key menu always have the same meaning ("help," "go up a level," "return to main menu," etc.) and the other function keys execute different ASYST words in each menu. In addition, some particularly useful words are available by function key presses in more than one menu. In addition, context-sensitive help is available within any of the function key menus by pressing "F9." Depending on which menu is active, this causes the appropriate help file to be read from disk.

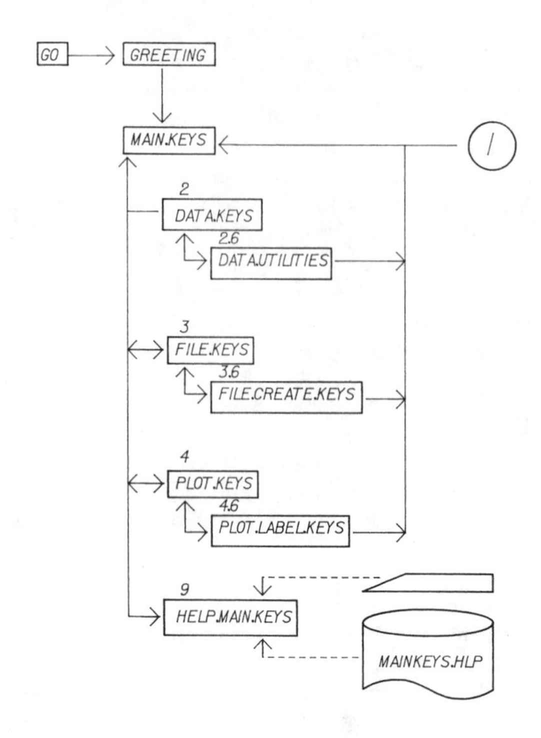

Figure 9.2 System major software modules.

The second level of user interface is the "expert level." Most of the words which constitute the data acquisition system can be executed at any time by an "<alt>" key combination. This saves the knowledgable user typing and having to navigate through the function key menus. On the other hand, the function keys are still active, and the knowledgable user may wish to use them at times also. A complete list of all meaningful "<alt>" key combinations is available at any time using "<alt> H" (see Fig. 9.4). An effort was made to use mnemonic combinations, but with only one letter, some compromises obviously had to be made. In order to impress the "<alt>" key combinations onto the mind of beginner users, the list of current meanings assigned to function keys (available by pressing "F8") also displays the corresponding "<alt>" key combinations.

A severe problem presented itself almost immediately when programming the user interface. It turns out that in ASYST it is easy to devise a scheme for descending a hierarchy of function key menus, but getting back up is not so easy! The problem can be illustrated with a simple example, which has the additional benefit of providing some of the flavor of ASYST programming. Recall that new words ("colon definitions") can contain only words which are already known to ASYST, either built-in or user-defined. One such word is FUNCTION.KEY.DOES, a built-in word that simply waits for a string and then attempts to execute it. In the following simple example, we assume that WORD1, WORD2, WORD3 and WORD4 are meaningful to ASYST (built-in or previously user-defined).

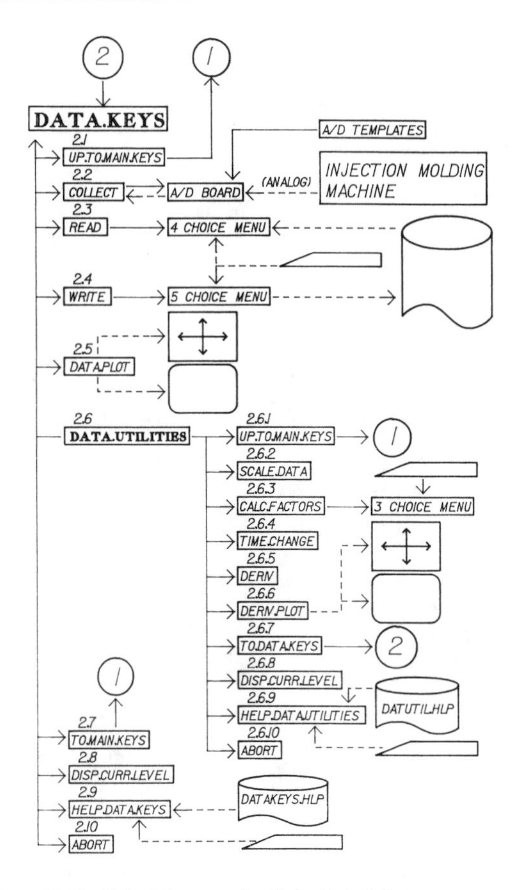

Figure 9.3 Detail of a Sub-Menu and a Sub-Sub-Menu.

```
Press <alt> + capitalized key within word definition name:
Array.readout        <alt> A       Scale.data            <alt> S
NOP                  <alt> B       Time.change           <alt> T
Close                <alt> C       create.file.menU      <alt> U
Data.plot            <alt> D       deriV                 <alt> V
graphics.rEadout     <alt> E       Write                 <alt> W
calc.Factors         <alt> F       create.neXt.file      <alt> X
Get.label.pos        <alt> G       deriv.plot (Y)        <alt> Y
Help.fkey            <alt> H       NOP                   <alt> Z
get.label (I)        <alt> I       NOP                   <alt> 1
put.label (J)        <alt> J       NOP                   <alt> 2
Collect (K)          <alt> K       NOP                   <alt> 3
file.Loop            <alt> L       NOP                   <alt> 4
proMpt               <alt> M       NOP                   <alt> 5
New.name             <alt> N       NOP                   <alt> 6
Open                 <alt> O       NOP                   <alt> 7
Plot.toggle          <alt> P       NOP                   <alt> 8
NOP                  <alt> Q       NOP                   <alt> 9
Read                 <alt> R       NOP                   <alt> 0
NOP                  <alt> -       CREATE.CURR.FILE      <alt> =

HELP SCREEN OBTAINED WHEN <alt> H IS PRESSED
```

Figure 9.4 Help screen obtained when "<alt> H" is pressed.

```
: SUB-SUB-MENU-1
F1 FUNCTION.KEY.DOES     WORD1
F2 FUNCTION.KEY.DOES     WORD2
;

: SUB-SUB-MENU-2
F1 FUNCTION.KEY.DOES     WORD3
F2 FUNCTION.KEY.DOES     WORD4
;

: SUB-MENU-1
F1 FUNCTION.KEY.DOES     SUB-SUB-MENU-1
F2 FUNCTION.KEY.DOES     SUB-SUB-MENU-2
;

: SUB-MENU-2
F1 FUNCTION.KEY.DOES     SUB-SUB-MENU-3
F2 FUNCTION.KEY.DOES     SUB-SUB-MENU-4
;

: MAIN-MENU
F1 FUNCTION.KEY.DOES     SUB-MENU-1
F2 FUNCTION.KEY.DOES     SUB-MENU-2
;
```

The definition of the first new word SUB-SUB-MENU-1 begins with a colon. All of the strings encountered before the semi-colon which terminates the definition are meaningful ("F1, F2" etc. represent the function keys). Same thing for SUB-SUB-MENU-2. Next, we start the definition of SUB-MENU-1 which soon encounters the word SUB-SUB-MENU-1. This is all right, since that word was defined immediately before. This definition compiles fine, so does SUB-MENU-1, and finally MAIN-MENU. There is no problem with any of this since all words are meaningful when they are used within a colon definition. One can see already the inverted strategy used, wherein subsidiary words must be defined prior to the words which later use them.

At this point, one can type MAIN-MENU which assigns meanings to function keys 1 and 2. Pressing "F1" executes SUB-MENU-1, which re-assigns the function keys. Pressing "F1" again now executes SUB-SUB-MENU-1, which re-assigns them again. Pressing "F1" yet again executes WORD1. So we see that descending to SUB-SUB-MENU-1 was extremely easy, now how to get back?

The temptation is simply to assign the word MAIN-MENU to one of the function keys assigned during the definition of the first word, SUB-SUB-MENU, thus:

```
: SUB-SUB-MENU-1
F1 FUNCTION.KEY.DOES     WORD1
F2 FUNCTION.KEY.DOES     WORD2
F3 FUNCTION.KEY.DOES     MAIN-MENU
;
```

This fails during compilation, however, because MAIN-MENU is not yet defined. In the same way, none of the function keys in any of the menu definition words can contain the name of a menu higher in the hierarchy because those words will not yet be meaningful to ASYST.

The solution to this grievous dilemma is the magic word "EXEC." This built-in ASYST word takes any string and executes it. The following illustrates its use. First, we declare a string variable MOVE-TO-MENU which will contain the name of the menu word we will wish to execute.

```
25 STRING MOVE-TO-MENU
```

Then we can create a colon definition in which this variable is assigned the value "MAIN-MENU." Next, we assign this new word to function key 3 during the definition of SUB-SUB-MENU-1, where it will cause no problem, as it is meaningful to ASYST.

```
: UP-TO-MAIN-MENU
" MAIN-MENU "
MOVE-TO-MENU ":=
MOVE-TO-MENU "EXEC
;

: SUB-SUB-MENU-1
F1 FUNCTION.KEY.DOES      WORD1
F2 FUNCTION.KEY.DOES      WORD2
F3 FUNCTION.KEY.DOES      UP-TO-MAIN-MENU
;
```

Now we continue the definitions as before, finishing with MAIN-MENU. We can descend to SUB-SUB-MENU-1 by pressing "F1" twice, as before, but now pressing "F3" causes the word UP-TO-MAIN-MENU to place the value "MAIN- MENU" onto the stack, and the "EXEC" word executes it, re-defining the function keys to their original meanings so that we are back in the main menu.

This simple example demonstrates the strategy actually used in the system to navigate back up the hierarchy to higher level function key menus.

Data File Handling

One of the useful features of this system is the ability to store data on the hard disk of the Compaq 286 computer. Since data is taken fairly rapidly (each molding shot usually consumes 100 points, and shots can occur as little as 10 seconds apart), and the data for each shot is stored in a separate file, we do not have time to type in file names. Therefore, a fairly elaborate system was built to manipulate file names. The standard format of data file names is #-MOxx.ext, where # is a number, MO is the month, xx is the date, and ext is a three letter DOS extension. Since DOS file names can be up to 8 characters in length, using a two-letter abbreviation for the month allows one to create up to 999 data files. An example might be 129-AU27.DAT, which is data file number 129 created on August 27. A variety of words are defined in the system based around a parsing algorithm which finds the hyphen in the file name and converts any characters in front of it into the numeric equivalent. Thus, there are words to increment and decrement the current file name, create a file with the current name, create a file with an incremented name, create a series of files all at once with numbers in a specified range, etc.

The file template used creates files with 3 comment fields. The first is automatically stamped with the current time, date, and the number of data points in the arrays (usually 100). The second is reserved for user comments. The third automatically fills in the melt temperature and the mold temperature which is measured at the time of the creation of the file.

Adaptability and Portability

The result of issuing an A/D.IN command is obviously several integers, in our case ranging from 0 to 4095 since we have a 12 bit A/D converter. In order to convert these integers into physically meaningful units, conversion factors must be supplied. In general, we are interested in the position of either the screw or the clamp of our Battenfeld injection molding machine, so these two choices are hardwired into the program. However, since the system is meant to be general purpose, useable on any molding machine, we have provided a word which actually determines the scaling factors in place. The user is prompted to move the machine component to its highest position (e.g. maximum plate separation or screw retraction) and input the value that this corresponds to (e.g. 10 inches separation). An A/D.IN is performed, then the user is prompted to repeat the procedure at a lower position. Another A/D.IN is performed and the two values are used to calculate scaling factors appropriate to that particular machine. From then on, the word SCALE.DATA will scale the position data correctly.

For our injection molding machine, the voltages we monitor are proportional to the positions of various components (screw, clamp, carriage, ejector). However, we often want to examine the velocity of the component during the injection cycle. This is a simple matter of differentiating the position with respect to time. The information is available: position data in the array DATA and elapsed time information in the array E.TIMES. ASYST has a built-in word DIFFERENTIATE.DATA, which unfortunately does not work as we expected it to. After much experimentation with simple two-dimensional arrays for which the differentiation was obvious by inspection, we arrived at a procedure contained in the word DERIV which seems to work well.

Since the goal of a smoothly running injection molding system is a stable, fully automatic cycle, obviously the data acquisition system must be able to run fully automatically as well. This is achieved by using the position of the mold clamp to trigger the acquisition of screw position data. There are two critical clamp positions, the "trigger position" and the "reset position." When the clamp opens beyond the reset position, it effectively sends a message, "start taking data when the clamp closes to the trigger position." The clamp then closes, triggers an adjustable delay, and then screw position data is taken. This data is written into a file (using an incremented file name), and the system waits until the clamp opens beyond the reset position again. Thus, there is no user involvement at all when operating in this fully automatic mode. Consistent with the desire to make the system transportable from machine to machine, the system can "learn" the trigger and reset positions. The user is prompted to move the clamp to the trigger position, an A/D.IN is performed, then to the reset position and another A/D.IN is performed.

Another feature of this system, which is consistent with the emphasis on portability from one injection molding machine to another, is the implementation of temperature measurement. The temperatures are measured by thermocouples which produce a potential proportional to temperature. These millivolt potentials are amplified by Omega Omni-Amp thermocouple amplifiers. However, interpreting the voltages as temperatures requires either a look-up table or a nonlinear equation. We chose to fit temperatures over the range of interest (0 - 450° C) with a third order polynomial using the built-in word LEASTSQ.POLY.FIT. Subsequently, making a temperature measurement involves performing an A/D.IN and invoking the built-in word POLY[X]. The data for potential vs. temperature for a J-type thermocouple are already in the system but can easily be changed. Thus, changing thermocouple types simply requires changing 10 numbers in the TEMPS.FIT word, and re-LOADING the system. Note that we found it necessary to use double precision real numbers for the coefficients in order to achieve fits corresponding to precision less than 1 degree F.

Control of Data Acquisition

In any data acquisition system one must specify the frequency with which data is to be taken. In this system, rather than changing the CONVERSION.DELAY of the A/D.TEMPLATEs we use the SYNCHRONIZE word, which delays the next A/D conversion until a period of time called the SYNC.PERIOD has elapsed (this is built into ASYST, not something we created). The frequency of data acquisition is changed by a word which prompts the user for a new value for SYNC.PERIOD. The data then populates arrays explicitly in a loop. We chose this route rather than using a buffer array because we wanted to do some manipulation of the data on-the-fly as it populates the arrays containing the position and elapsed time information. The software overhead was not prohibitive, and we can take data every 5 milliseconds or so, if we want. In fact, we rarely go faster than 30 milliseconds between data points.

Besides the frequency of the data, the user is also interested in the total number of data points. Unfortunately, ASYST did not support dynamic sizing of arrays so we have arbitrarily chosen 100 points as the size of the arrays to fill. It is possible to change this by reloading the program into a blank copy of ASYST. At a certain point during the loading, the user is asked to input the size of the data arrays, or accept the default of 100 with a carriage return. The number he provides is then used in the array variable declarations. Unfortunately, reloading the entire program is quite time-consuming (about 5 minutes) so in practice we rarely change the length of the data arrays. With the release of ASYST Version 2.0, both difficulties have been addressed. For one thing, dynamic sizing of arrays is now possible through "tokens," though this enhancement has not yet been incorporated into our system. Also, the new version loads and compiles far more quickly, so that re-loading to change the size of data arrays is not as prohibitive as before.

9.2.5 Examples of Application

In this section we will show some examples of our data acquisition system in action and discuss the utility of the gathered information. These applications will demonstrate both fundamental machine behavior as observed by purging to ambient, and mold-specific behavior, first in a spiral flow mold, and then in a three-plate mold with pin gates. Of particular interest will be the influence of mold features on machine behavior.

Purging to Ambient

The first application of the system was to measure the velocity with which the screw comes forward during injection. This was necessary since the velocity setting on our Battenfeld machine with Unilog 3000 closed-loop controller is input in terms of percent of maximum velocity, from 1% to 99%. One naturally wonders 99% of what!? This information is not provided in the operators manuals. Furthermore, our particular machine is equipped with an accumulator for high speed injection, so it is of great interest to determine what is the quantitative effect of using the accumulator.

Figure 9.5 shows the velocity as a function of screw position for shots purged to ambient (i.e. no resistance to flow), using a hydraulic pressure of 2000 PSI and various velocity settings. As expected, the screw accelerates to a constant velocity in a relatively short distance at low speeds, but takes a longer distance at higher speeds. With the accumulator inactive speeds are in the 0 - 3 in/sec range. Figure 9.6 shows similar plots, with the accumulator active. We see that speeds are higher, as expected, in the 0 - 12 in/sec range. Interestingly, the screw now takes considerably longer to reach its final velocity. For example, at 80% without the accumulator, the screw reached 2.7 in/sec in about 74ms. At 40% with the accumulator active, the screw stabilizes at 5.5 in/sec only after about 145ms! For this reason, the length of the stroke was increased to 2 inches for the 40% and 80% measurements.

An additional interesting observation is that the screw starts to slow down before reaching 0 inches, i.e. it does not simply bottom out at full speed. The point at which the screw starts to slow down increases with velocity. For example, at 80% with no accumulator, the screw starts to slow down at .1 inches, and at 80% with the accumulator active, it is already slowing down by .3 inches.

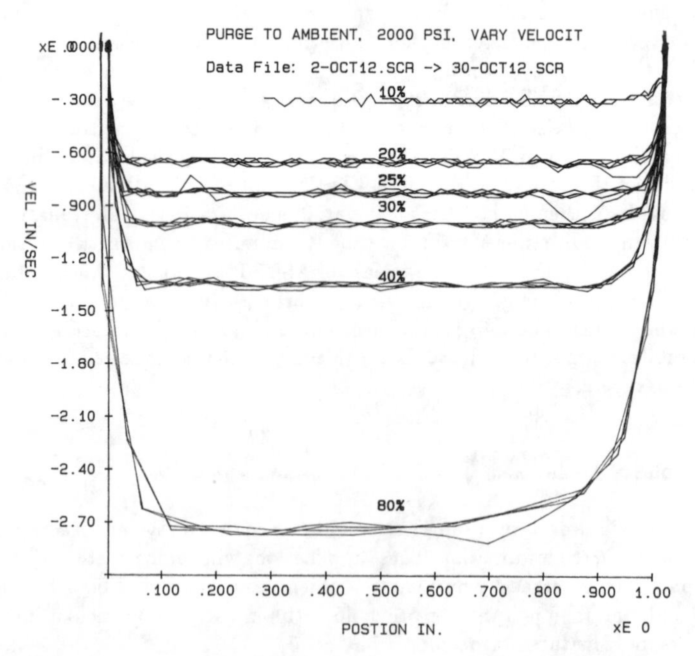

Figure 9.5 Screw velocity vs. screw position while purging to ambient, 2000 PSI, accumulator inactive.

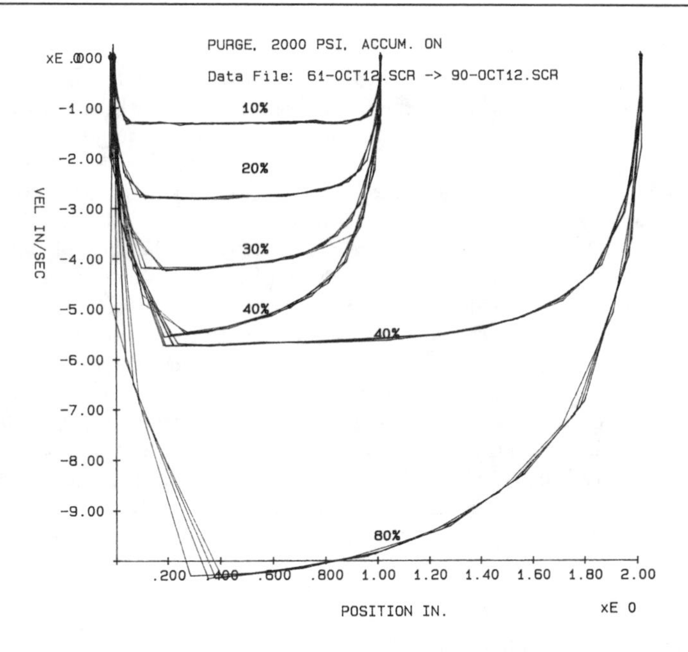

Figure 9.6 Screw velocity vs. screw position while purging to ambient, 2000 PSI, accumulator active.

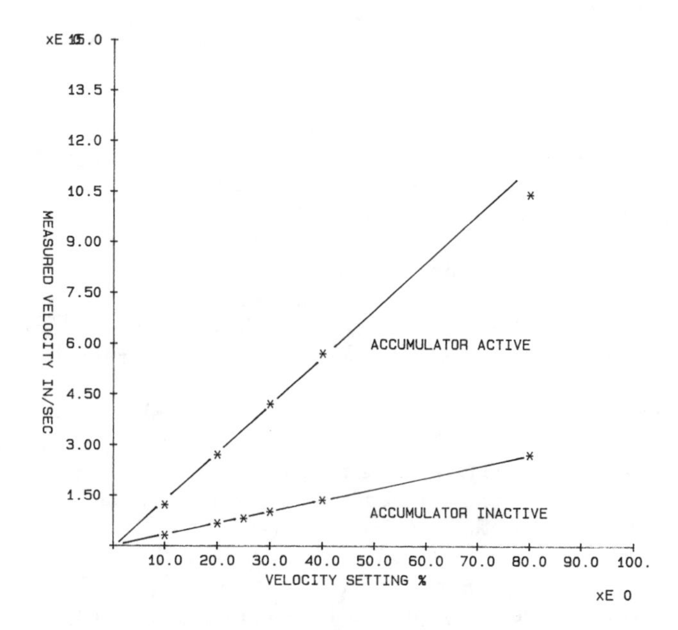

Figure 9.7 Measured velocity vs. % setting from purging to ambient.

Figure 9.7 shows measured velocity as a function of the set percentage with the accumulator both inactive and active. We observe that with the accumulator inactive, all of the points lie on a straight line which intersects the origin, an extremely satisfying result. The slope of this line is 0.03375. Thus, with the accumulator inactive, one can obtain the actual injection velocity by multiplying the set value (in percent) by 0.03375. With the accumulator on, all of the points again lie on a straight line intersecting the origin, with the exception of the highest percent velocity, 80%. The slope in this case is .14.

Figures 9.5-9.7 show that activating the accumulator has the effect of multiplying the injection velocity by a factor of about 4.1, throughout the lower half of the speed range. However, at higher speeds, there does not appear to be enough hydraulic pressure to achieve the desired velocity. In addition, the fraction of the injection stroke required to accelerate and decelerate the screw is surprisingly large, especially with the accumulator active. Therefore, control of the screw at high velocities and short travels will not be particularly good.

This conclusion was reached by observing shots which were purged to ambient, the most favorable situation possible for the machine, since the resistance to injection is nil. Next, we examined the behavior for spiral flow mold, which is more demanding of the machine, since the resistance to injection is non-zero and increases throughout the shot.

Spiral Flow Molding

The spiral flow configuration is an interesting situation halfway between purging to ambient and making real parts, since the resistance to injection starts low and builds monotonically, until finally the resistance to flow exceeds the hydraulic pressure available to drive the screw forward. In a well controlled machine, one would hope to see velocity to be fairly independent of position until nearly the end of the shot, as the servo-valve compensates for increased resistance to injection by supplying more hydraulic pressure to the screw.

Figure 9.8 shows velocity vs. position traces for high density polyethylene injected into a spiral flow mold, for two values of hydraulic pressure (1500 and 2000 PSI) and a variety of velocities, with the accumulator inactive. Note that the nominal hydraulic pressure cited is the pressure available upstream from the servo-valve, and thus, the maximum available to the screw. The pressure actually applied to the screw is determined by the servo-valve in response to conditions which occur during the shot. Oddly, this latter pressure, which is the more relevant one, cannot be monitored by the operator. Our data acquisition system at present does not monitor it either, but this is a planned enhancement.

The traces in Figure 9.8 show that the velocity achieved at each setting is the same as the expected value based on shots purged to ambient. This behavior is independent of the nominal hydraulic pressure, demonstrating the assertion in the previous paragraph that this value is the maximum available to the servo-valve, and the servo-valve simply uses as much as it needs. We observe that although the initial velocity was correct, the velocity measured as the shot progresses droped continuously for all but the lowest velocity (10%, .3 in/sec). There can be two explanations for this: either the servo-valve is not responding rapidly enough to changing pressure demands, or the valve is completely open and the maximum pressure is insufficient to maintain the desired velocity further into the shot. Two observations lead us to believe that the latter explanation was correct. First, note that the position at the end of injection (i.e. the length of the spiral) is independent of the injection velocity but is greater for higher hydraulic pressure. Second, and more persuasive, is the observation that at a setting of 20%, both pressures give the correct initial velocity of .6 in/sec, but the velocity decreases more rapidly when using 1500 PSI than when using 2000 PSI.

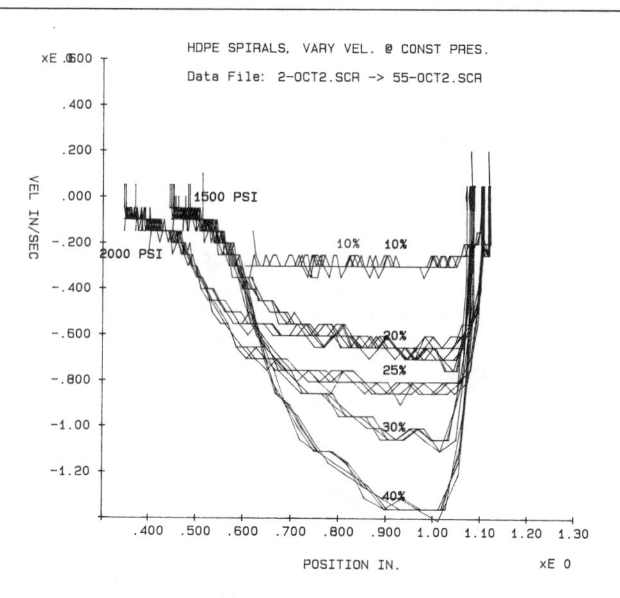

Figure 9.8 Screw velocity vs. screw position, spiral flow molding of HDPE, varying % setting at constant hydraulic pressure.

The conclusion from these observations is that, surprisingly, insufficient hydraulic pressure is available to maintain a constant injection velocity for all but the slowest injection speed. This may in a way be an unfair test, since the flow length for a typical part does not increase as rapidly as the shot progresses. Nevertheless, it is an extremely revealing test of the machine's performance.

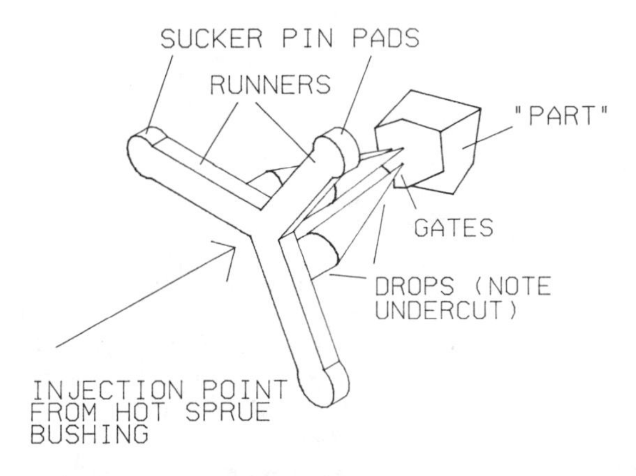

Figure 9.9 View of injection molded part and runner system.

Performance with a Real Part

Figure 9.9 shows a schematic of a real part made in a three-plate mold using pin gates. The actual part cannot be shown, but it is about the same size relative to the runner system as the irregular object shown, i.e. quite small.

Figure 9.10 shows traces for screw position vs. injection time. We observe that the screw travel during the filling of the runner system is quite reproducible, but the position vs. time when traversing the gates is not. We also find that the cushion is reproducible (.28 inch) except for one part which did not fill because the previous part failed to eject. Finally, we find that the bounce-back upon release of follow-up pressure is quite reproducible. Note that the final screw position after bounce-back (.42 inch) is actually farther back than the point at which the gates are encountered on the way in (.39 inch). This is due to the small size of the part and the compression of the melt during injection.

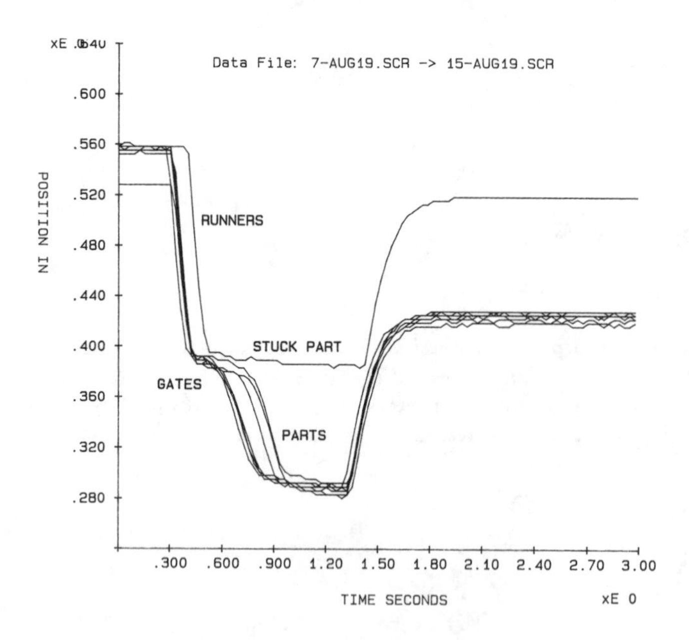

Figure 9.10 Screw position vs. time for injection of real part.

Figures 9.11-9.13 show velocity vs. position traces for the same part made at velocity settings of 40, 60, and 80%. We see immediately the rather disheartening fact that on none of the shots did the screw actually achieve a constant velocity at any point. On the other hand, the velocity/position behavior seems to be quite reproducible. One has to ask the question, does it matter if the velocity is not constant, or is reproducibility sufficient to make good quality parts?

Examining Figures 9.11-9.13 in more detail, we see that at the lowest velocity setting, 40%, the measured peak velocity of 1.2 in/sec is about right based on the measurements described above. However, the measured peak velocities of 1.4 in/sec and 1.6 in/sec observed

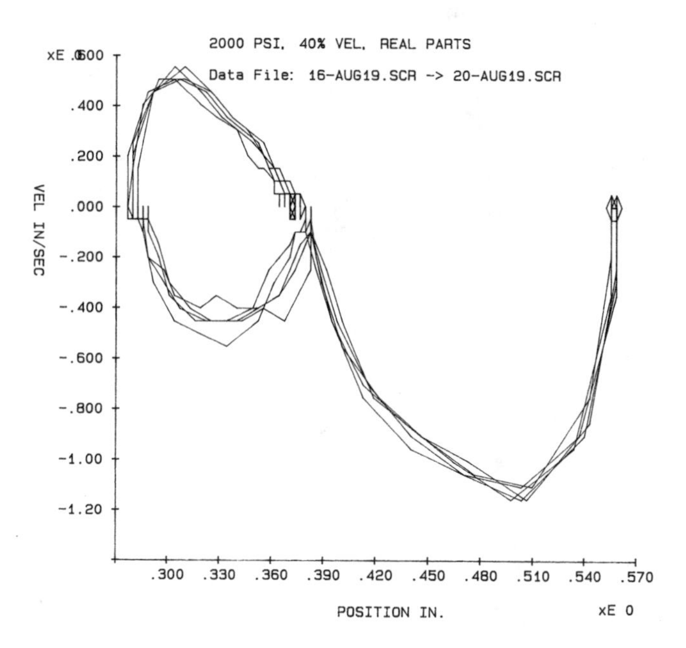

Figure 9.11 Screw velocity vs. screw position for injection of real part at 40% velocity setting.

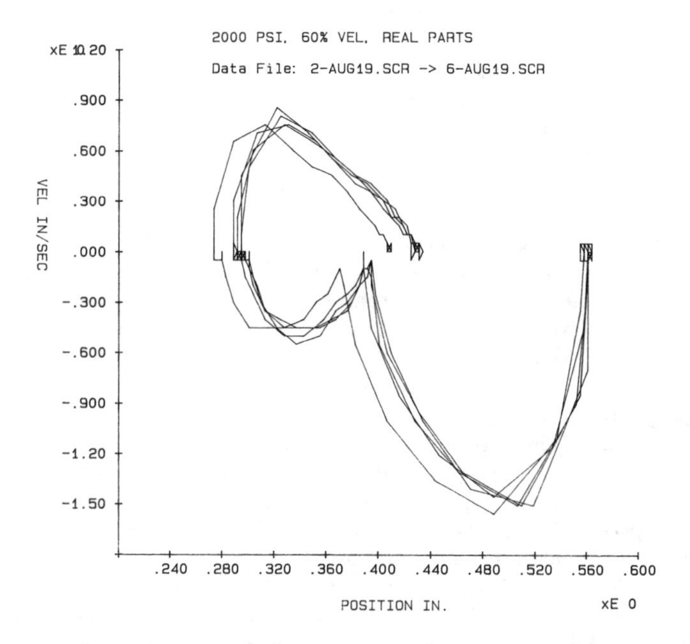

Figure 9.12 Screw velocity vs. screw position for injection of real part at 60% velocity setting.

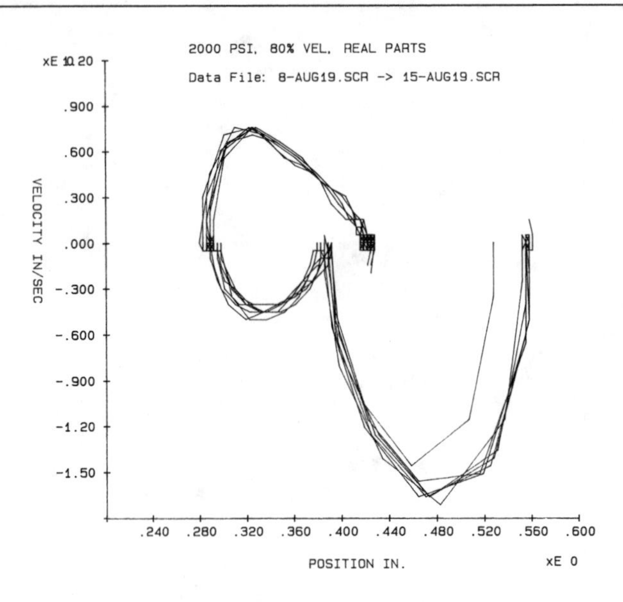

Figure 9.13 Screw velocity vs. screw position for injection of real part at 80% velocity setting.

at 60% and 80% settings are too low. The velocity peaks are achieved rapidly, in about 0.05 - 0.1 inch of travel, then velocity starts to decay almost immediately. By the time the advancing melt encounters the gates at about 0.38 inches, the screw has come to a virtual halt. Finally, the servo-valve reacts to provide more pressure to the screw and the part is filled. It is disappointing to note that the actual part filling occurred at 0.45 in/sec for all three velocity settings, due to the throttling effect of the gates. The lesson from these three figures are that not only does the operator not get what he wants (i.e. constant injection at three different velocities), but he does not even have the ability to influence the filling speed in the part itself, only in the runner system!

This example might also be construed as an unfairly difficult test, as the part is small, the gates are of extremely small cross-section, injection was attempted at relatively high velocity, and the machine is too large for the part (out of a total screw travel of 4 inches, we use the last 1/2 inch or so). Nevertheless, it is a real-world situation that actually did arise in our laboratory, and we were shocked to learn that we had no control over the part filling. Clearly the servo-valve in our machine was unequal to this task.

Another factor in the experiments depicted in Figures 9.11-9.13 was the fact that we were using a high viscosity melt. In order to reduce the viscosity of the melt, we mixed in 33% by weight of a liquid crystalline polymer [14] in the form of dry blended pellets. The difference in behavior, shown in Figure 9.14, was quite astounding. The injection velocity at 80% is relatively flat, between 1.8 - 2.0 in/sec over most of the shot and the velocity is hardly affected at all when the gates are encountered at a position of 0.38 inches. In fact, the viscosity is so low when using the LCP additive that the screw continues to move forward after the part is filled at 0.25 inches and no bounce-back is observed. No doubt significant leakage past the screw flights was occurring during these shots. The LCP blend example illustrates the value of being able to observe the machine's behavior directly and in detail.

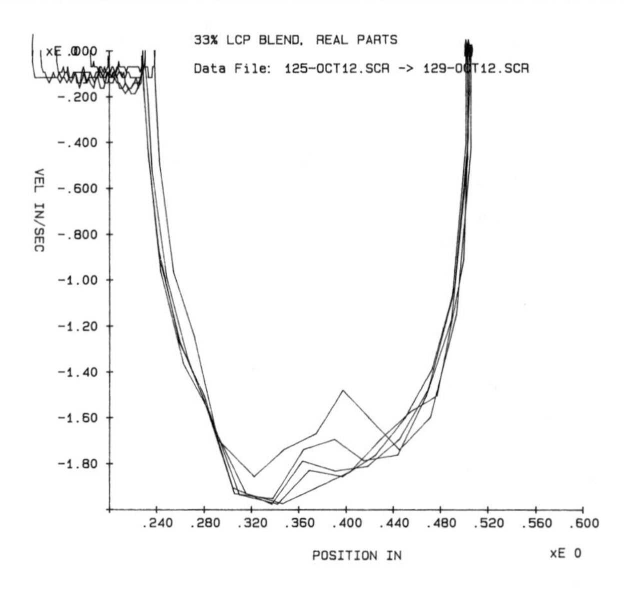

Figure 9.14 Screw velocity vs. screw position for injection of real part using 33% LCP blend for reduced viscosity velocity setting.

9.3 SUMMARY AND CONCLUSIONS

Most laboratory injection molding machines are instrumented to some degree. Researchers have used strip chart recorders, data loggers, voltmeters, etc. for many years. The data one can take are similar: temperatures, pressures (hydraulic and melt), and displacements. The advantage of using a dedicated microcomputer is that one can do real-time data manipulation (i.e. differentiation of position to obtain velocity) and that one can store the data for each shot for later analysis and comparison. In principle, most of the capabilities of the system described here could be duplicated with lots of chart paper and lots of time-consuming, mind-numbing work. In practice, no one would do it. Having the computer do the work enables and encourages the researcher to examine the process much more closely.

With the advent of CIM (computer integrated manufacturing), JIT (just in time delivery), SPC (statistical process control), and other applications of information-age technology to manufacturing, process control inevitably becomes much more stringent. It is not at all far-fetched to examine manufacturing processes in similar detail, on a shot-by-shot basis in the case of injection molding. In fact, some sophisticated controllers already perform such functions, e.g. the Battenfeld Unilog 4000 with "injection actual-value graphics," which compares each shot with a stored ideal curve, and flags out-of-spec shots. Naturally, a dedicated system which is hardwired into a sophisticated controller is much less flexible than the system described here.

Our system is unusual that it uses extremely mundane off-the-shelf hardware which can be purchased at low cost. The most "exotic" component is the analog-to-digital conversion board, and even there the system allows great flexibility in the choice of the board. On the other hand, the programming language/environment truly is exotic. ASYST is an extremely specialized and idiosyncratic language which happens to be ideal for this application.

Another data acquisition system for injection molding which we are aware of has been assembled by Kalyon et al. [15]. Their system takes the opposite approach of using rather sophisticated hardware (Hewlett Packard A/D converter and microcomputer) but programming in HP-BASIC. The merit of their approach is that the language is easy to work with and very powerful in some respects. HP-BASIC has built-in HPIB commands for interfacing to the hardware, commands for touch-screen input, and graphics to build a dramatic and easy to use user interface.

These and many other one-of-a-kind data acquisition systems can provide extremely useful observations of actual machine behavior. We are continuing to enhance the system and welcome any inquiries from persons who might be interested in duplicating it. One thing we have learned at this point is that the velocity changes are far from the step functions that manufacturers seem to imply and that many molders would like to believe. It makes one question the utility of some sophisticated controllers which allow as many as 10 velocity changes during a shot. Is this meaningful? Does it favorably impact part quality? Is it merely advertising and techno-babble?

We have also learned that in many situations the velocity of injection is limited by the hydraulic pressure generated by the machine, a maximum of 2000 PSI in our case, under surprisingly non-demanding conditions. A sophisticated closed-loop controller is of no value if the servo-valve is running wide open rather than modulating the pressure applied to the screw.

The lesson is that it pays to maintain a healthy scepticism with regard to the performance of your machine. One must resist the temptation to simply "set it and forget it," to have faith in the black box. Calibrations can drift, valves can stick, the machine's logic might not conform to the operator's logic, there are a lot of things to go wrong in this complex dance of mechanics, hydraulics, and electronics. It pays to make measurements and monitor performance to realize the promise of a sophisticated machine, for both research and manufacturing.

REFERENCES

1. Ajay, S., PhD thesis, Carnegie-Mellon University (1978).
2. Hieber, C.A., Wang, V.W., Wang, K.K., Chung, B., *SPE Tech. Papers*, *30*, 769 (1987).
3. Sandall, D.J., Campbell, G.A., *SPE Tech. Papers*, *28*, 789 (1985).
4. Burghoff, G., Menges, G., *SPE Tech. Papers*, *28*, 341 (1982).
5. Paulsen, D.C., "Theory, Practice, and Control of Injection Molding," Paulsen Seminar Programs Inc., Southington CT.
6. "A/D Converter Manual," Interactive Structures Inc. (1981).
7. "Data Acquisition Handbook," Taurus Computer Products, Manchester NH, p. 2-22 (1984).
8. Osinski, J.S, *SPE Tech. Papers*, *30*, 979 (1984).
9. Electronic Design, Sept. 29, 1983.
10. Coates, P.D., Sivakumar, A.I., Johnson, A.F, Master's theses, University of Bradford, England (1983).
11. IBM PS/2 Data sheet.
12. Apple MacIntosh and Apple+ Data Sheet.
13. Hewlett Packard PC Data Sheet.

14. Kiss, G.D., *Polym. Eng. Sci.*, *27*, 410 (1987).
15. Kalyon, D.M., Dey, S., Wagner, A., presented at Polymer Processing Society Meeting, Buffalo NY (1987).

10. SUBJECT INDEX

A

A/D
board 281-282
converter 280, 283, 290, 300
expansion board 285
ABAQUS 28
Adaptive meshing 168
Arrhenius-type temperature dependence 206, 212-213
ASYST 278, 282-286, 289-291, 299
package 283
Axisymmetric
contraction flow 4
finite element implementation 123

B

Band matrix 189
Barrel film 3
Birefringence 5, 6, 248-249, 257, 264, 270-274
BKZ-model 164
Bleeder ply molding 9
Blow molding 2, 9, 20, 118, 120-123, 127, 138, 140, 142, 144, 157-161, 163-164, 167-168
of a rectangular box 158
Boger fluid 6
Boundary
condition 3, 4, 20-22, 24-26, 28, 30-31, 34, 56-58, 62-63, 65, 84-90, 103-105, 107-109, 111-112, 121-122, 177, 181, 183, 188-189, 198, 207-208, 210, 213-215, 217, 219-220, 224-225, 230, 248, 253, 254, 270, 274
condition in surface boiling 25
element method 9
integral method 8, 10
Bulk temperature 91, 96-97, 226

C

Calendering 2, 8
of Phan-Thien-Tanner viscoelastic fluid 8
of viscoelastic fluid 8
Carreau model 90, 208
Cauchy
problem 21-22
stresses 136
Center gated disk 249-250, 253-254, 257-272, 274
Circular-tube 207-208, 211, 232
Claxton model 208

Closed-loop controller 292, 300
Coat hanger die 4
Coating 2, 7-8
Coextrusion 4, 7-8
Compressible
    flow 180, 186-187, 191, 198
    molding 9
    packing phase 191
Compression molding 206
Computer interface 278, 280
Constitutive equation in finite elasticity 21
Continuity equation 176, 180, 182, 190, 210, 218, 226
Control-volume 186-187, 189-190, 206, 208-209, 218-223, 225, 227-232, 241-242
    finite-element method 206, 208, 218-219, 221, 223, 225, 228, 241
    method 209, 222, 227-228, 230-232, 242
    technique 190, 227, 242
Convective redistribution 4
Conveying region 2-3
Cooling/solidification phase 177-178, 187, 192
Counter-rotating twin-screw
    extruder 80
    machine 48
Crank-Nicholson scheme 181
Criminale-Ericksen-Filbey constitutive equation 7
Crystallization temperature 174
Curing kinetic model 206-207, 211-212
Current signal 280
Curtiss-Bird model 6

D

Data acquisition software 281
Data acquisition system 278-281, 285-286, 290-291, 294, 300
Deborah number 6
Degree of cure 206-208, 212, 226
Design methodology 159
Die design 2-3
Differential
    constitutive equation 6
    scanning calorimeter 206
Diffusion 4, 8, 177
Digital signal 278-279
Direct
    approximate-analytic method 23
    formulation 20, 26
    numerical method 23
Dispersive mixing model 5
Drag flow 2, 7, 67, 91, 95, 98

E

Energy equation 84, 86, 88-89, 183-184, 197, 206, 208-213, 224, 241, 250, 257, 274
Entry and exit flows 26
Equation of state 174, 176-177, 180-184, 187-190, 196-197, 202-203
Euler
  finite-difference scheme 182
  Lagrange equation 32
Expansion flow 4
Explicit Eulerian scheme 254
Extrudate swell 4, 31
Extrusion 2-4, 6-7, 20, 26, 28, 31, 38-39, 41-42, 45-46, 52, 54, 63, 78-80, 84, 89, 92, 98, 111, 118, 164, 249
  blow molding 118
  die design 3

F

Fiber
  orientation 4, 10
  spinning 5-6
Field signals 278-280
Filling phase 174-175, 177, 191, 198
Film blowing 2, 5-7
Finite
  control-volume integration approach 186
  difference 2-4, 81-82, 88-89, 92, 174-175, 177-178, 180, 186, 206, 216, 248, 254-255
  difference method 2-3, 78, 81, 174, 206, 208, 216, 224, 241, 255
  element method 62-63, 78, 88, 118, 121-123, 126, 145, 154, 158, 206, 207-209, 218-219, 221, 223, 225, 228, 232, 241, 249
  element model 81, 89, 111, 122, 133, 147, 151, 154, 168
First normal stress difference in the strip cavity 268-269
Five-zone model 3
Flow
  analysis network 3, 26, 207
  analysis network method 3, 26
  path 207
Forming a deep cylinder 154
4-noded element 127-128
Free inflation 121, 144-147, 150, 153
Frozen-in birefringence 5, 249, 270, 273-274
  in fibers 5
Frozen layer 248-249, 257, 259

G

Galerkin
finite-element formulation 216
finite-element method 207-209, 216-218, 222, 225, 227-232, 241-242
method 90, 186, 208, 222, 227-232, 242
Gamma function 33
Generalized
Hele-Shaw flow 207
Newtonian fluid 20, 182, 196, 248
Geometric instabilities 167
Glass transition temperature 133-135, 174, 183, 198, 257-258, 270, 274
Global
flowchart 180-181, 188
load vector 130
matrix equation 90
stiffness matrix 90, 130
Green's
deformation tensor 125, 129, 134
theorem 219
Griffith Number 82, 86, 91, 95, 102

H

Half-disk 207
Hard disk 279, 281, 289
Heat
conduction 2, 22-23, 25, 174, 177-178, 183-184, 209, 211, 248, 274
convection 2
of crystallization 197
Hele-Shaw creeping flow 174, 176
approximation 176
Hyperelastic material 122, 134, 136
model 122, 136

I

Ill-posed problems 22
Impermeable boundary region 215, 217
Induction time 206-209, 212-214, 226, 233, 235, 240-242
Inelastic modeling 10
Injection-blow molding 20, 118
Injection molding 2, 10, 25, 42, 160, 174, 176, 196-198, 202, 206-209, 212-213, 216, 240, 248-249, 256, 274, 278-279, 281, 290, 299-300
of rubber 206
of thermoplastics 10, 196, 207, 209, 212
Injection stretch-blow molding 160-161, 163

Intermeshing co-rotating twin-screw extruder 5, 40-42, 44-45, 47-48, 50-52, 55, 59, 71
Internal mixer 5
Inverse
    formulation 3, 20, 23, 26, 34
    heat conduction problem 25
    Laplace transform 24
    Stefan problem 25
Isothermal flow 6-7, 80, 90, 174, 206, 248, 270, 274

J

Jacobian matrix 124-125

K

Kinetics of crystallization 177, 197
Kneading disc 38, 42-49, 51, 53-55, 64-71
    element 38, 42-49, 51, 53-55, 64-71
    mixer 44-47

L

Lagrangian multiplier 90
Laplace's equation 27
Lay flat assumption 179
Left-handed screw element 64
Leonov model 248-249
Linear
    interpolation 90, 127, 129, 182
    matrix 189
Longitudinal velocity 253, 259-261
Lubrication approximation 3, 5, 7-8, 20, 26

M

Main-processor 89
Malaxator 38
Maxwell model 5-6
Mean-stream method 21
Melt
    conveying zone 79
    front 175, 198, 206-209, 211, 215, 217, 225-226, 228-229, 235-236, 238, 241, 248-249, 253, 257-259, 264-265, 269-270, 272
    pool 3
    spinning 2, 5-6

spinning of viscoelastic fluid 5
Melting
region 2-3
zone 3
Metering
region 2-3
zone 2-3
Method of iteration regularization 23
Microprocessor 281
Mixing 2, 4-5, 39-40, 43, 45, 48, 52-53, 78-79, 81-82, 87, 89, 90, 103-105, 107-108, 110-112
in a single-screw extruder 79
parameter 78, 90, 103-105, 107, 110-112
region 78, 81, 87, 103-104, 107-108, 111
zone 111-112
Modeling
of coextrusion 4
of extrusion 26
Modified
Cross model 209, 212, 233
Mooney-Rivlin form 152
Modular twin screw machine 46, 51-52
Mold
cooling system 10
heating analysis 9
Mooney-Rivlin
constants 148
formulation 134-135, 137
model 149
Multifilament spinning 6
Multilayer extrusion 4
Multimanifold vane flat die 4

N

Neo-Hookean rubber constitutive relationship 121
Neumann boundary condition 4
Neumann stability criteria 255
Newton-Raphson method 89
Non
-circular channel 207
-isothermal induction time 207, 209, 213, 241
-isothermal simulation 2
-Newtonian flow in the screw section 59
Nonlinear
elastic behavior 133-134
finite elasticity 134
Nonuniformity of the mold temperature 10

O

Ogden
expansion 140, 142-144, 165-166
formulation 134, 136
material138, 152, 159
model 134, 136, 142, 144, 156
Oldroyd fluid B 6
1-D element 127-128

P

Packing phase 177, 179, 185-187, 191-193, 196
Parison 10, 118, 120-121, 157-159, 164, 167
inflation 121
Peclet Number 82, 86, 102
Penalty function formulation 4
Physical instabilities 118, 167, 174
Planar flow 4
Plastic preform 163
Plug-assist thermoforming 118-119
Post-filling stage 187, 206, 211, 239, 241
Power-law
fluid 5, 7-8, 59, 62, 64, 71, 79
index 7, 78, 82, 84, 86, 104-105, 107, 109, 110, 112
model 5, 7, 90, 212
Pre-processor 89
Pressure 2, 3, 5, 8, 79-80, 83, 86-87, 90-01, 95-96, 98, 103-105, 107-112, 118, 121, 128-129, 131, 133, 144, 148, 154, 159-160, 165-167, 174-180, 182-187, 189-198, 202, 206-210, 214-219, 221-224, 227-232, 235-237, 241-242, 248, 249, 270, 278, 292, 294-296, 298-300
dependence on viscosity 202
drop optimization 31
time evolution 191
Primitive variables formulation 4
Process control 4, 120, 299

Q

Quadratic
interpolation 90
triangular element 90

R

Rate of crystallization 197

Reaction injection molding 10
Reactive injection-molding 206, 208, 233
Rectangular cavity 186, 191, 207
Regularized
algebraic method 23
numerical method 23
Residence time distribution 78-80, 91, 98-102, 112
Residual load vector 131-133
Resistance signal 280
Reverse Polish notation 283
Reynolds
' equation 26-28, 30
number 26-28, 30
Right-handed screw element 45, 52, 54

S

Scale up of extrusion dies 4
Screw
characteristic curve 56-58, 60-61, 63-64, 66-67
film 3
flight film 3
push-out experiment 3
shank film 3
Second order fluid 21
2nd Piola-Kirchhoff stress tensor 126, 129-130, 134-136
Self-wiping screw 51, 56, 62
Shear
-free squeezing flow 9
-induced extension 4
rate in the strip cavity 259, 263-264, 266
Sheet molding compound 9
Shrinkage 10, 174-175, 190, 195-196
Single-screw extruder 54, 78-79, 81-83, 88, 111-112
Singular element 229, 231-232
Singularity 209, 229, 231-232, 242
Slip effect 4
Slit die 249
Solid
bed 2-3
bed conveying region 2
conveying zone 79
Solidification phase 174, 177-178, 183, 187, 192
Spencer-Dillon model 28
Spencer-Gilmore equation of state 176
Spin coating deposition 8
Spiral mandrel die 4
Staggered meshes 180, 187, 191

Stefan problem 25
Stokes-Windsor intermeshing co-rotating twin-screw extruder 42
Straight vacuum forming 118
Strain
distribution 79
energy function 134-136, 146, 150, 165-166
Stress
-induced dispersion 4
-optical coefficient 257
relaxation 6, 248, 270
Stretch-blow molding 118, 160-161, 163
Strip cavity 248-250, 253-254, 256-274
Subcontrol volume 219-220, 223-224

T

Tadmor melting model 3
Tait's equation 183, 190, 196-197, 202-203
of state 183, 190, 196
Tangent stiffness matrix 131, 133
Temperature
at the instant of fill 237
in the cavity 257
Thermoforming 2, 10, 118-123, 127, 138, 140, 142, 144-145, 147, 150-151, 154-157, 159, 164, 166-168
crystallizing polyethylene terephthalate 10
of a rectangular box 156
Thermosetting resin 4
Thin-cavity approximation 179
Three
-dimensional finite element 3
-layer model 3
Translation
region 78, 81, 107-109
zone 78, 91, 111
Transverse velocity 249, 253, 257, 259, 262-263
Twin-screw extruder 5, 40-42, 44-45, 47-48, 50-52, 55, 59, 61, 71, 78-82, 87-89, 103-104, 106, 108-109, 111
Two
-dimensional finite element simulation 9
-dimensional isothermal extrusion 3
-lobed screw 53
-node tubular elements 208, 220, 232, 241
-phase flow 7, 186
-phase model 7
2-noded element 127

U

Unsymmetric case of calendering 8
User interface 282, 284-286, 300

V

Velocity vectors at the centroid 226, 237
Venturi-type device 52
Viscoelasticity 4, 7-8, 164, 248
    constitutive equation 249, 274
    effects in compression molding 9
    modeling of injection molding 10, 248
Viscous heating 177, 181, 184-185, 188, 194, 210-211, 224, 227, 248, 250-251, 257, 274
Voltage signals 279-281
Volterra integral equation 24
Vortex patterns 8
Vulcanization 206-213, 234, 241

W

White-Metzner viscoelastic equation 248
Wire-coating 7-8
    coextrusion 7-8

Y

Yield stress 4